# THE FUTURE OF ENERGY USE

# THE FUTURE OF ENERGY USE

Phil O'Keefe, Geoff O'Brien and Nicola Pearsall

London • Washington, DC

First published in 2010 by Earthscan



Earthscan Ltd, Dunstan House, 14a St Cross Street, London EC1N 8XA, UK
Earthscan LLC, 1616 P Street, NW, Washington, DC 20036, USA
Earthscan publishes in association with the International Institute for Environment and Development

For more information on Earthscan publications, see www.earthscan.co.uk or write to earthinfo@earthscan.co.uk

ISBN: 978-1-84407-504-1 hardback
ISBN: 978-1-84407-505-8 paperback

Typeset by JS Typesetting Ltd, Porthcawl, Mid Glamorgan
Cover design by Hugh Adams

A catalogue record for this book is available from the British Library

Library of Congress Cataloging-in-Publication Data

O'Keefe, Philip.
The future of energy use / Phil O'Keefe, Geoff O'Brien, and Nicola Pearsall. – 2nd ed.
p. cm.
Rev. ed. of: The future of energy use / Robert Hill, Phil O'Keefe, and Colin Snape, 1995.
Includes bibliographical references and index.
ISBN 978-1-84407-504-1 (hardback) – ISBN 978-1-84407-505-8 (pbk.) 1. Power resources. 2. Energy policy. I. O'Brien, Geoff. II. Pearsall, Nicola. III. Hill, R. (Robert), 1937- Future of energy use. IV. Title.
TJ163.2.H53 2010
333.79–dc22

2009020025

At Earthscan we strive to minimize our environmental impacts and carbon footprint through reducing waste, recycling and offsetting our $CO_2$ emissions, including those created through publication of this book. For more details of our environmental policy, see www.earthscan.co.uk.

Printed and bound in the UK by TJ International, an ISO 14001 accredited company. The paper used is FSC certified and the inks are vegetable based.

# Contents

# List of Figures, Tables and Boxes

## Figures

# Tables

## Boxes

# Acknowledgements

First, we would like to acknowledge the former co-authors of the first edition of *The Future of Energy Use*, namely the late Professor Robert Hill and the now retired Dr Colin Snape. At Northumbria, they helped build a tradition of studying science and society that has resonated through this institution, from undergraduate to postdoctoral level, for many generations. Secondly, we would like to thank the generous reviewers of the first edition whose encouragement pushed us collectively to embark on a second edition. This brings us to an important third acknowledgement that requires just a little explanation.

We had initially thought that, as nothing significant had happened in terms of new technologies or even the deployment of existing technologies, it would be a simple matter of updating figures and diagrams with additional commentary on improved efficiency. How wrong we were. The policy debate on energy has entirely changed in the last 20 years and, despite the fact that there remain few new commitments on the ground to new energy practice, the framework is entirely new. It is driven by a commitment to stop despoiling the global commons through continued accelerating increases in anthropogenic greenhouse gases which, in turn, suggest the twin pillars of energy policy are efficiency and renewables. This, however, poses problems since the energy system, especially the electricity and transport sectors, with their respective emphasis on existing fuel-technology combinations of coal or nuclear with large transmission and oil product with road network, have so much embedded capital that it is difficult to move towards a sustainable future. As we have struggled with this conundrum, Earthscan have provided substantial support despite considerable delay on our part in delivery. A heroine of this book is Claire Lamont, supported by Jonathan Sinclair Wilson.

Our fourth acknowledgement is to Zaina Gadema who not only helped with the editing but provided research assistance and input to this book. How she did this, while juggling graduate school, other jobs and raising a family, is beyond what we could ask of ordinary folk. A large debt of gratitude is owed. Fifthly, there were others who, at various times, provided research input including Leanne Wilson and Joanne Rose, from a base in ETC UK.

Finally, we acknowledge the generous generation of students who have pushed us in argument, who showed us that it was necessary to say that all politicians, of whatever political view, and all policy makers, of whatever institutional bent, have essentially ducked the issue of an energy secure future in an environmentally secure space. As we continue our work, and our necessary lobbying, the final acknowledgement is the largest of all.

*Phil O'Keefe, Nicola Pearsall and Geoff O'Brien*
*Northumbria University, April 2009*

# Foreword

## Energy futures

The energy future of the global commons faces two direct physical challenges that can ironically only be met by generating political will. The first challenge is to address accelerating climate change and increased climatic variability, a situation created by anthropogenic expansion of greenhouse emissions. In addressing this challenge it must be emphasized that there are subsets of problems namely:

- How do we gauge and miss the environmental tipping points by ensuring that the carbon equivalents are kept below 350 parts per million (ppm)?
- What are the type and scale of technologies that will allow us to deliver a 350ppm goal recognizing that all technologies change relationships to the environment and simultaneously change people's relationships to other people?
- How can we deliver this sustainable energy world within and between generations of people, without starting from a policy position that recognizes energy and by implication carbon inequalities within and between nation states?

The second challenge is to build the hydrogen – non fossil and non nuclear fuel – future while recognizing that current technical configurations of the transport and electrical systems work against this. The dominance of hydrocarbons, particularly oil-based products for private transport and the dominance of hydrocarbons and nuclear power in the large scale generation and transmission system distort current systems. Neither system can be said to reflect 'market prices' since both systems are heavily reliant on direct and indirect subsidies together with an inherent tendency to expand gigantism because of embedded capital such as roads and transmission systems. Obtaining end use fuel–technology combinations that are efficient must be the driving goal as we move to the hydrogen economy. That will again require a willingness to address energy poverty in both developed and developing worlds, together with a willingness to develop local scale supply solutions owned by the community.

As we end this book it is worth reflecting on the 30 years of work each of us has done around energy and environment research. While we would never claim that there has been little research ongoing over the last 30 years, the impact on energy futures remains marginal. Phil O'Keefe finds that, despite better computer modelling, there is little commitment to international and national energy planning and, instead with financial liberalization of commercial energy, the last 30 years has lived on the mantra of market delivery. Furthermore, with reference to biomass supply solutions and appropriate technology for local consumption, he sees a decline in commitment to delivery with efforts devolving, lacking support to non-governmental organizations (NGOs) and community groups. Geoff O'Brien, after 15 years work in the oil industry where he developed a portfolio of environmental interventions, is concerned about the continuing inability to make energy efficiency a core function of energy futures, coupled with the governance issue that sees private and public institutions act as oligopolies against local energy solutions. Nicola Pearsall continues to be concerned as both a theoretical and bench scientist that while renewables, particularly her own specialism of 'photovoltaic generation' have continued to progress, the absence of substantive research and development monies compared with conventional generation, and the lack of drive to large-scale commercialization, have significantly reduced the impact of renewables.

Little progress in 30 years. Who is to blame? Well, in one sense all of us who did not keep energy on the agenda. But in the key sense, it is politicians and policy makers of whatever political hue who refused to think through the numbers and safeguard present and future generations. In a situation of globalization ('One world many places; many worlds one place') the new energy economy must be generated, governed by the core principles of democratic participation and a constrained pluralism. This implies that the underlying choices of energy futures lies not simply in a discussion of rival technical claims but to human rights, bioethics, security and justice, responsible finance and a commitment to global relevant sustainable science. Any takers?

*Geoff O'Brien*
*Phil O'Keefe*
*Nicola Pearsall*

# List of Acronyms and Abbreviations

| | |
|---|---|
| AAU | assigned amount unit |
| ABWR | advanced boiling water reactor |
| AC | alternating current |
| ACE | Association for the Conservation of Energy |
| ACEA | EU Automobile Manufacturers Association |
| AGR | advanced gas cooled reactor |
| ANWR | Arctic National Wildlife Refuge |
| ATM | air-traffic management |
| BE | binding energy |
| BIPV | building integrated photovoltaics |
| BNFL | British Nuclear Fuels Ltd |
| Bq | becquerel |
| BTL | biomass to liquid |
| Btu | British thermal unit |
| BWB | blended wing-body |
| BWR | boiling water reactor |
| CBA | cost–benefit analysis |
| CCGT | combined cycle gas turbine |
| CCS | carbon capture and storage |
| CDCF | Community Development Carbon Fund (World Bank) |
| CDM | clean development mechanism |
| CdTe | cadmium telluride |
| CEN | European Committee for Standardisation |
| CER | certified emission units |
| CERs | carbon emission reductions |
| CHP | combined heat and power |
| CIGS | copper indium gallium diselenide |
| CIP | Competiveness and Innovation Programme (EU) |
| COM | Communication from the Commission (EU) |
| COP | Coefficient of Performance |
| COP | Conference of Parties |
| CSLF | Carbon Sequestration Leadership Forum |
| CSP | concentrating solar power |
| CTL | coal to liquid |
| GTL | gas to liquid |
| CUM | cubic metres |
| DC | direct current |
| DCF | Discounted Cash Flow |

| | |
|---|---|
| DLR | *Deutsches Zentrum für Luft- und Raumfahrt* (German Aerospace Centre) |
| DOE | Department of Energy (US) |
| (D-T) | deuterium/tritium mixture |
| EASE | Enabling Access to Sustainable Energy |
| ECBM | Enhanced Coal Bed Methane (recovery) |
| EEW | Energy Efficiency Watch |
| EIA (US) | Energy Information Agency (US Department of Energy) |
| EIA | environmental impact assessment |
| EIT | economies in transition |
| ENS | European Nuclear Society |
| EOR | enhanced oil recovery |
| EPBD | Energy Performance of Buildings Directive (EU) |
| EPC | Energy Performance Certificate |
| ERU | emission reduction unit |
| ESCOs | energy service companies |
| ESD | Energy for Sustainable Development |
| EST | Energy Savings Trust |
| ETS | Emissions Trading Scheme |
| ETSAP | Energy Technology Systems Analysis Programme |
| EU | European Union |
| EuP | energy-using products |
| EPR | European Pressurized Water reactor |
| EUR | European Utility Requirements |
| EW | exempt waste |
| FAME | fatty acid methyl ester |
| FAO | Food and Agriculture Organization (UN) |
| FBR | Fast Breeder Reactors |
| FIT | feed-in tariff system |
| GDP | gross domestic product |
| GEF | global environmental facility |
| GHGs | greenhouse gases |
| GIF | Generation IV International Forum |
| GNP | gross national product |
| GPS | global positioning satellites |
| GTI | Great Transition Initiative |
| GTL | gas to liquid |
| Gwe | Gigawatts |
| GWP | Global Warming Potential |
| HFCs | hydroflourocarbons |
| HHI | Hirschmann–Herfindahl Index |
| HLW | high level waste |
| HVAC | heating, ventilation and air conditioning |
| IAEA | International Atomic Energy Agency |
| ICE | internal combustion engine |
| ICF | inertial confinement fusion |
| ICT | Information and Communications Technology |
| IEA | International Energy Agency |
| IEE | Intelligent Energy Europe (EU) |

| | |
|---|---|
| ILW | intermediate level waste |
| IPCC | Inter-governmental Panel on Climate Change |
| ISO | International Organisation for Standardisation |
| ITER | International Thermonuclear Experimental Reactor |
| ITF | International Transport Forum |
| JET | Joint European Torus |
| JI | Joint Implementation |
| KWDP | Kenyan Woodfuel Development Programme |
| LED | light emitting diode |
| LILW | low and intermediate level waste |
| LILW-LL | long lived waste |
| LILW-SL | short lived waste |
| LLCC | least lifecycle cost |
| LMFBR | Liquid Metal Fast Breeder |
| LNG | Liquified Natural Gas |
| LNT | linear, no-threshold |
| MDGs | Millennium Development Goals |
| MEPS | Minimum Energy Performance Standards |
| MFE | magnetic confinement fusion |
| MOX | mixed oxide |
| $MtCO_2$ | million tonnes of $CO_2$ |
| mtoe | million tonnes oil equivalent |
| MW | megawatt |
| MWh | megawatt hour |
| NAPAs | National Adaptation Programmes of Action |
| NEI | Nuclear Energy Institute |
| NGO | non-governmental organization |
| NPPs | nuclear power plants |
| NPT | Nuclear Nonproliferation Treaty |
| NPV | net present value |
| NRC | Nuclear Regulatory Commission (US) |
| NREL | National Renewable Energy Laboratory (US) |
| ODS | Ozone Depleting Substances |
| OEs | Operational Entities |
| OECD | Organisation for Economic Co-operation and Development |
| OGZEB | Off-Grid ZEB |
| OJ | Official Journal of the European Communities |
| OPEC | Organisation of Petroleum Exporting Countries |
| OTEC | ocean thermal energy conversion |
| OWC | Oscillating Water Column |
| PDHU | Pimlico District Heating Undertaking |
| PEM | polymer electrolyte membrane |
| PFC | perfluorocarbons |
| PHEV | Plug-In Hybrid Electric Vehicle |
| PJ | pico joules |
| PMBR | pebble bed modular reactor |
| ppm | parts per million |
| PV | photovoltaic |

| | |
|---|---|
| PWR | pressurized water reactor |
| RAR | 'Reasonably Assured Resources' (NEA) |
| RBMK | Reactor Bolshoy Moshchnosty Kanalny |
| RCEP | Royal Commission on Environmental Pollution |
| RD&D | Research, Development and Deployment |
| RES-E | renewable energy source for electricity |
| RMU | removal unit |
| SAP | Standard Assessment Procedure |
| SCORE | Stove for Cooking, Refrigeration and Electricity |
| SD | Sustainable Development |
| SES | Single European Sky |
| Sherpa | Sustainable Heat and Energy Research for Heat Pump Application |
| SIDS | Small Island Developing States |
| SRES | Special Report on Emissions Scenarios |
| SRTP | social rate of time preference |
| STPR | social time preference rate |
| Sv | sievert |
| SWI | Shannon–Wiener Index |
| t HM | tonnes of heavy metal |
| TPES | total primary energy supplies |
| TORCH | The Other Report on Chernobyl |
| UK | United Kingdom |
| UKCIP | UK Climate Impacts Programme |
| UNCED | United Nations Conference on Environment and Development |
| UNCSD | United Nations Commission on Sustainable Development |
| UNEP | United Nations Environment Programme |
| UNFCCC | United Nations Framework Convention on Climate Change |
| UN/ISDR | United Nations International Strategy for Disaster Reduction |
| USGS | US Geological Survey |
| USNRC | United States Nuclear Regulatory Commission |
| VSD | variable speed drive |
| WACC | Weighted Average Cost of Capital |
| WBCSD | World Business Council for Sustainable Development |
| WEC | World Energy Council |
| WIGs | wing-in-ground effect vehicles |
| WISE | World Information Service on Energy (Paris) |
| WNA | World Nuclear Association |
| WSSD | World Summit on Sustainable Development |
| WWF | World Wide Fund for Nature |
| ZEB | zero energy building |

# 1

# The Changing Energy Landscape

## Introduction

Why another book on energy, particularly from academics at Northumbria University, when past outputs have been, generally, very favourably received? The short answer is that the energy future for which we need to plan will have to be undertaken in an exceptionally different political economy of the 21st century. It may seem a contradictory starting point but, in general, the developed world and, as a consequence, the developing world, does not have a robust energy policy framework. Typically energy policy conclusions are drawn from environmental policy, such as the focus on renewables, yet environmental policy itself is not largely drawn from environmental considerations but from the requirements of market competition policy. In short, environmental policies, and the regulations and standards that follow from them are commonly drawn up to minimize unfair competition rather than address environmental concerns. Quite simply the market, however imperfect or distorted, rules.

Energy itself is fundamental to social development. Simultaneously, energy is central to one of the greatest environmental challenges humanity faces: climate change. Solutions to climate change will have a significant and lasting impact on the future of energy use. But climate change is not the only problem that the energy system faces. There are growing concerns about energy security, with real fears that geopolitical disruption could place many of the world's economies in jeopardy. On top of this is the problem of the longevity of existing fossil fuel supplies. While coal appears to be plentiful, oil and gas have increasingly limited lifetimes. In 2008, there were rapid fluctuations in the price of energy supplies, with oil peaking near to US$150 a barrel at one point and falling to less than US$50 towards the end of 2008. The knock-on effects of this were increases in the cost of basic food stuffs as well as manufactured products, showing that energy is a fundamental component of our lifestyles. Some believe these price fluctuations signal that we have reached what is termed 'peak oil'; where more energy oil reserves are used than are produced. Trying to think through energy futures with the pressing problems of climate change and issues of energy resource and energy price security is particularly difficult without an energy policy framework, since the market itself gives very mixed messages.

There are lessons to be learned from the recent oil price instabilities: the International Energy Agency (IEA) in its *World Energy Outlook* for 2008 says prices could rise to as high as US$200 a barrel by 2030. The era of cheap oil is over. However, as the IEA points out, the reasons for rising prices lie not in the shortage of energy resources but a lack of investment in energy infrastructure. The IEA argues that such investment in infrastructure (for example, exploration and refinery capacity) will amount to approximately US$26 trillion by 2030. The paradox is that if we continue to rely upon and invest in fossil fuels

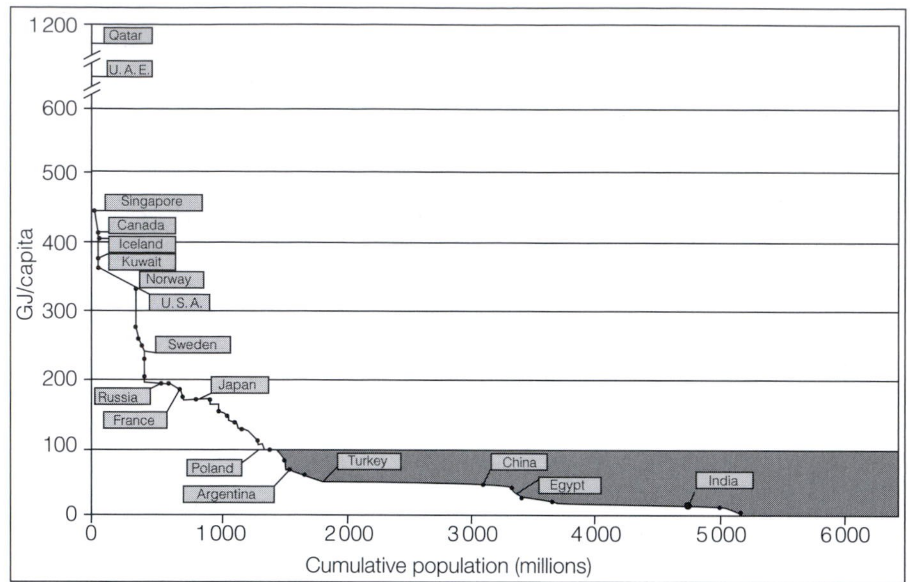


*Note:* Area between dashed line and data points is 500EJ/year and represents everyone below Poland today achieving this same energy usage of 100EJ per capita.

*Source:* World Energy Council (WEC), 2007

**Figure 1.1** Energy per capita as a function of cumulative population

without investing in measures to address the carbon problem then the climate consequences could be severe (IEA, 2008).

Renewable technologies are the obvious answer to the carbon problem but they only provide a fraction of the world's energy capacity. Some 80 per cent of global energy is produced from fossil fuels, while some 13 per cent is produced by traditional biomass, with the rest being produced by a mixture of nuclear, hydropower and other renewable technologies.

In summary, the major influences on the future of energy use will be mitigating climate change using low-carbon and renewable energy sources, improving energy security and ensuring a smooth transition into new forms of energy technology use. Each on its own will be difficult and, collectively, represent a major challenge to the global community. In this book we explore the issues around these topics as well as look at the current trends in the energy system and some of the proposals that have been made to deal with the carbon problem. We also touch briefly on global climate agreements as these will have a profound effect on the future of energy use.

## Energy and gross domestic product

There has always been a close link between social development and energy use. In brief, this has meant that as economies have grown there has been a corresponding growth in energy demand. This close coupling between economic growth and energy demand has been a feature of societies up until recently, where, for instance, focused efforts in Japan to decouple energy use and economic growth have become increasingly prominent. Energy intensity is something which governments are looking at quite seriously as one of the tools to be used in reducing greenhouse gas emissions. However, the picture is not quite as simple as that. Energy demand is heavily influenced by prevailing climate conditions, lifestyles and income, as can be seen in Figure 1.1. Two striking observations about the graph are the high consumption in oil-rich countries and the relatively low consumption in India and China, the most rapidly industrializing countries.

Many of those countries above 100EJ per capita are what are commonly termed as developed world countries. Typically it is this collective of countries that are members of the Organisation for Economic Co-operation and Development (OECD); 30 in total, which make up the developed world. It is important to note, however, that other countries, such as Russia and Kuwait, for example, though not members of the OECD, still have high energy use per capita, mainly due to prevailing climatic conditions. Kuwait, for instance, has a typically constant hot, dry climate, whilst Russia on average has a particularly cold climate.

In the near future other factors will significantly shape the demand for energy. These include a growing world population (estimated to rise by at least another one billion by 2025 and perhaps, eventually stabilizing at 9 billion people), and the rapidly rising percentage of the world's population moving from rural to urban areas. Many rural to urban migrants, particularly in Asia, quickly emulate Western lifestyles as incomes increase (greater per capita income), in turn, expanding the number of those joining the middle classes. This scenario is almost certainly true when considering the future despite recent global economic decline. Together, these demographic changes are likely to increase demand for scarce energy resources. This will run parallel to emerging constraints on new production partially through the control exercised by state-run companies and partially by climate concerns. This implies that the future of energy is unlikely to be determined by market forces. Climate concerns, energy security and technological developments are the most likely determinants of the future of energy.

Studies focusing on the likely mix of energy resources suggest that for some years to come the supply side will be dominated by fossil fuels. The IEA (2008) predict that there are sufficient oil and gas reserves to meet current and projected demand for the next 40 years. Coal reserves are sufficient for the next few hundred years. The issue is not lack of reserves but whether or not ways can be found to use these supplies in ways that do not jeopardize future generations.

## Global energy resources

There is considerable debate about global energy resources and their longevity. Coupled with this debate is the ongoing and increasing apprehension that if we do not find clean ways of using fossil fuels or renewable technologies, severe impacts upon the climate system will become more frequent and increasingly difficult to overcome. For many OECD nations, the paramount concern lies with security of supply as many of the existing or proven reserves are in areas regarded as geopolitically unstable, which could in turn jeopardize supplies. Figures 1.2, 1.3 and 1.4 show proven oil reserves, consumption and production for 2007 (BP, 2008). The Middle East is currently the most geopolitically unstable region of the world, in part because of wars generated by developed countries.

From these figures it is relatively straightforward to calculate the expected longevity of current reserves at the daily consumption rate for 2007. This gives a value of:

> (1208.2 thousand million barrels/83,719 thousand barrels per day)/365 = 38 years.

This figure will change if, as expected, demand for oil increases, particularly in the industrializing countries of India and China. However, there are still considerable reserves of non-conventional oil such as tar-sand, shale and heavy oils that could be exploited to meet demand. These are discussed in more detail in Chapter 5.

The case for natural gas is even more promising in terms of resource longevity. Figures 1.5, 1.6 and 1.7 show proven gas reserves, consumption and production for 2007 (BP, 2008).

The longevity of proven natural gas supplies can be calculated from the reserves and consumption figures (expressed in cubic metres) as:

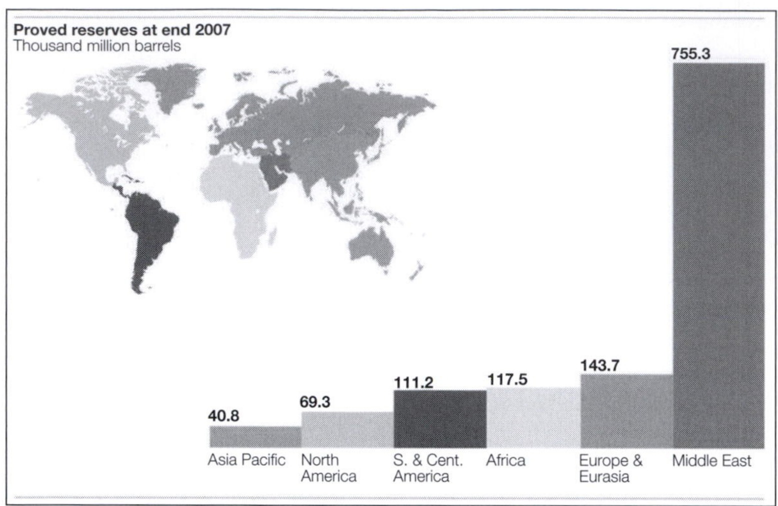


*Source:* BP, 2008

**Figure 1.2** Proven oil reserves by area

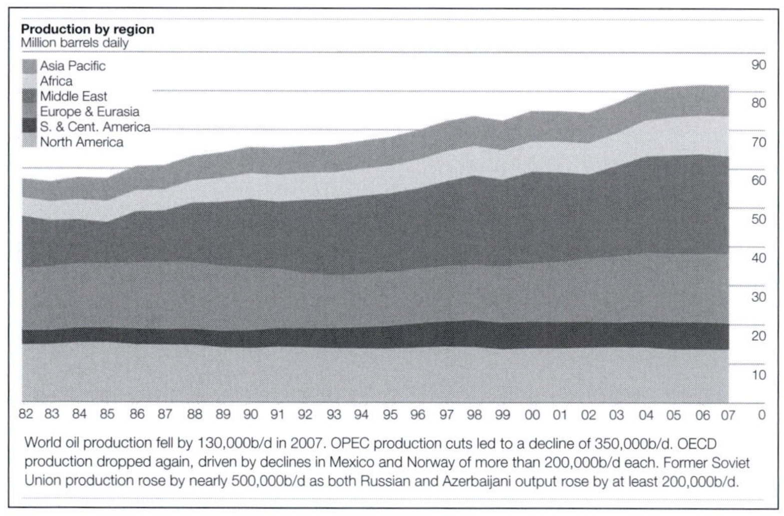


*Source:* BP, 2008

**Figure 1.3** Oil production by region

(181.6 trillion/2.8 billion per day)/365 = 175 years.

Again, this value will change as consumption rates as expected, increase. In contrast to oil there are no other types of gas resource that can be exploited through conventional exploration techniques. Although other methods of producing gas are available, such as the gasification of coal seams. This is discussed in more detail in Chapter 5.

Resources such as coal, (see Figure 1.8) are relatively plentiful, for which the lifetime of current reserves is thought to be in excess of 150 years (BP, 2008).

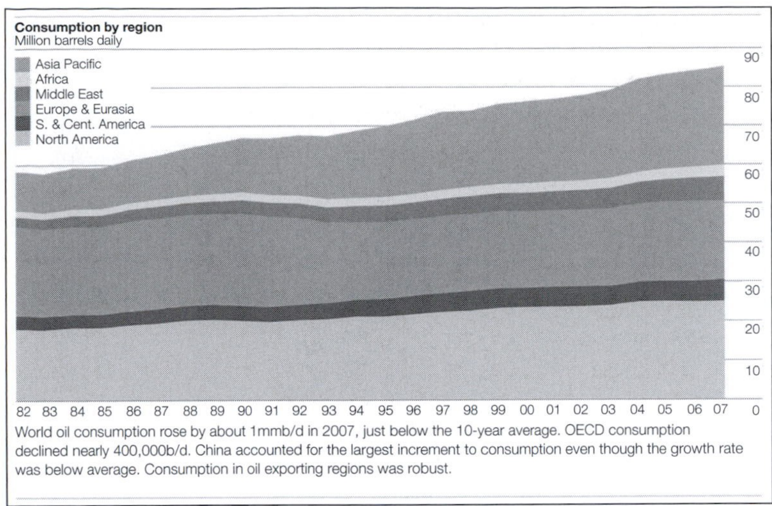


*Source:* BP, 2008

**Figure 1.4** Oil consumption by region

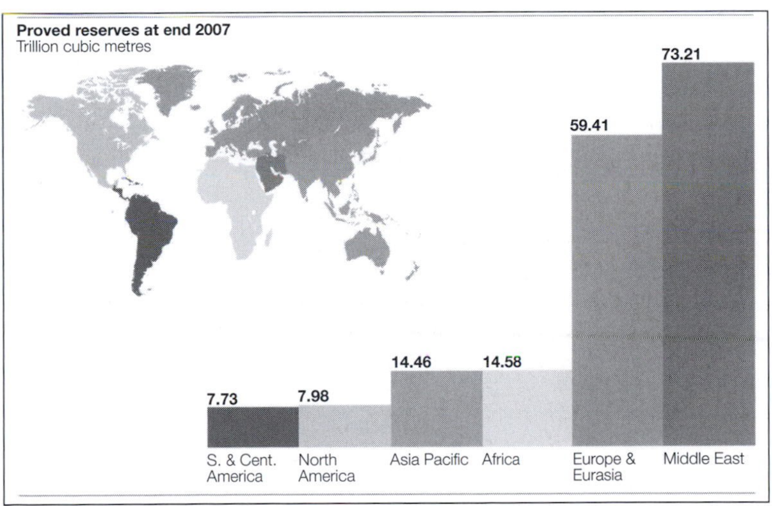


*Source:* BP, 2008

**Figure 1.5** Natural gas reserves by area

In terms of fossil fuels it can be argued that sufficient resources are available to meet current demand. If demand continues to accelerate at the current rate, then it is likely that a resource constraint in terms of availability may become a contentious issue. This is illustrated best by inspecting resource and consumption maps, which clearly show that the distribution of resources does not match consumption locations. This is likely to be increasingly problematic in terms of energy security.

Other forms of energy supply that produce a significant amount of power are nuclear and hydroelectric power, as shown in Figures 1.9

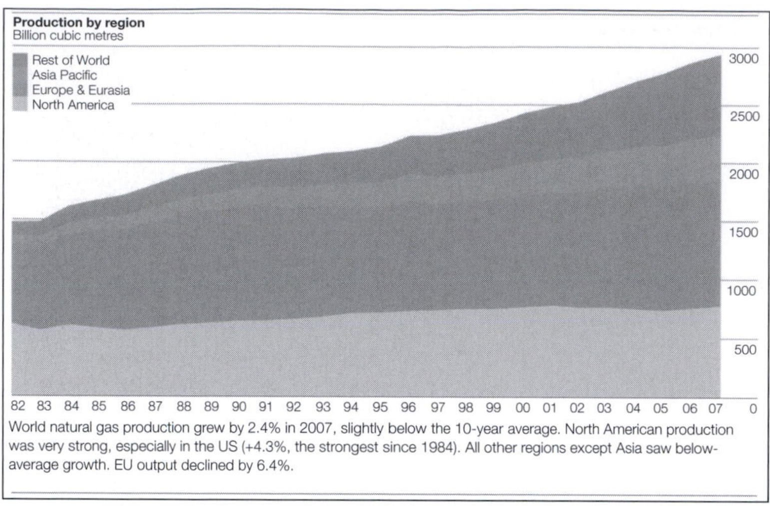


*Source:* BP, 2008

**Figure 1.6** Natural gas production by region

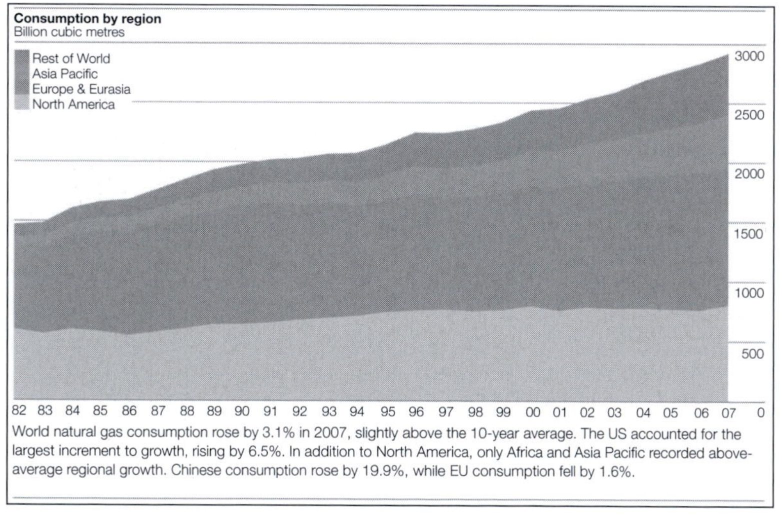


*Source:* BP, 2008

**Figure 1.7** Natural gas consumption by region

and 1.10 (BP, 2008). Figure 1.11 (again from BP (2008)) shows world energy consumption by fuel type. What is clear from all the figures shown is that demand for energy has increased over given reporting periods. It is apparent that demand for energy is predicted to increase. The speed at which demand will grow is uncertain and will depend upon a range of factors. Determining future energy demand is difficult and in many ways is more of an art than a science. Ways in which energy futures, using examples from developing countries, are predicted are discussed in more detail in Chapter 3.

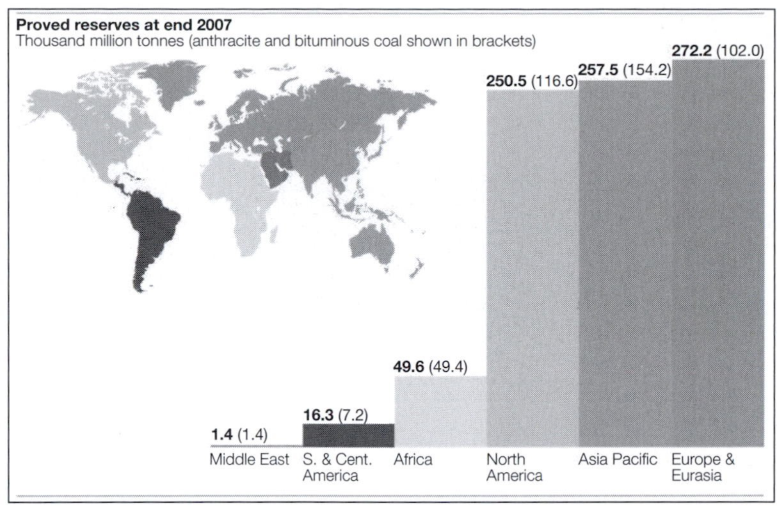


*Source:* BP, 2008

**Figure 1.8** Proven coal reserves by area

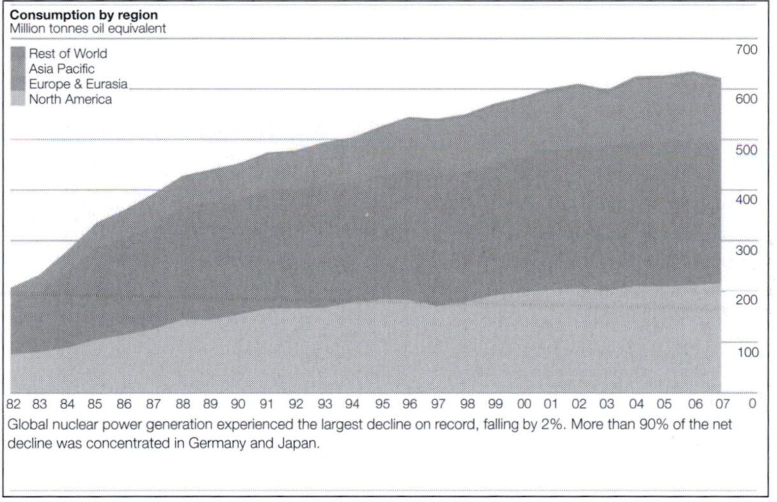


*Source:* BNP, 2008

**Figure 1.9** Nuclear power by area

## Energy futures

Although there are a number of different energy predictions as discussed in Chapter 2, Figures 1.12 to 1.16 give a sense of what many believe is likely to happen in the future. In general, predictions are determined on no interventions other than those in a place and of a time at which the prediction was made. In this sense, projections represent what may happen based upon existing agreements and actions and/or non-actions. For

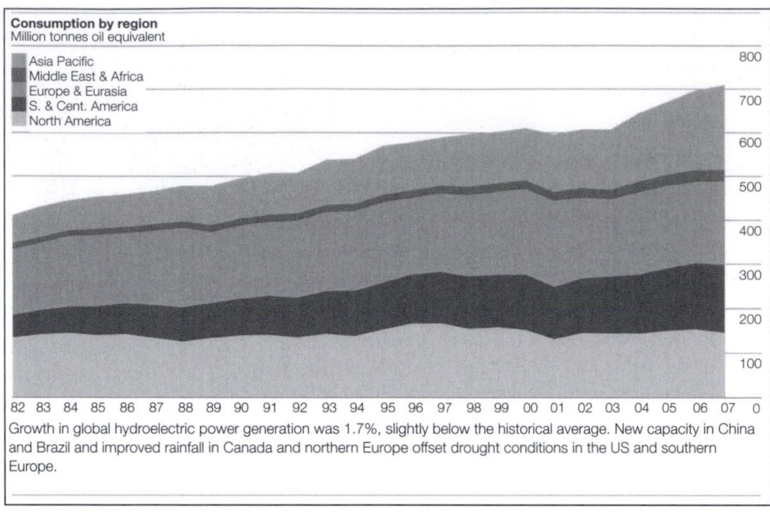


*Source:* BP, 2008

**Figure 1.10** Hydroelectric power by area

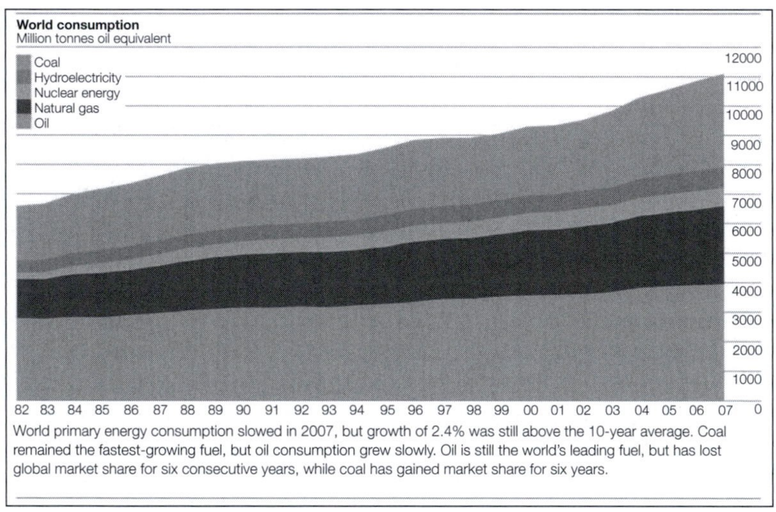


*Note:* For further information relating to energy resources, definitions, and conversion factors see appendices.

*Source:* BP, 2008

**Figure 1.11** World energy consumption by type

instance, one major inaction can be said to be the relatively weak international effort to meaningfully address or reverse climate change. In addition, in order to appeal to a wider audience, especially market forces, projections are commonly packaged in a Business-as-Usual or Baseline case narrative. Figure 1.12 shows the prediction by the Energy Information Agency (EIA (US), part of the US Department of Energy) for use by fuel types up to 2030.

Of note in this prediction is the relatively high growth in the use of coal as a primary form

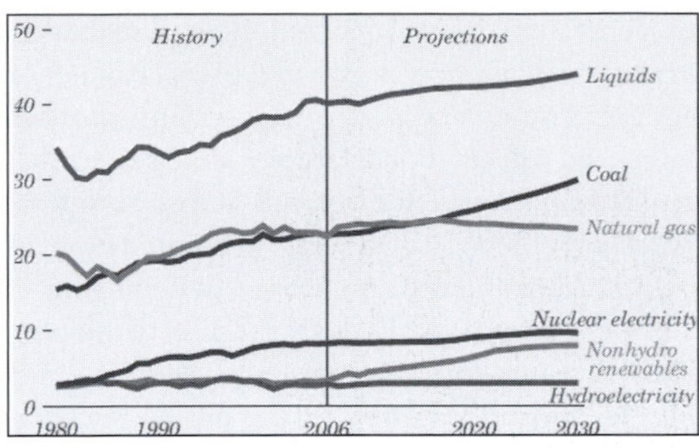


*Source:* EIA (US)/DOE, 2008

**Figure 1.12** Energy consumption by fuel type (quadrillion Btus)

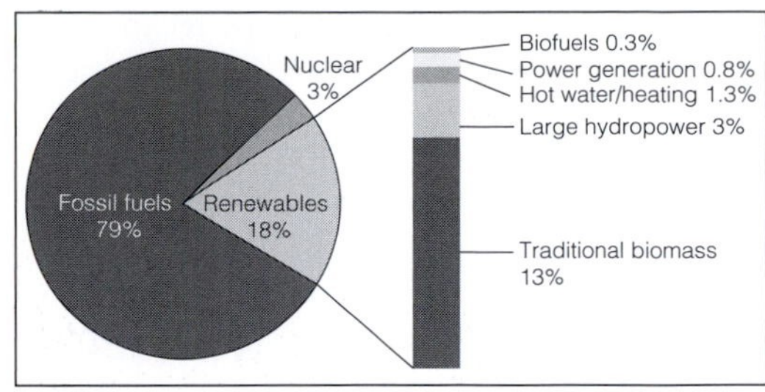


*Source:* Martinot, 2008

**Figure 1.13** Global share of energy production by source type

of energy supply. Figure 1.8 shows that the Asia Pacific area has the highest levels of coal reserves and industrializing nations such as China and India are likely to use indigenous resources to drive economic growth. Other principal growth areas are likely to lie within the realms of liquid fuels (oil and related products and liquefied natural gas) and renewables (excluding hydroelectricity where most of the major opportunities for large hydroelectric power generation have been exploited). It is likely that reliance on nuclear power will not change significantly. This is despite renewed interest in many countries in developing nuclear capacity. However, much of this will replace existing nuclear stock, which is approaching the end of its service life. It seems quite probable that growth of conventional fuels may outstrip the growth of renewable capacity. In reality modern renewable capacity such as photovoltaics (PVs) and wind power account for only a small fraction of overall energy use, as shown in Figure 1.13.

Although renewables account for 18 per cent of global energy supplies, the bulk of this consists of traditional biomass. This type of fuel is made up from sources such as wood-fuel, charcoal, animal dung and crop residues that are typically used in poorer countries. A significant number of people depend on these sources to provide energy for cooking, heating and lighting. Chapter 3 discusses traditional biomass fuels in greater detail. In total, low carbon sources, excluding traditional biomass but including nuclear, account for less that 8 per cent of the global energy resource.

Slow progress on the development and implementation of renewable capacity and the uneven distribution of energy supplies threatens the stability of the global atmospheric commons and the economic stability of those countries that rely heavily on imported supplies. The need for energy policy makers to develop and implement approaches that both reconcile increasing demand for energy and growing climate concerns represents an extremely challenging backdrop for the future of energy. The essential elements of energy policy in this vein will need to be shaped to address the necessity of attaining the international goals of sustainable development, climate change and the MDGs (Millennium Development Goals). This is a challenge on a number of fronts: technological, social, political and economic. In the developed world the emphasis is on technical solutions. While there is logic to this approach in that the developed world is both technology dependent and innovative, one of the problems with energy is the timescales and cost involved. Of the US$26 trillion of investment needed by 2030, half of that is required to replace existing infrastructure (IEA, 2008). With the credit crunch and the financial problems of 2008 facing the banking sector, it is uncertain whether sufficient levels of funding will be available for refurbishment and for the development of new and more efficient technologies. There is a danger that investment in new technologies may well be delayed. Box 1.1 highlights some of the problems associated with bringing about long-term change in energy technologies.

## Vulnerability of energy systems

Vulnerability of an energy system has typically been expressed in terms of its susceptibility to accidents through technical failure or operator errors. However, it is increasingly recognized that a range of threats and hazards can adversely affect a given energy system. Energy systems (defined here as the resource base, transformation technologies and delivery infrastructure that provides end-user services) are vulnerable in a number of ways; including system complexity, instrumental disruptions, hazards and geopolitical disruptions. Today, policy makers are shifting their focus to promoting energy security. Energy security is defined as an uninterruptible supply of energy, in terms of quantities required to meet demand at affordable prices (WEC, 2008) on a 24/7 basis. Promoting energy security requires a much broader approach to the range of vulnerabilities inherent in large scale interconnected systems.

Energy systems are progressively more complex and interconnected. Complexity, or tight coupling, in technological systems generates a number of vulnerabilities. No matter how effective conventional safety devices are in technological systems, accidents are inevitable, with catastrophic potential. Examples of systems that have catastrophic potential include nuclear power plants and weapons systems, recombinant DNA production and ships carrying highly toxic or explosive cargoes. Tight coupling can result in errors, either in design or operation and potentially leads to accidents (Perrow, 1999). Lovins and Lovins (1982) explain the vulnerability of electrical power systems as the 'unintended side effect of the nature and organization of highly centralized technologies' (Lovins and Lovins, 1982, p2).

Interconnected electrical systems are complex spatially dispersed systems reliant on a series of generators interconnected to a distribution grid to deliver power to end users. Failure or compromise of one part of the system can have knock-on effects. For example, an incident that resulted in a series of blackouts across Europe in November 2006 was due to a combination of cold weather and the concurrent switching off of a power cable to allow the passage of a shipping vessel across a riverway, which caused a sudden hike in demand and stress on the overall electrical network. In turn, this led to parts of the system temporarily failing, leading to many blackouts over much of the energy system. This combination led to the system collapsing like 'a house of cards' (Willsher and McMahon, 2006). The cascading effect whereupon a system shuts down is a feature of interconnected systems. Here, an incident triggers a series of outages across the network that ultimately can lead to collapse. Contributing factors that influence the rate and frequency of blackouts include the steady increase in electric loading and economic pressures to maximize grid usage that, in turn, adds a number of stresses to the system (Dobson et al, 2007).

Deregulation, driven by competition for lower prices and efficiencies, requires complex interaction between an increasing number of agents in different markets (for instance, energy, capacity, ancillary services and transmission rights) and in multiple timeframes (futures, day ahead and real time). An electrical power system is a single entity and requires real time coordination. This is even more problematic when markets operate across national borders where systems and operating procedures may differ (Watts, 2003). Close coordination between different systems operators is needed to maintain network integrity at times of high demand or system disruption. A significant blackout that took place in Italy in 2003 has been attributed to a lack of coordination between national operators following the loss of interconnection capacity in Switzerland. Blackouts are likely to become more frequent as electrical systems are exposed to increasing numbers of severe weather events driven by climate change. The fragile nature of electrical supply infrastructure such as pylons and switching stations will always be a physical risk. The IPCC predicts increasingly

## Box 1.1 Timing is everything

Current technologies cannot replace traditional energy architectures on the scale needed.

New energy technologies will probably not be commercially viable and widespread by 2025. The current generation of biofuels is too expensive, threatens food prices and releases carbon. Other biomass or chemical resources may be more promising, such as those based on high-growth algae or agricultural waste products, especially cellulosic biomass, but remain in their infancy in terms of viability.

The development of clean coal technologies and carbon capture and storage are gaining momentum and, if cost-competitive by 2025, would enable coal to generate more electricity in a carbon-constrained regulatory environment. But the size of the carbon capture and storage problem in a 'Clean Coal' scenario is beyond current technology.

Long-lasting hydrogen fuel cells have potential, but they are at least a decade away from commercial production. Enormous infrastructure investment is needed to support a 'hydrogen economy'.

The adoption lag is real – it takes about 25 years in the energy sector for a new production technology to become widely adopted, mainly because of the need for new infrastructure to handle major innovation. For energy, in particular, massive and sustained infrastructure investments made for almost 150 years encompass production, transportation, refining, marketing and retail activities.

Gas is attractive. Despite its widespread availability since the 1970s, gas continues to lag behind oil. An example of this contradiction can be seen in the transport sector. Due to higher technical and investment requirements for producing and transporting gas, oil-based fuels remain dominant.

Simply meeting baseline energy demand over the next two decades is estimated to require more than $3 trillion of investment in traditional hydrocarbons.

A transition to a new energy system by 2025 should not be ruled out if improved renewable generation sources (photovoltaic and wind) and improvements in battery technology are increasingly viable.

Decentralized and autonomous approaches would mean lower infrastructure costs such as stationary fuel cells powering homes and offices, recharging plug-in hybrid vehicles and selling energy back to the grid.

Energy conversion schemes (hydrogen generation for automotive fuel cells from household electricity) could avoid the need to develop a complex hydrogen transportation infrastructure.

*Source:* Adapted from National Intelligence Council, 2008

frequent and severe weather events that as a direct causal effect will undoubtedly have a detrimental impact on energy infrastructures.

The 1998 ice storms in Quebec damaged 350 power lines and some 16,000 structures collapsed (Figure 1.14). Due to an unforeseen and expectedly deadly heat wave in Europe in 2003, a number of nuclear reactors in France were shut down due to a lack of cooling water as river flows dropped (United Nations Environment Programme (UNEP), 2004). The UK floods of 2007 highlighted with great clarity the vulnerability of physical infrastructure. For example, emergency action was urgently required in order to prevent an electrical sub-station from being flooded that would have left 500,000 people without power.

The vulnerability of an energy system to the physical disruption to imported supplies is strongly linked to its dependence on those imported supplies. Each energy source to some extent supplies a captive market (for example, oil in the transport sector) and uses different logistic

*Source:* Cohen, 2003

**Figure 1.14** Damaged high voltage towers in Quebec

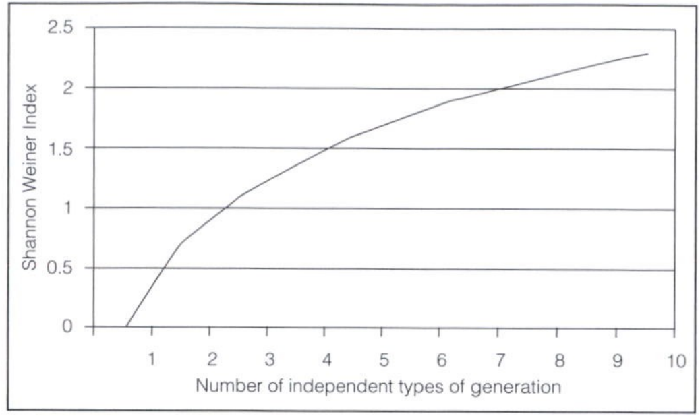


*Source:* Grubb et al, 2006, p4052

**Figure 1.15** Shannon–Wiener Index for generation types

systems for delivery. Oil has the highest energy vulnerability in Europe as it is significantly reliant on imports, a substantial volume of which come from regions considered to have a high geopolitical risk.

There are two ways of assessing vulnerability to imports. The first measures the reliance on the sources of supply, for instance, relying upon a few places to supply primary energy. The second looks at the vulnerability of production, for example over reliance on a technology type that may use one source of fuel such as gas. The first assessment uses the Hirschmann–Herfindahl Index (HHI), which is an indicator of energy import dependence. HHI is the sum of the squared market shares held by various suppliers. In short, it measures the concentration of supplies from particular places. If, for example, imported supplies were sourced from a few places such as Saudi Arabia or the Gulf States, then this would indicate a high supplier concentration leading to increased vulnerability. Scores produced by the index determine the level of vulnerability. Scores between 8000 and 10, 000 indicate high vulnerability. Scores of less than 1600 indicate a diversified supply of sources, which should mean reduced vulnerability.

Another method for assessing vulnerability is by evaluating the diversification of supply using the Shannon–Wiener Index (SWI). This index calculates the mean square of the proportions of the total energy supplied by different sources and gives a measure of the diversification of the energy mix of a given sector. The minimum value of the index would be zero, which indicates reliance on a single source. This would imply a high level of vulnerability. As shown in Figure 1.15 the index can also be used to assess the vulnerability of generation capacity.

The higher the value of the index the less vulnerable the system would be to a disruption to a single component of the generation mix (Grubb et al, 2006). In general it is reasonable to assume that a high level of dependency on imports from a small number of suppliers and over reliance on a small number of generator types is likely to amplify the vulnerability of a given system. The more diverse the supply base and type of production capacity (coal, nuclear, gas, renewable energies, etc.), the lower the vulnerability.

Vulnerability is multi-dimensional and a study conducted by the World Energy Council into the vulnerability of the European energy system suggests a range of factors that influence vulnerability (WEC, 2008).

At the macro-economic level:

- Energy dependence/energy independence determined by the HHI index.
- Energy intensity – a measure of the industrial structure that, for example, reflects the number of energy intensive industries in the economy.

This has steadily declined in Europe with the decrease in traditional industries such as steel making and the increase in value-added and knowledge-based industries.

- Net energy import bill – a rise in energy costs has a detrimental economic impact. This has been a real problem in 2008 when oil and gas prices fluctuated widely. Price volatility can severely affect revenue as more is generally spent on energy leading to inflation and interest rate increases (although this has been offset by the global financial slowdown in 2008), and greater reliance on trade as the need for energy imports rise.
- Carbon content of total primary energy supplies (TPES) – increasing expenditure on imported fossil fuels could impact the development of renewable capacity. This could hamper progress in meeting the Kyoto targets. Rising concerns about global climate change will make greenhouse gases, and particularly $CO_2$ emissions become more and more costly.
- Currency exchange rate – a fluctuating currency could exacerbate the energy import bill.

At the micro-economic level:

- For the consumer, vulnerability is characterized by the risk of supply disruption and associated price increases. A recent example of a physical supply disruption event occurred during the summer of 2005 in the US caused by hurricanes Rita and Katrina, which not only destroyed oil and gas production rigs in the Gulf of Mexico, but also damaged several refineries. The European Union (EU) has a strategic oil reserve of 90 days but there is no similar facility for natural gas.
- Electricity is more problematic as there is no means of storage. Electrical systems are vulnerable to massive interruptions as outlined previously and, because of the way in which systems have developed and the costs of long distance transmission, less than 10 per cent of electricity is traded across borders compared with some 60 per cent of natural gas. Additionally economic pressures to maximize grid-use can add additional stress to grid infrastructure. Although these factors are important at the macro-economic level, there is a need to consider mechanisms for stabilizing prices in order to protect certain areas of the industrial sector from extreme price fluctuations.

At the technological level:

- Development of an integrated and well-functioning electricity market in Europe (an aim of EU policy) will necessitate sufficient generation to meet demand, an adequate infrastructure to deliver the power and robust technical and administrative operational procedures. The EU Commission has set out plans for greater connectivity between EU member states (Communication from the Commission, 2007). When these will be realized is unknown. In the interim, it is predicted that between 2006 and 2030 electricity consumption will grow at 1.5 per cent per year. However much of the existing generator capacity is ageing (Figure 1.16). To meet the expected growth in demand, additional capacity of 265GW installed by 2030 giving an overall capacity of 843GW will be required. This will entail considerable investment. Stricter environmental rules, political decisions, higher costs due to the ETS or falling profits due to lower efficiencies could mean that investment needs may well be double that of adding additional capacity with some 520GW of installed capacity required by 2030.

At the social level:

- Fuel poverty, generally defined as being when more than 10 per cent of household income is spent on fuel, is an issue that can affect many households. The Fuel Poverty Concept is an interaction between poorly insulated housing and inefficient in-house energy systems, low-income households and high-energy service prices. Although there are a number of schemes designed to help those in fuel poverty, the costs to government and the distress

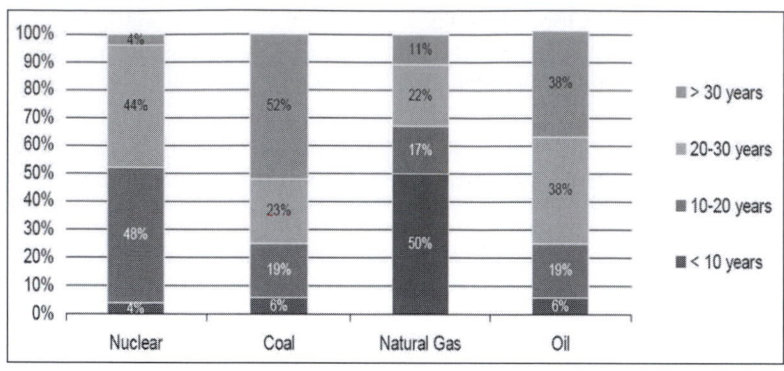


*Source:* WEC, 2008, p60

**Figure 1.16** Age structure of installed capacity in the EU

felt by those dependent on welfare schemes are important factors for public policy.

At the geopolitical level:

- Disruptions to the supply of primary fuels due to geopolitical events or terrorist attacks will exacerbate vulnerability. *Globally, some 60 per cent of energy supplies are transported by ship, vulnerable to both extreme weather events, accidents and increasing piracy. By 2030, some 70 per cent of EU energy supplies will be imported, often from places regarded as politically unstable. Of equal concern is disruption from potential terrorist attacks. Small, highly motivated groups could cause severe damage to port facilities and further disruption by attacking critical chokepoints such as the Straits of Hormuz and the Suez Canal. Disruptions of this kind would have a catastrophic impact on energy markets* (Kröger, 2006; Homer-Dixon, 2002). Land-based supply infrastructure is equally vulnerable to terrorist attacks.

In response to concerns about the vulnerability of the energy system and energy security, the EU Strategic Energy Review highlights the growing dependency on gas imports from Russia. The EU currently imports 61 per cent of its gas consumption with some 42 per cent of those imports from Russia. By 2020 gas imports are expected to grow to 73 per cent of consumption. The EU plans to reduce this level of vulnerability by diversifying its supply base through the construction of two major gas pipeline projects – Nabucco and South Stream – to deliver gas to southern Europe from central Asia and Russia, respectively (Figure 1.17).

The Strategic Energy Review also highlights the need to strengthen and diversify the electricity grid. It calls for a North Sea offshore grid to be developed that would link up national electricity grids in northwestern Europe and plug in numerous planned offshore wind farms. This, along with the Mediterranean Ring and the Baltic Interconnection Project, would form the building blocks of a European 'supergrid'.

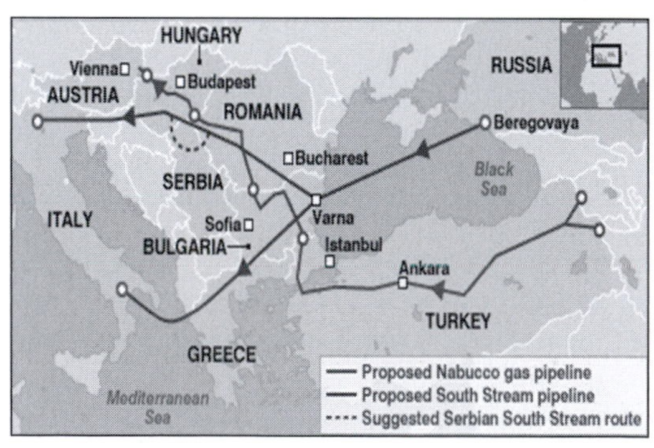


*Source:* BBC, 2008

**Figure 1.17** Gas pipelines

A Mediterranean energy ring – interconnecting electricity and gas networks is viewed as essential in developing the region's vast solar and wind energy potential. The review highlights:

- infrastructure needs and diversifying energy supplies;
- external energy relations to secure access to supplies;
- oil and gas stocks and crisis response mechanisms;
- energy efficiency;
- making best use of the EU's indigenous energy resources.

*Source:* COM, 2008

## Global climate policy

As mentioned earlier, climate policy is one of the major determinants of energy policy both now and for the future. Climate change has been described as one of the greatest threats to humankind. Climate change is a greater threat than global terrorism according to Sir David King. However, there are immediate concerns related to climate variability that sees an increasing frequency in the occurrence of extreme weather events. For example, the European heat wave in 2003, Hurricane Katrina in 2005 and the UK floods of 2007. But it should be noted that there are a significant number of more commonplace events, such as droughts, that are having severe impacts on livelihoods across the globe.

Historically, the United Nations Framework Convention on Climate Change (UNFCCC) is the first temporal marker in terms of attempting to globally address climate change; adverse consequences of which derive from a world reliant on carbon intensive energy. Inception of the UNFCCC took place during the 1992 United Nations Conference on Environment and Development (UNCED), more commonly known as the Rio or Earth summit before being enacted in March 1994 (UNFCCC, 2002), yet debate continued as to the implementation and interpretation of the Convention (Najam et al, 2003).

The original objective of the UNFCCC began with the principal aim of stabilizing greenhouse gas concentrations to a level that would prevent anthropogenic interference with the climate system. UNFCCC Article 2 states that stabilization must be achieved in a timeframe that would enable:

- ecosystems to adapt naturally to climate change;
- food production to continue unthreatened; and
- economic development to proceed in a sustainable manner.

The principle of the UNFCCC recognizes that:

1 scientific uncertainty is an insufficient argument for not taking precautionary measures;
2 countries have 'common but differentiating objectives'; and
3 industrial countries with larger historical contributions of greenhouse gases must take the lead in addressing the problem.

Negotiations ensued for a legally binding protocol to meet the '*common but differentiated responsibilities*' (UNFCCC, 2007), as set out by the UNFCCC over quantified emission reduction objectives, with the UNFCCC adopting the Kyoto Protocol in December 1997 (Dunn, 2002; Ison et al, 2002; Najam et al, 2003).

Countries that ratified this Protocol committed industrial nations (if listed as an Annex I country – these are the 36 industrialized countries and economies in transition listed in Annex I of UNFCCC) to collectively reduce emissions of greenhouse gases (GHGs) by 5 per cent from 1990 levels between 2008–2012 via a range of flexible mechanisms (Dunn, 2002; Ison et al, 2002; Najam et al, 2003). Slow uptake to the Protocol by Annex I countries led to flexibility mechanisms being constantly revised (Agarwal et al, 2001 cited by Najam et al, 2003). Developing countries expressed concern that such flexibility was a departure from sustainable development (Najam et al, 2003).

Significant changes, not least the collapse of the Soviet Union, the resultant new salient world order and globalization, brought with them a host of challenges to addressing climate change and energy policy issues in terms of applicability and practicability, particularly the difficulty of embracing the multiple interests of indigenous peoples, transnational corporations, governments, their agencies and practitioners (Sneddon et al, 2006).

Although developed countries agreed to lead a reduction in greenhouse gases through the acceptance of emission reduction targets in 1992 (including the US), (Ison et al, 2002) and Kyoto 1997, very few countries reduced emissions in line with Kyoto targets (Table 1.1).

Following a series of measures to update flexibility mechanisms within the Kyoto Protocol, a plan of action and timeline for finalizing the Protocol's specific policy objectives was agreed in 1998 (UNFCCC, 2002).

Mitigative measures, primarily technological and neo-liberal market-based emission reduction strategies, therein, capital dependent, predominantly shaped the formation of Kyoto. Conversely, southern states, those most vulnerable and little equipped to deal with the impacts of climate change were increasingly marginalized. Commodification of GHGs via the global introduction of caps, national targets and a lucrative environmental trading regime left little in terms of adaptation instruments, deepening North–South

**Table 1.1** Total greenhouse gas emissions, percentage change from 1990 to 2004 (relative to 1990), Annex I countries

| | Total GHG emissions without LULUCF (Tg/million tonnes $CO_2$ equivalent) | | | Changes in emissions (%) | | Emission reduction target under the Kyoto Protocol[a, b] |
|---|---|---|---|---|---|---|
| Party | 1990 | 2000 | 2004 | 1990–2004 | 2000–2004 | (%) |
| Australia | 423.1 | 504.2 | 529.2 | 25.1 | 5.0 | —[c] |
| Austria | 78.9 | 81.3 | 91.3 | 15.7 | 12.4 | –8 (–13) |
| Belarus | 127.4 | 69.8 | 74.4 | –41.6 | 6.6 | No target yet |
| Belgium | 145.8 | 147.4 | 147.9 | 1.4 | 0.3 | –8 (–7.5) |
| Bulgaria | 132.3 | 64.3 | 67.5 | –49.0 | 5.1 | –8 |
| Canada | 598.9 | 725.0 | 758.1 | 26.6 | 4.6 | –6 |
| Croatia | 31.1 | 25.3 | 29.4 | –5.4 | 16.5 | —[c] |
| Czech Republic | 196.2 | 149.2 | 147.1 | –25.0 | –1.4 | –8 |
| Denmark | 70.4 | 69.6 | 69.6 | –1.1 | 0.1 | –8 (–21) |
| Estonia | 43.5 | 19.7 | 21.3 | –51.0 | 8.4 | –8 |
| European Community | 4252.5 | 4129.3 | 4228.0 | –0.6 | 2.4 | –8 |
| Finland | 71.1 | 70.0 | 81.4 | 14.5 | 16.4 | –8(0) |
| France | 567.1 | 561.4 | 562.6 | –0.8 | 0.2 | –8(0) |
| Germany | 1226.3 | 1022.8 | 1015.3 | –17.2 | –0.7 | –8(–21) |
| Greece | 108.7 | 131.8 | 137.6 | 26.6 | 4.5 | –8(+25) |
| Hungary | 123.1 | 81.9 | 83.9 | –31.8 | 2.5 | –6 |
| Iceland | 3.28 | 3.54 | 3.11 | –5.0 | –12.2 | +10 |
| Ireland | 55.6 | 68.7 | 68.5 | 23.1 | –0.4 | –8(+13) |
| Italy | 519.6 | 554.6 | 582.5 | 12.1 | 5.0 | –8(–6.5) |
| Japan | 1272.1 | 1345.5 | 1355.2 | 6.5 | 0.7 | –6 |
| Latvia | 25.9 | 9.9 | 10.7 | –58.5 | 8.2 | –8 |

| | | | | | | |
|---|---|---|---|---|---|---|
| Liechtenstein | 0.229 | 0.256 | 0.271 | 18.5 | 6.0 | –8 |
| Lithuania | 50.9 | 20.8 | 20.2 | –60.4 | –3.1 | –8 |
| Luxembourg | 12.7 | 9.7 | 12.7 | 0.3 | 31.3 | –8(–28) |
| Monaco | 0.108 | 0.117 | 0.104 | –3.1 | –11.0 | –8 |
| Netherlands | 213.0 | 214.4 | 218.1 | 2.4 | 1.7 | –8(–6) |
| New Zealand | 61.9 | 70.3 | 75.1 | 21.3 | 6.8 | 0 |
| Norway | 49.8 | 53.5 | 54.9 | 10.3 | 2.7 | +1 |
| Poland | 564.4 | 386.2 | 388.1 | –31.2 | 0.5 | –6 |
| Portugal | 60.0 | 82.2 | 84.5 | 41.0 | 2.9 | –8(+27) |
| Romania | 262.3 | 131.8 | 154.6 | –41.0 | 17.3 | –8 |
| Russian Federation | 2974.9 | 1944.8 | 2024.2 | –32.0 | 4.1 | 0 |
| Slovakia | 73.4 | 49.4 | 51.0 | –30.4 | 3.3 | –8 |
| Slovenia | 20.2 | 18.8 | 20.1 | –0.8 | 6.6 | –8 |
| Spain | 287.2 | 384.2 | 427.9 | 49.0 | 11.4 | –8(+15) |
| Sweden | 72.4 | 68.4 | 69.9 | –3.5 | 2.1 | –8(+4) |
| Switzerland | 52.8 | 51.7 | 53.0 | 0.4 | 2.6 | –8 |
| Turkey | 170.2 | 278.9 | 293.8 | 72.6 | 5.3 | —[c] |
| Ukraine | 925.4 | 395.1 | 413.4 | –55.3 | 4.6 | 0 |
| UK | 776.1 | 672.2 | 665.3 | –14.3 | –1.0 | –8(12.5) |
| US | 6103.3 | 6975.9 | 7067.6 | 15.8 | 1.3 | –[c] |
| Annex I EIT Parties | 5551.0 | 3366.9 | 3506.0 | –36.8 | 4.1 | — |
| Annex I non-EIT Parties to the Convention | 18,551.5 | 17,514.6 | 17,931.6 | –3.3 | 2.4 | — |
| Annex I Kyoto Protocol Parties | 11,823.8 | 9730.3 | 10,011.5 | –15.3 | 2.9 | –5 |

*Notes:*

[a] The national reduction targets as per the 'burden-sharing' agreement of the European Community are shown in percentages.

[b] The national reduction targets relate to the first commitment period under the Kyoto Protocol, which is from 2008 to 2012.

[c] A party to the Climate Change Convention but not party to the Kyoto Protocol.

Base year data (under the Climate Change Convention) are used here instead of 1990 data (as per COP decision 9/CP.2 and 11/CP.4) for Bulgaria (1998), Hungary (average of 1985–1987), Poland (1988), Romania (1989) and Slovenia (1986).

LULUCF = Land use, land use change and forestry; EIT = Economies in transition, Tg/million tonnes $CO_2$ equivalent = Emission reductions from voluntary programmes will generally be expressed in million metric tonnes of $CO_2$ equivalent. For the purposes of national greenhouse gas inventories, emissions are expressed as teragrams of $CO_2$ equivalent (Tg $CO_2$ Eq). One teragram is equal to $10^{12}$ grams, or one million metric tonnes (EPA, 2005).

*Source:* Adapted from: UNFCCC, 2006

tensions in the global climate regime (Agarwal et al, 1999 cited by Najam et al, 2003).

For example, the introduction of CDMs (clean development mechanisms – a mechanism whereby developed countries can offset GHGs by mutually agreeing to implement or update cleaner technologies in developing countries (Ison et al, 2002)) are seen largely as high risk, involving high transaction costs and being difficult to implement (IETA, 2007) with the added complexity that, possibly, CDMs could serve to increase GHGs by creating an incentive to indiscriminately develop inappropriate new technologies in developing countries (Ison et al, 2002). On the other hand, CDMs if used effectively are viewed as a potential valuable policy instrument in which to encourage financial support to promote low carbon development in developing countries (Stern, 2006).

In 2001, in a symbolic gesture by COP 6 (Sixth Session of the UNFCCC Conference of the Parties, COP 6, The Hague, The Netherlands,

13–24 November, 2000, (UNFCCC, 2007)), a range of voluntary funding initiatives to counterbalance the needs of least developed countries and Small Island Developing States (SIDS) were agreed to promote capacity building, technology transfer and assistance in climate change adaptation. These were poorly funded, voluntary, managed by the controversial global environmental facility (GEF) and increasingly attached as environmental pre-conditions to aid development programmes, further exacerbating Southern issues (Ison et al, 2002; Najam et al, 2003).

Kyoto was well designed with a flexible framework but its focus in the negotiating process was driven by persuading Annex I countries to ratify the protocol through a series of mitigative measures, signalling a departure from the initial calls for sustainable development integration within global climate policy. Southern preferences for a negotiated formula (based on the long-term objectives of the UNFCCC) encompassing clear linkages between climate change and sustainable development were largely sidelined, further marginalizing developing countries (Banuri and Sagar, 1999, cited by Najam et al, 2003).

Unlike the internationally cohesive and successful outcomes of the Montreal Protocol, progress over the next three years of negotiations for implementation of the Kyoto Protocol waned. It appeared largely ineffective in achieving any real difference in terms of slowing the rate of climate change through political governance (Dunn, 2002; Najam et al, 2003). Throughout the climate negotiation process the least developed countries had been generally reactive in their environmental negotiations with the North as needs to address climate change evolved (Najam et al, 2003).

Additionally, critics argued that the Kyoto target requirement of a 5 per cent reduction in GHGs was inadequate, particularly in light of scientific opinion as presented by the IPCC II, Second Assessment Report, which calls for at least 50–80 per cent reductions in GHGs (Najam et al, 2003).

The main sticking point was that the US, arguably the world's largest polluter (Table 1.2) refused to ratify the protocol (Najam et al, 2003, Middleton and O'Keefe, 2003). President Bush (Junior) reinforced US unilateralism with its transparent scepticism of IPCC findings, insistence that ratification would inevitably lead to a catastrophic economic downturn and complaints that Kyoto targets were unfairly biased in favour of developing countries, especially China and India, comparing their total GHG emissions on a par to the US (by conveniently ignoring per capita rates of emissions) (Middleton and O'Keefe, 2003). See Table 1.2 and Figure 1.18.

Disagreement in late 2000 between the US and the EU led to US withdrawal from the negotiating process in March 2001 (Dunn,

**Table 1.2** Per capita emissions

| Country | Metric tonnes (millions) | Population (millions) | Tonnes of emissions (per person) |
|---|---|---|---|
| Total G7 nations | 9061 | 688 | 13.2 |
| US | 5302 | 273 | 19.4 |
| Canada | 409 | 31 | 13.2 |
| Germany | 861 | 82 | 10.5 |
| UK | 557 | 59 | 9.4 |
| Japan | 1168 | 126 | 9.3 |
| Italy | 403 | 58 | 6.9 |
| France | 362 | 59 | 6.1 |
| Rest of the world | 13,269 | 5209 | 2.6 |
| World total | 22,690 | 5897 | 3.8 |

*Source:* Based on Foster, 2002. Figures from World Bank, 2000/2001, cited by Middleton and O'Keefe, 2003

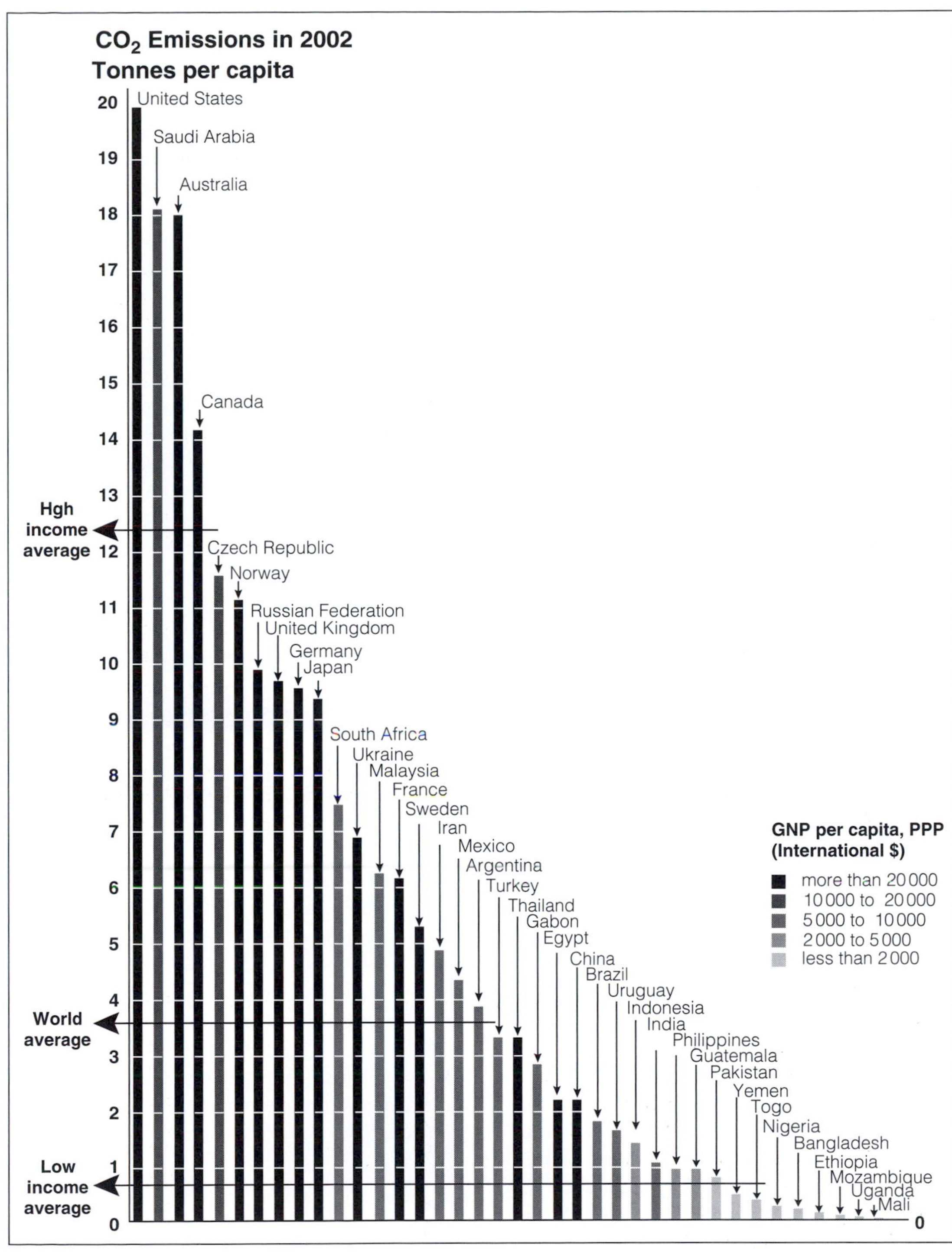


*Source:* UNEP, 2005

**Figure 1.18** World $CO_2$ emissions

2002; UNFCCC, 2002), only serving to galvanize Europe, Canada, Japan and other industrial nations into resolving points of contention in Bonn, Germany and Marrakech, Morocco during the same year (Dunn, 2002).

The EU, having adopted and harnessed the *Precautionary Principle* (COM, 2000) within climate change policy, notably steered the push to bring Kyoto into force arguing the case for mandatory reductions in GHG emissions. The US's unfalteringly market driven position had little changed from that held at debates at the time of Rio in 1992 (Dunn, 2002).

The 1992 Rio Earth Summit clearly placed climate change as an undisputed interest of both Northern and Southern nations (Najam et al, 2002). Slow uptake of UNCED issues within global political regimes, endless negotiating processes and inadequate implementation of policy acted as progenitors for increased public pressure, which ultimately brought about the World Summit on Sustainable Development (WSSD). This was a ten-year review of progress towards the global commitment to sustainable development since Rio, 1992, called by the UN General Assembly and known as The World Summit on Sustainable Development (WSSD) or Rio + 10. It was held in Johannesburg, 2002 (Ison et al, 2002), albeit in a climate of political and economic instability (Rechkemmer, 2006).

## WSSD, Johannesburg, 2002

The WSSD was established primarily as a review process of UNCED policy implementation. By the time WSSD in Johannesburg was held in 2002, climate change had moved to the forefront of global environmental political discourse. Indeed, a survey by Najam et al in 2002 (Najam et al, 2003) confirmed that after poverty, climate change was the next most important issue for experts and practitioners in 71 countries (Najam et al, 2003; Rechkemmer, 2006).

Globalization, an open market driven global economy and the US's persistent unilateral pursuit of its own best interests and the unprecedented events in the US on September 11, 2001 only served to further resolve the US unilateralist agenda (Middleton and O'Keefe, 2003; Rechkemmer, 2006). September 11 acted as a catalyst to do so, all the while undermining multilateral and collective global governance, particularly in the context of climate change (Middleton and O'Keefe, 2003).

By 2007, a seminal report known as the Stern Review of 2006 (herein referred to as the Review) together with the fourth assessment report of the IPCC (AR4) in 2007 reignited the climate change debate. Together, these reports firmly asserted the role of economics and policy formation in line with energy and climate as essential inexorably linked elements and they outlined the urgency of achieving a tougher carbon emission reduction trajectory to minimize future adverse economic and environmental consequences. The summary of the AR4 avoids specific policy recommendations but does summarize a range of possibilities for the consideration of policy makers. Table 1.3 outlines the series of these AR4 options.

Although nuclear power is briefly suggested as a possible carbon neutral solution, the AR4 recognizes associated problems of potential negative external weapons proliferation, safety and waste constraints. Arguably more optimistic is recognition of the importance of forest-related mitigation activities:

> *About 65% of the total mitigation potential (up to 100 US$/t$Co_2$-eq) is located in the tropics and about 50% of the total could be achieved by reducing emissions from deforestation.* (IPCC, 2007, p14).

However, the most profound of findings lie with the firm assertion that the costs of early action far outweigh the costs of inaction:

> *It is often more cost-effective to invest in end-use energy efficiency improvement than in increasing energy supply to satisfy demand for energy services. Efficiency improvement has a positive effect on energy security, local and*

**Table 1.3** Possible policies, measures and instruments for environmental protection

| Sector | Policies,[a] measures and instruments shown to be environmentally effective | Key constraints or opportunities |
|---|---|---|
| Energy supply | Reduction of fossil fuel subsidies<br>Taxes or carbon charges on fossil fuels<br>Feed-in tariffs for renewable energy technologies<br>Renewable energy obligations<br>Producer subsidies | Resistance by vested interests may make them difficult to implement<br>May be appropriate to create markets for low emissions technologies |
| Transport | Mandatory fuel economy, biofuel blending and $CO_2$ standards for road transport<br>Taxes on vehicle purchase, registration, use and motor fuels, road and parking pricing<br>Influence mobility needs through land use regulations, and infrastructure planning<br>Investment in attractive public transport facilities and non-motorized forms of transport | Partial coverage of vehicle fleet may limit effectiveness<br>Effectiveness may drop with higher incomes<br>Particularly appropriate for countries that are building up their transportation systems |
| Buildings | Appliance standards and labelling<br>Building codes and certification<br>Demand-side management programmes<br>Public sector leadership programmes, including procurement<br>Incentives for energy service companies (ESCOs) | Periodic revision of standards needed<br>Attractive for new buildings. Enforcement can be difficult<br>Need for regulations so that utilities may profit<br>Government purchasing can expand demand for energy-efficient products<br>Success factor: Access to third party financing |
| Industry | Provision of benchmark information<br>Performance standards<br>Subsidies, tax credits<br>Tradable permits<br>Voluntary agreements | May be appropriate to stimulate technology uptake<br>Stability of national policy important in view of international competitiveness<br>Predictable allocation mechanisms and stable price signals important for investments<br>Success factors include: clear targets, a baseline scenario, third party involvement in design and review and formal provisions of monitoring, close cooperation between government and industry |
| Agriculture | Financial incentives and regulations for improved land management, maintaining soil carbon content, efficient use of fertilizers and irrigation | May encourage synergy with sustainable development and with reducing vulnerability to climate change, thereby overcoming barriers to implementation |
| Forestry/forests | Financial incentives (national and international) to increase forest area, to reduce deforestation and to maintain and manage forests<br>Land use regulation and enforcement | Constraints include lack of investment capital and land tenure issues. Can help poverty alleviation |
| Waste management | Financial incentives for improved waste and wastewater management<br>Renewable energy incentives or obligations<br>Waste management negotiations | May stimulate technology diffusion<br>Local availability of low-cost fuel<br>Most effectively applied at national level with enforcement strategies |

*Note:* [a] Public Research, Development and Deployment (RD&D) investment in low emissions technologies have proven to be effective in all sectors.

*Source:* Adapted from IPCC, 2007

> *regional air pollution abatement, and employment.*
>
> *Renewable energy generally has a positive effect in energy security, employment and on air quality. Given costs relative to other supply options, renewable electricity, which accounted for 18% of electricity supply in 2005, can have a 30–35% share of the total electricity supply in 2030 as carbon prices up to 50 US\$/t$CO_2$-eq.* (IPCC, 2007, p13)

Similarly of note in the Review is the lengthy exploration of economic costs directly associated with climate change as well as the costs and benefits of action to reverse or reduce its negative impacts. The benefits of early and firm action are strongly advocated as considerably outweighing future costs (as does the later publication of the AR4). Through the use of economic models, a range of possibilities to tackle the problem of an energy dependent world is given and the danger of inaction is again highlighted as are the environmental and economic impacts of climate change. Chiefly, an annual investment of 1 per cent of global gross domestic product (GDP) within the next 10–20 years is called for to avoid the worst effects of climate change. Failure to do so, based on its economic models could mean that when taking into account the overall costs and risks of climate change, inaction could cost at least 5 per cent of global GDP each year, from 2006 onwards. Moreover, according to the Review's worst case scenario, inaction could result in the estimated risk of global GDP output declining by up to 20 per cent (Stern, 2006).

The Review received a mixed critical response from the national and international community. Some disputed methods, especially of calculation (with particular reference to the discount rate used) but many agreed with its main conclusions (Tol and Yohe, 2006; Nordhaus, 2007). Undoubtedly, this Review, together with the AR4, succeeded in propelling the economics of climate change into the economic, political, scientific and public spheres at a time when climate change issues simultaneously began to appear more regularly across a host of different media platforms.

Although the climate convention had been agreed more than a decade previously, anthropogenic emissions, concurrent with public concern, increased more rapidly than ever before. The need to establish a post-Kyoto agreement including greater targets for the most significantly polluting countries, such as the US, by and large the most recalcitrant state, soon became central to formulating a space within which the world's most powerful industrialized countries could, in the near future, demonstrate leadership in significant GHG reduction. However, tensions between the US and the rest of the world are but a fraction of the overall picture.

The US's overtly hostile stance towards rapidly industrializing economies such as China and India in the context of the climate change debate means that urgent reconciliation of complex factors including access, ownership and capacity to use natural resources for development are fraught with difficulty and are distinctly dichotomously entrenched around the 'wants' of the global North and 'needs' of the global South. For instance, cumulatively, industrialized countries account for almost 80 per cent of the world's emissions. Yet, access to resources such as coal in China, which are readily, widely and cheaply available, are carbon intensive but essential for rapid economic development. Do industrialized countries have a right to argue against energy produced cheaply, particularly as these countries have already attained a level of economic development via the same or similar means? Evidently, this jars against the paradoxical need to curb emission growth in tandem with rapid industrialization or development in general.

In essence, developing countries generally and rightly perceive that industrialized 'Northern' countries have developed, achieving higher standards of living over the past two centuries through considerable land use change, heavy industry, manufacturing and technological development, reliant on carbon intensive energy use. Indeed, those now pressing for emission reduction have consumed most of any 'ecological space' available. These are challenges that will face those seeking harmony between future energy requirement and use, and the carving out of a constructive set of achievable future international agreements centred on drastic emission reduction. Murmurs of hope have been made.

Recognition of the adverse role of anthropogenic emissions by President Bush (Junior) in 2007 signals a definite change in tone from the US and the election of President Obama will accelerate that change. However, the heterogeneity of the climate and energy problem is immense, requiring an immense cohesive international response. As advocated by both the Stern and IPCC reports, prevention of 'dangerous climate change' requires a total re-think and drastic reduction in emissions within 10–20 years. Dealing with a post-Kyoto world that increasingly threatens a greater number of adverse consequences (many of which are likely to lead to augmented intensity and variability of extreme weather events), due to our ongoing consumption and reliance on energy, essentially necessitates a harmonious, tenable and pragmatic range of solutions. The advent of a new route delineating a path towards emission reduction onwards from the ending of the Kyoto Protocol in 2012 began with negotiations in Bali.

## The Bali Roadmap

Post-2012 negotiations opened with the establishment of the so-called Bali Roadmap that was the beginning of a negotiation process to produce a post Kyoto agreement to be ratified in Copenhagen in 2009. As such, the specific focus of Bali negotiations was to develop a roadmap that would facilitate negotiation of an international agreement on climate for the period after 2012, being the first commitment period in which the Kyoto Protocol expires.

The Bali conference held during December 2007 realized three key outcomes. First, developing countries that had previously resisted proposals to tackle emission growth under a climate treaty joined the negotiating table, offering mitigation plans, although these will largely depend upon the extent to which industrialized countries address emission growth. Second, delegates, having recognized the urgent need for deforestation prevention to reduce emissions, reached a consensus on the need to feature deforestation as a valid climate change issue. Third, adaptation firmly moved to centre stage as delegates shifted from the foci of mitigation (often viewed upon as a techno-centric, capital intensive and a developed world possibility) to other ways of addressing emission growth that would both encourage wider formulation of appropriate and more people centred policies and enable greater uptake of such policies for the benefit of livelihood and food security, especially pertinent to developing countries.

On the road from Bali to Copenhagen there was essentially a staging point, the Poznan negotiations. In one sense Poznan did not produce significant forward movement. It occurred in and around the US 2008 presidential elections, when a significant democratic victory was celebrated, but at a time when no indication of the real content of the US position in addressing climate change was made. Problems in agreeing emission reductions continue. For instance, though many developing countries have pushed Annex I countries for a mid-term goal emission reduction of 25–40 per cent (with reference to 1990 levels) by 2020, maintaining that this is crucial to any long-term goal commitment, some industrialized countries including Japan, Canada and Australia rendered the target unfeasible under current conditions (Xinhuanet, 2008).

Thus the advent of a new route delineating a definite path towards policy formation for emission reduction onwards from the ending of the Kyoto Protocol in 2012 began with negotiations in Bali, recently with more negotiations in Poznan and continues in Copenhagen (December 2009). Trading mechanisms are likely to feature throughout such negotiations.

## Trading mechanisms

There are two difficult trading mechanisms that have to be resolved: namely the issue of global carbon trading, including the CDMs, and the

**Table 1.4** Key mitigation technologies and practices by sector

| Sector | Key mitigation technologies and practices currently commercially available | Key mitigation technologies and practices projected to be commercialized |
|---|---|---|
| Energy supply | Improved supply and distribution efficiency; fuel switching from coal to gas; nuclear power; renewable heat and power (hydropower, solar, wind, geothermal and bioenergy); combined heat and power; early applications of Carbon Capture and Storage (CCS, e.g. storage of removed $CO_2$ from natural gas). | CCS for gas, biomass and coal-fired electricity generating facilities; advanced nuclear power; advanced renewable energy, including tidal and wave energy, concentrating solar and solar PV. |
| Transport | More fuel efficient vehicles; hybrid vehicles; cleaner diesel vehicles; biofuels; modal shifts from road transport to rail and public transport systems; non-motorized transport (cycling, walking); land-use and transport planning. | Second generation biofuels; higher efficiency aircraft; advanced electric and hybrid vehicles with more powerful and reliable batteries. |
| Buildings | Efficient lighting and daylighting; more efficient electrical appliances and heating and cooling devices; improved cook stoves, improved insulation; passive and active solar design for heating and cooling; alternative refrigeration fluids, recovery and recycle of fluorinated gases. | Integrated design of commercial buildings including technologies, such as intelligent meters that provide feedback and control; solar PV integrated in buildings. |
| Industry | More efficient end use electrical equipment; heat and power recovery; material recycling and substitution; control of non-$CO_2$ gas emissions; and a wide array of process-specific technologies. | Advanced energy efficiency; CCS for cement, ammonia, and iron manufacture; inert electrodes for aluminium manufacture. |
| Agriculture | Improved crop and grazing land management to increase soil carbon storage; restoration of cultivated peaty soils and degraded lands; improved rice cultivation techniques and livestock and manure management to reduce $CH_4$ emissions; improved nitrogen fertilizer application techniques to reduce $N_2O$ emissions; dedicated energy crops to replace fossil fuel use; improved energy efficiency. | Improvement of crop yields. |
| Forestry/forests | Afforestation; reforestation; forest management; reduced deforestation; harvested wood product management; use of forestry products for bioenergy to replace fossil fuel use. | Tree species improvement to increase biomass productivity and carbon sequestration. Improved remote sensing technologies for analysis of vegetation/soil carbon sequestration potential and mapping land use change. |
| Waste management | Landfill methane recovery; waste incineration with energy recovery; composting of organic waste; controlled waste water treatment; recycling and waste minimization. | Biocovers and biofilters to optimize $CH_4$ oxidation. |

*Source:* Adapted from IPCC, 2007, p10

issue of compensation for non-deforestation. In addition, the issue of adaptation funding remains problematic for developing countries where the costs of adaptation clearly outweigh any compensation mechanism at the moment. In general, however, Copenhagen looks as if the emphasis will remain on mitigation, that is on the search for new and renewable technologies to address energy needs in a lower carbon future. In essence this is the challenge for the future of energy. Carbon resources of oil, gas and coal are already 'captured' by geological process; renewable resources have yet to be captured because they require harvesting and storage before use.

Existing infrastructures favour captured carbon not renewable harvesting, so it is not simply investment in resources but in infrastructures to facilitate end use that must be the focus for investment. This emphasis on end use broadly raises the three large sectors, currently carbon-based, that require new strategic thinking. The first is transport, highly dependent on an oil resource and infrastructure but where there seems to be little policy or purpose to move from private to public transport systems except for commuting in larger cities. The challenge here is not simply one of switching fuels but a broader one of designing habitat.

The second large end use sector is household energy use where there are two broad requirements, namely for space heating, especially in the temperate regions where most developed countries are situated, and end use devices associated with the use of electricity. Again the challenge for the household sector is in one sense, particularly for space heating, a challenge beyond the energy sector. Urban redesign for efficient heat retention in buildings in winter and associated cooling opportunities in summer is a challenge that has largely been undelivered with the possible exception of Scandinavia. In terms of end use technologies that are electricity dependent, there is a challenge to match load demand to end use performance but again, with major exceptions of specific devices such as navigation aids, generation is not associated with a specific technology.

The third sector is industrial demand where there is an urgent need to pursue energy efficiency initiatives so that economic growth is decoupled from energy growth; together with the search for a range of renewables to support industrial production, there is a challenge for the industrial sector, which is largely in the private sector, to show initiative by using the market to deliver a lower carbon future.

In many senses, these decisions will all have to be made but are necessarily too late. This leads to a constant dilemma of whether investment should be at least cost – which would favour the existing system – or to maximize benefits – which would favour a renewable and efficient future. As decisions on energy investment taken today would see little production capacity in place in the next ten years (2020) and that investment would then last for some 50 years (2070), it seems inevitable that market forces drive towards least cost solutions. Ironically, this brings into play proven expensive technologies such as nuclear power because it can provide base load through existing transmission and distribution infrastructure. This is probably not a solution that many people would favour from a social and technological perspective but energy policy decisions are made in the context of existing grounded capital infrastructure. Quite simply, in determining the future of energy, politicians and planners have to consider whether it is the determinism of financial markets or broader technical and social issues that underlie the choice of energy system. Hopefully this book will, again, allow an exploration of what could be possible for a lower carbon and an energy secure future.

## Mitigation and adaptation

To address the climate challenge, there is an emphasis on mitigation and adaptation strategies. Mitigation strategies are essentially ones that are technologically focused and imply substantial capital investment; adaptation strategies are essentially livelihood focused and imply recurrent expenditure rather than the one-off investment associated with mitigation. Fundamental differences in north–south perceptions and values of the environment and economic development prevail, as international relations in terms of global climate change regime remain dominated by north–south dichotomies of mitigation versus adaptation (the latter driven by the need to prioritize maintaining livelihoods over economic development). In general, the emphasis from the developed world has been on mitigation strategies, where the diffusion of new technology is seen as the best option to address climate change whereas the adaptation emphasis has been much stronger in the developing world.

Any mitigation strategy has to acknowledge that there is a resource constraint, particularly with reference to oil and gas and that, even without that resource constraint, supplies are vulnerable to geopolitical disruption. Mitigation implies a transition but in the transition itself there are considerable regulatory and acceptance barriers. These barriers are reinforced by the fact that over the last 50 years the global energy system has migrated from small distribution systems to large-scale systems that have a commercial life span of 40 years plus. Moving away from such centralized systems remains a challenge.

Ironically central to this challenge is the fact that fossil fuels, especially coal and lignite remain an option, despite them compounding the problem of accelerated climate change. Nuclear power remains an option although there continues to be considerable public concern about the deployment of this resource, not least because the issue of high level nuclear waste has not been resolved. Both existing fossil fuel and nuclear power are essentially large-scale and somewhat inflexible, useful for base-load generation but unable to address the issue of peak demand. IPCC 4 addressed key mitigation technologies by sector. This is detailed in Table 1.4. On the energy supply side, the emphasis is on carbon capture and sequestration but in most other sectors the emphasis is on a surge for high efficiency technologies.

Mitigation is necessary to stabilize the concentration of greenhouse gases in the atmosphere, the stronger the mitigation effort, the larger and quicker will be the impact and stabilization. In all IPCC scenarios the strongest emphasis is on the role of efficiency coupled with an emphasis on renewables, nuclear and carbon capture and sequestration. However, most commentators on these scenarios do not realize that all scenarios agree on projections to 2030, only diverging after that point. In simple terms, there is common agreement that climate change is already with us.

As this book is published, a new post-Kyoto Agreement is to be signed in Copenhagen, although developing countries have offered more in emissions trading than developed countries, where the US, Canada, Japan and Australia have been particularly cautious about future commitments. There is hope that a new treaty can see a 20 per cent reduction in emissions by 2020. This follows the agreement signed between EU members in late 2008. Central to the establishment of a successful post-Kyoto settlement will be a strong emissions trading market within which the Clean Development Mechanism (CDM) – where developed countries can buy emission rights from developing countries – will flourish. There are still issues, however, around the creation of adaptation funds, particularly with respect to the size of funds not adequately addressing the challenge of adaptation in developing countries.

Part of the problem of determining energy futures is the powerful role of monopoly and oligopoly suppliers in determining choice. One needs look no further than Gazprom where Russia's position as a monopoly supplier of gas to Europe through the Ukraine has produced a cold winter for many European states dependent on imports. It is this mixture of economic and political muscle that gives rise to such resource confrontations where the legalities of market position are abandoned because the politics is more powerful than the law, but there are significant antecedents to this recent demonstration of Russia's political economy of energy, not least the operations of BP, Shell and Exxon in attempting to sway political leadership in the Middle East in the early part of the 20th century. Over time, however, the oil and gas oligopoly, once known as the seven sisters has been consolidated which again produces interesting anomalies. Despite, for example, Shell, BP and Exxon still leading the

Forbes List, a position generated by their dominance in the oil and gas industry, they proclaim via advertising that they are taking a leading role in promoting renewable energy; this marketing position is aptly described as 'greenwash' for it is but a small proportion of their overall budgets which continue to be derived from carbon-based fuels. Furthermore, depending on the state–market axis, monopolies can also exist with the state. Nowhere is this more true than in France where EDF has difficulty in defining its position against European common market requirements of competition and where it promotes intensification of nuclear generation not only in France but in the more liberalized energy economies such as the UK. There is much to be explored in the political economy of energy resource and infrastructure.

One of the most disturbing observations to make about the future of energy is that, largely, most developed country governments do not share the debate with their own citizens. This is particularly the case in the UK where since 1979 – the last major oil shock before that of the early 21st century – there has been little formal debate on the energy resource and infrastructure mix. There have been a number of ad hoc changes, such as the rush for gas to generate electricity, but there has been no coherent analysis of the energy investment framework. As such, existing monopolies and oligopolies continue to direct an energy future that remains carbon heavy.

Key determinants of energy policy are continuing to be driven by geo-political considerations and energy security, which are simultaneously increasingly anchored by environmental concerns, particularly the issue of climate change. These individual but generally inseparable factors in turn influence the shaping and evolving nature of energy technologies and system design. One of the most significant influences on the future of energy will be the approach adopted by the US in light of a newly elected president and his promise to consider the climate change challenge. Whether a Cap and Trade approach (as outlined in the Kyoto Protocol) will be adopted, therein realizing the European ETS scheme, or whether a more straightforward method such as a carbon tax will be adopted, is yet to be revealed. While there is great enthusiasm for a market-based approach to energy systems, there are a number of cautionary voices arguing the need for a robust market-based mechanism in order to create a price signal that strongly infers the need to reduce fossil fuel consumption.

In this opening chapter, the problems surrounding the future of energy use have been touched upon. Specific challenges including the drive for efficiency, as well as an uncertain future for the present dominance of conventional fuels and potential shift to nuclear power, are discussed in more detail in the following chapters. This book concludes by considering the range of possible energy futures.

## References


Agarwal, A., Narain, S., Sharma, A. and Imchen, A. (2001) *Green Politics: Global Environmental Negotiation-2: Poles Apart*, New Delhi: Centre for Science and Environment

BBC (2008) EU seeks to expand energy grids, BBC News Online 13 November 2008. Available at: http://news.bbc.co.uk/1/hi/world/europe/7727028.stm

BP (2008) BP Statistical Review of World Energy June 2008. Available at: www.bp.com/liveassets/bp_internet/globalbp/globalbp_uk_english/reports_and_publications/statistical_energy_review_2008/STAGING/local_assets/downloads/pdf/statistical_review_of_world_energy_full_review_2008.pdf

Cohen, S. J. (2003) Climate Change Impacts and Adaptation: Role of Extreme Events, Presentation at the Environment Canada Scenarios Workshop, Victoria, 16–17 October 2003. Available at: http://www.cics.uvic.ca/scenarios/pdf/2003extremes/cohen.pps

COM (2000) Communication from the Commission on the Precautionary Principle. Available at: http://ec.europa.eu/dgs/health_consumer/library/pub/pub07_en.pdf

COM (2007) Communication from the Commission to the European Council and the European Parliament –- An energy policy for Europe, COM/2007/0001 final {SEC(2007) 12}. Available at: http://eur-lex.europa.eu/LexUriServ/site/en/com/2007/com2007_0001en01.pdf

COM (2008) Communication from the Commission to the European Parliament, the Council, the European Economic and Social Committee and the Committee of the Regions, Brussels, COM (2008) 744/3, Second Strategic Energy Review: An EU Energy Security and Solidarity Action Plan. Available at: http://ec.europa.eu/energy/strategies/2008/doc/2008_11_ser2/strategic_energy_review_communication.pdf

Dobson, I., Carreras, B. A., Lync, V. E. and Newman D. E. (2007) Complex Systems Analysis of Series of Blackouts: Cascading Failure, Critical Points, and Self-organization. Available at: http://eceserv0.ece.wisc.edu/%7Edobson/PAPERS/dobsonCHAOS07.pdf

Dunn, S. (2002) *Reading the Weathervane: Climate Policy From Rio to Johannesburg*, Worldwatch Paper 160, Washington, DC: Worldwatch Institute

EIA (DOE) (2008) Annual Energy Outlook: With projections to 2030, EIA/DOE, Washington DC, USA. Available at: www.eia.doe.gov/oiaf/aeo/pdf/0383(2008).pdf

EPA (2005) Emission Facts, EPA, USA. Available at: www.epa.gov/OMS/climate/420f05002.pdf

Grubb, M., Butler, L. and Twomey, B. (2006) 'Diversity and security in UK electricity generation: The influence of low-carbon objectives', *Energy Policy*, vol. 34, pp4050–4062

Homer-Dixon, T. (2002), The Rise of Complex Terrorism, Global Policy Forum. Available at: www.globalpolicy.org/wtc/terrorism/2002/0115complex.htm

IEA (2008) World Energy Outlook: Executive Summary, IEA, Paris, France. Available at: www.iea.org/Textbase/npsum/WEO2008SUM.pdf

IETA (2007) Available at: www.ieta.org/ieta/ww/pages/index.php?IdSitePage=618

IPCC (2007) Summary for Policymakers In: *Climate Change 2007: Mitigation*. Contribution of Working Group III to the Fourth Assessment Report of the Intergovernmental Panel on Climate Change [B. Metz, O. R. Davidson, P. R. Bosch, R. Dave, and L. A. Meyer (eds)], Cambridge and New York: Cambridge University Press. Available at: www.ipcc.ch/pdf/assessment-report/ar4/wg3/ar4-wg3-spm.pdf

Ison, S., Peake, S. and Wall, S. (2002) *Environmental Issues and Policies*, 1st edn, Harlow, Essex: Pearson Education Limited

Kröger, W. (2006) Issues of Secure Energy Supply, Latsis Symposium 2006, Research Frontiers in Energy Science and Technology. Energy and Reliability. Available at: www.lsa.ethz.ch/docs/061012_Latsis_WK.pdf

Lovins, A. B. and Lovins, L. H. (1982) Brittle Power: Energy Strategy for National Security, Amherst, NH: Brick House Publishing. Available at: http://reactor-core.org/downloads/Brittle-Power-Parts123.pdf

Martinot, E. (2008) Renewables 2007 Global Status Report, REN21, Paris: REN21 Secretariat and Washington, DC: Worldwatch Institute. Available at: www.ren21.net/pdf/RE2007_Global_Status_Report.pdf

Middleton, N. and O'Keefe, P. (2003) *Rio Plus Ten – Politics, Poverty and the Environment*, London: Pluto Press

Najam, A., Huq, S. and Sokona, Y. (2003) 'Climate negotiations beyond Kyoto: Developing countries concerns and interests', *Climate Policy*, Vol. 3, pp221–231

Najam, A., Poling, J. M., Yamagishi, N., Straub, D. G., Sarno, J., DeRitter, S. M. and Kim, E. M. (2002) 'From Rio to Johannesburg: Progress and prospects', *Environment*, vol. 44, pp26–38

National Intelligence Council (2008) Global Trends 2025: A Transformed World, Washington, DC: National Intelligence Council. Available at: www.dni.gov/nic/PDF_2025/2025_Global_Trends_Final_Report.pdf

Nordhaus, W. D. (2007) 'A review of the Stern Review on the Economics of Climate', *Journal of Economic Literature*, vol. 45, no. 3, pp686–702

Perrow, C. (1999) *Normal Accidents: Living with High-Risk Technologies*, Princeton, NJ: Princeton University Press

Rechkemmer, A. (2006) International Environmental Governance – Issues, Achievements and Perspectives, UNU Institute for Environment and Human Security, pp19, 21, 27, 31–34, 44–45, 48

Sneddon, C., Howarth, R. B. and Norgaard, R. B. (2006) 'Sustainable development in a post-Brundtland world', *Ecological Economics*, vol. 57, pp253–268

Stern, N. (2006) The Economics of Climate Change. Available at: www.hm-treasury.gov.uk/sternreview_index.htm

Tol, R. S. J. and Yohe, G. (2006) 'A Review of the Stern Review', *World Economics*, vol. 7, no. 4, pp233–250

UNEP (2004) Impacts of summer 2003 heat wave in Europe, Environment Alert Bulletin. Available at: www.grid.unep.ch/product/publication/download/ew_heat_wave.en.pdf

UNEP (2005) Vital Climate Change Graphics, UNEP. Available at: www.grida.no/_res/site/file/publications/vital-climate_change_update.pdf

UNFCCC (2002) *A Guide to the Climate Change Convention Process*, Preliminary 2nd edn, Climate Change, Copenhagen: UNFCCC

UNFCCC (2006) Changes in GHG emissions from 1990 to 2004 for Annex I Parties. Available at: http://unfccc.int/files/essential_background/background_publications_htmlpdf/application/pdf/ghg_table_06.pdf

UNFCCC (2007) National Adaptation Programmes of Action (NAPAs). Available at: http://unfccc.int/national_reports/napa/items/2719.php

Watts, D. (2003) Security and Vulnerability in Electric Power Systems, NAPS 2003, 35th North American Power Symposium, University of Missouri-Rolla in Rolla, Missouri, 20–21 October 2003, pp559–566. Available at: www2.ing.puc.cl/power/paperspdf/PaperECE723v39Format.pdf

WEC (2007) Deciding the Future: Energy Policy Scenarios to 2050, World Energy Council. Available at: www.worldenergy.org/documents/scenarios_study_online_1.pdf

WEC (2008) Europe's Vulnerability to Energy Crises, World Energy Council, UK. Available at: www.worldenergy.org/documents/finalvulnerabilityofeurope2008.pdf

Willsher, K. and McMahon, B. (2006) Millions blacked out across Europe as cold snap triggers power surge, Guardian Unlimited. Available at: www.guardian.co.uk/germany/article/0,,1940415,00.html

Xinhuanet (2008) UN softens tone on Poznan outcome as climate meeting draws to end, China News 12 December 2008. Available at: http://news.xinhuanet.com/english/2008-12/12/content_10491839.htm

# 2

# Cost of Energy and Scenario Planning

This chapter provides an overview of the cost of energy and scenario planning. These broad, often overlapping themes, underpin wider aspects of energy security and energy planning. In exploring the cost of energy and scenario planning, complexities arise including aspects related to the modern day issue of climate change and drive for sustainability. For management decisions to effectively address the multi-faceted nature of an increasingly globalized economy, national and international energy markets must harness approaches that adopt multi-perspective and holistic planning. Ultimately, addressing the energy conundrum is a 21st century challenge. This challenge is associated specifically with the urgent need to decarbonize economies without compromising economic development, all the while striving towards energy and environmental security. This means that revisiting the question of cost is an imperative for all stakeholders. 'What is the cost of energy?' is as pertinent a question as ever. In this chapter we will examine and explain why several answers are misleading and unsatisfactory.

The cost of energy, for most of us, is represented in our electricity, gas or petrol bills. These, however, show only the prices charged by the companies to the consumer and usually have a fairly tenuous relationship to the costs of energy. Government tax policies, tax credits for investment or for research and development, and energy policies all have a great influence on prices, as do company policies on rates of return and on marketing.

However, estimating the real cost of energy from any given source is necessary if decisions on new investment are to be made. Investments may be in the supply of energy, its distribution or in the technologies for converting the distributed energy into the energy services demanded by the customer. Most public controversy surrounds the construction of new electricity generation plants and the electricity industry will be used as the main example in this chapter. It is chosen not only because of the controversy over nuclear, coal, gas or renewable sources of supply, but also because it is the most intensively analysed of the energy industries and the issues are more clearly understood. Other energy sectors, such as space heating, use a larger fraction of the nation's delivered energy and transport is becoming the largest polluter of the environment. But the multiplicity of buildings and of vehicles, each with their different characteristics, makes them difficult examples to use.

## The costs of electricity generation

In the global village, energy services are not there to meet everyone's need; 2 billion people rely on biomass fuels and 1.6 billion have no access to electricity. Lack of access to quality energy services, particularly electricity, entrenches poverty and constrains the delivery of social capital, impacting especially on women and female children. Experiences around the globe show there is no single or unique way of achieving electrification from both a technological perspective and a financial one. The range of technologies is constantly expanding but the sustainability of particular schemes is becoming increasingly complex; for example, the first generation of hydroelectricity schemes in Africa are now facing significant problems of siltation, which lowers their power capacity.

Recent reviews by the World Bank identify four categories of investment to access electricity. These are grid-based peri-urban and urban electrification, grid-based rural electrification, off-grid rural electrification and the generation of rural electricity funds totalling US$486,000,000 for financial years 2003–2005 (ESMAP, 2007). In further work, the Bank tried to assess power generation technologies under a size range of 50W to 500MW. These were organized into three distinct electricity configurations; off-grid, mini-grid and grid. Table 2.1 outlines the technologies that were examined.

The findings of this review concluded that renewable energy is more economical than conventional generation for off-grid applications of less than 5kW. It went on to conclude that several renewable energy technologies, especially biogas, are potentially the least cost for mini-grid generation where there were isolated loads of between 5kW and 500kW hours. However, conventional

**Table 2.1** Generation technology options and configurations

| Generating types | Life span (Year) | Off-grid | | Mini-grid | | Grid-connected | | | |
|---|---|---|---|---|---|---|---|---|---|
| | | | | | | Base load | | Peak | |
| | | Capacity | CF (%) | Capacity | CF(%) | Capacity | CF(%) | Capacity | CF(%) |
| Solar-PV | 20<br>25 | 50W<br>300W | 20 | 25kW | 20 | 5MW | 20 | | |
| Wind | 20 | 300W | 25 | 100kW | 30 | 10MW<br>100MW | 30 | | |
| PV-wind hybrids | 20 | 300W | 25 | 100kW | 30 | | | | |
| Solar thermal with storage | 30 | | | | | 30MW | 50 | | |
| Solar thermal without storage | 30 | | | | | 30MW | 20 | | |
| Geothermal binary | 20 | | | 200kW | 70 | | | | |
| Geothermal binary | 30 | | | | | 20MW | 90 | | |
| Geothermal flash | 30 | | | | | 50MW | 90 | | |
| Biomass gasifier | 20 | | | | | | | | |
| Biomass steam | 20 | | | | | | | | |
| MSW/landfill gas | 20 | | | | | | | | |

| | | | | | | | | | |
|---|---|---|---|---|---|---|---|---|---|
| Biogas | 20 | | | 60kW | 80 | | | | |
| Pico/microhydro | 5 | 300W | 30 | | | | | | |
| | 15 | 1kW | 30 | | | | | | |
| | 30 | | | 100kW | 30 | | | | |
| Mini hydro | 30 | | | | | 5MW | 45 | | |
| Large hydro | 40 | | | | | 100MW | 50 | | |
| Pumped storage hydro | 40 | | | | | | | 150MW | 10 |
| Diesel/gasoline generator | 10 | 300, 1kW | 30 | 30 | | | | | |
| | 20 | | | 100kW | 80 | 5MW | 80 | 5MW | 10 |
| Microturbines | 20 | | | 150kW | 80 | | | | |
| Fuel cells | 20 | | | | 80 | 5MW | 80 | | |
| Oil/gas combined turbines | 25 | | | | | | | 150MW | 10 |
| Oil/gas combined cycle | 25 | | | | | 300MW | 80 | | |
| Coal steam subcritical sub, SC, USC | 30 | | | | | 300MW | 80 | | |
| | 30 | | | | | 500MW | 80 | | |
| Coal IGCC | 30 | | | | | 300MW | 80 | | |
| | 30 | | | | | 500MW | 80 | | |
| Coal AFB | 30 | | | | | 300MW | 80 | | |
| | 30 | | | | | 500MW | 80 | | |
| Oil steam | 30 | | | | | 300MW | 80 | | |

*Source:* Adapted from ESMAP, 2007, p28

power generation technologies which included combined gas cycle gas turbines and coal and oil fired steam turbines remain most economical for large grid connected applications despite projected increases in carbon energy prices. Two new coal fired plant technologies are attracting considerable attention including the integrated gasification combined cycle power station which can use either coal or lignite for stations of up to 400MW. The review ends by noting that

> *new technologies are becoming more and technologically mature, uncertainty in fuel and other inputs is creating increasing risk regarding future electricity costs, and old assumptions about economies of scale in generation may be breaking down. (ESMAP, 2007, p33)*

## External costs of energy

Energy is a central function of society and economies. As such, adverse impacts on an energy dependent world, where energy resource distribution is uneven and unequal, have costs but the costs of damages caused are not integrated into the energy pricing system. Many believe that a mechanism should be developed to reflect external costs within the price paid for energy services. This concept, borrowed from the field of welfare economics, is often termed 'damage cost externalities' or external costs; more usually termed externalities. The concept aims to ensure that prices reflect the total costs of an activity, incorporating the cost of damages caused by employing taxes, subsidies or other economic instruments. This internalization of external costs is intended as a strategy to rebalance social and

## Box 2.1 Externalities

Externalities are defined as the unintentional side effects of an activity affecting people other than those directly involved with a given activity. A negative externality is one that creates side effects that could be harmful to either the general public directly or through the environment, such as pollution generated from burning fossil fuels to produce electricity. A positive externality, on the other hand, is an unpaid benefit that extends beyond those directly initiating the activity such as the development of a public park.

Traditionally, both negative and positive externalities are considered as forms of market failure – when a free market does not allocate resources efficiently. Arthur Pigou, a British economist best known for his work in welfare economics, argues that the existence of externalities justifies government intervention through legislation or regulation. Pigovian taxation philosophy promotes taxation of negative externalities, essentially, activities associated with detrimental impacts. The Pigovian tax therefore shifts the emphasis from the subsidization of negative externalities, advocating the subsidization of positive externalities, that is, activities which create benefits in order to further positively incentivize associated activities.

Many economists believe Pigovian taxation on pollution is a preferable, effective and more efficient means of dealing with pollution as an externality, than government-imposed regulatory standards. Taxes leave the decision of how to deal with pollution to individual sources by assessing and setting a fee or 'tax' on the amount of pollution generated. Consequently, in theory, under the Pigovian tax approach, a source looking to maximize profit will need to consider/account for stipulated targets and taxation costs related to the reduction and/or control of pollution emissions in order to operate at least-cost.

Other economists believe that the most efficient solution to externalities is to include them within the cost for those engaged in the activity, that is, to internalize any given externality. This means that externalities are not necessarily seen as market failures which can, in turn, weaken the case for government intervention. Many externalities can be internalized via the creation of well-defined property rights. Economist Ronald Coase showed that taxes and subsidies were typically not necessary as long as involved parties could successfully negotiate voluntary agreements. According to Coase's theorem, it does not matter who has ownership, so long as property rights exist and free trade is possible.

Another method for controlling negative externalities associated with energy production, loosely related to property rights, is the 'Cap and Trade' system. The Cap and Trade system sets maximum emission levels for a given group of sources over a specific time period that can be traded, bought and sold, or banked for future use. Over time, the cap is lowered and in theory this should encourage more efficient processes so that additional profits can be realized by selling allowances to less efficient producers.

environmental dimensions with the purely economic, accordingly leading to greater environmental sustainability. However, finding the most appropriate and effective mechanisms for dealing with externalities is complex. Including external costs is just one mechanism – see Box 2.1.

All energy technologies will have some sort of environmental impact and this will vary according to the type of primary energy resource used. For example, it is generally agreed that when considered comparatively, fossil fuels fare worst in terms of the range and severity of their direct adverse impacts. Despite this given known, it should be recognized that all forms of energy generation are associated with detrimental impacts, even, for example, wind power. For instance, when considering its whole life cycle, negative impacts include the construction, installation and decommissioning of a wind turbine. Concerns can arise regarding perceived impacts upon amenity values (usually aesthetic impacts) for onshore systems, and concerns about ecological impacts on marine systems for offshore wind generation. Finding an agreed and consistent method capable of calculating these values is problematic but important. If an agreed method, with the capacity to calculate whole life costs of energy generation were established, certain 'hot spots' and/or 'burdens', could be identified and utilized to address potential implications for investment decisions in energy generation capacity and associated infrastructure.

In the 1990s, the European Union (EU) Commission launched ExternE, a major research programme to provide a scientific basis for the quantification of energy related externalities and to give guidance supporting the design of internalization measures. The programme used a bottom-up impact pathway assessment in which environmental benefits and costs are estimated by following the pathway from single source emissions via changes of air, soil and water quality to physical impacts, such as increased emissions. ExternE argues that a bottom-up approach is needed as external costs are highly site dependent. A reference scenario based on background is then used as a comparison to a scenario where the additional emissions from the activity are introduced. An analysis of pathway dispersion to different receptors is used to establish the dose rate and exposure to derive differences in physical impacts on public health, crops and building material. The final stage in the analysis evaluates the impacts in monetary terms. For damage to marketed items such as crops and materials, market prices are used to evaluate damage. Drawing from welfare theory, non-marketed goods such as damage to human health and amenity loss are evaluated on the basis of the willingness-to-pay or willingness-to-accept approach that is based on individual preferences as illustrated in Box 2.2.

The ExternE study found that externalities ranged from 40 billion euro to 70 billion euro for fossil fuel and nuclear in 2003. The study highlighted, that if included in energy prices, identified externalities would double the cost of producing electricity from coal or oil and increase the cost of electricity production from gas by 30 per cent (European Commission, 2003). Table 2.2 details a comprehensive list of external costs associated with electricity production for different technology options in different EU member states. Unsurprisingly, fossil fuels are most costly. Energy costs rose sharply in 2008. The cost of energy in 2008 ranged from 8 to 18 pence per kWh, dependent on the supplier and tariff regime, proving that externalities from fossil fuels continue to have a substantial impact upon price.

Since the original work of ExternE, a number of related studies have been undertaken to develop more accurate methods of quantifying the social and environmental damage resulting from energy consumption. Some of these studies are outlined here:

### *NEWEXT (New Elements for the Assessment of External Costs from Energy Technologies)*

NEWEXT revisited some of the major uncertainties in external cost data. Major uncertainties in the current external cost data result from uncertainties in the monetary valuation of mortality effects. This study also addressed some omissions, such as impacts on ecosystems due to acidification, eutrofication and global warming. The existing accounting framework was also criticized for not taking into account the contamination of water and soil. It addressed the unbalanced treatment of severe accidents, as the current framework is very much focused on accidents in the nuclear fuel chain, while neglecting severe accidents from other energy sources. The principal areas of research encompassed:

- monetary valuation of increased mortality risks from air pollution;
- monetary valuation of ecological and $CO_2$ impacts based on preferences revealed in political negotiations (standard-price approach);
- assessment of environmental impacts and resulting externalities from multi-compartment (air/water/soil) impact pathways;

## Box 2.2 Willingness-to-pay and willingness-to-accept

These are methods for placing a monetary value on non monetary goods. There are two approaches: stated preference or willingness-to-pay and revealed preference or willingness-to-accept.

**Stated preference:** this tries to establish what individuals are willing to pay for a particular environmental good such as a landscape. Surveys estimate mean willingness to pay for a defined environmental asset and aggregate for the whole population. One problem with this is that as ecosystem services are not fully captured in commercial markets, or adequately defined in terms comparable with economic services and manufactured capital, they are often given too little weight in policy decisions.

**Revealed preference:** two methods can be used to determine value:

1 *Travel costs* have been used to value National Parks in the US by surveying visitors to determine drive time, socio-economic status, frequency of visits, and so forth. The derived data are used to calculate the value of the amenity.
2 *Hedonic pricing* is used to estimate economic values for ecosystem or environmental services that directly affect market prices. It is most commonly applied to variations in housing prices that reflect the value of local environmental attributes, meaning that people will pay more for a property in an area of high environmental quality as compared to a similar property in an area with a lower environmental quality.

*Source:* O'Riordan, 2001

**Table 2.2** External costs for electricity production in the EU for existing technologies[a] (in € cent per kWh[b])

| Country | Coal and lignite | Peat | Oil | Gas | Nuclear | Biomass | Hydro | PV | Wind |
|---|---|---|---|---|---|---|---|---|---|
| AT | | | | 1–3 | | 2–3 | 0.1 | | |
| BE | 4–15 | | | 1–2 | 0.5 | | | | |
| DE | 3–6 | | 5–8 | 1–2 | 0.2 | 3 | | 0.6 | 0.05 |
| DK | 4–7 | | 2–3 | | 1 | | | | 0.1 |
| ES | 5–8 | | 1–2 | | | 3–5[c] | | | 0.2 |
| FI | 2–4 | 2–5 | | | | 1 | | | |
| FR | 7–10 | | 8–11 | 2–4 | 0.3 | 1 | 1 | | |
| GR | 5–8 | | 3–5 | 1 | | 0–0.8 | 1 | | 0.25 |
| IE | 6–8 | 3–4 | | | | | | | |
| IT | | | 3–6 | 2–3 | | | 0.3 | | |
| NL | 3–4 | | | 1–2 | 0.7 | 0.5 | | | |
| NO | | | | 1–2 | | 0.2 | 0.2 | | 0–0.25 |
| PT | 4–7 | | | 1–2 | | 1–2 | 0.03 | | |
| SE | 2–4 | | | | | 0.3 | 0–0.07 | | |
| UK | 4–7 | | 3–5 | 1–2 | 1 | | | | 0.15 |

*Notes:* [a] Global warming is valued with a range of damage cost estimates from €18 to €46 per ton of $CO_2$
[b] Sub-total of quantifiable externalities (such as global warming, public health, occupational health material damage)
[c] Biomass co-fired with lignites

*Source:* Adapted from the European Commission, 2003, p13

- assessment of externalities from major accidents in non nuclear fuel chains;
- methodology testing and revision of external cost estimates;
- dissemination.

For further detail see the NEWTEXT Website (www.ier.uni-stuttgart.de/forschung/projektwebsites/newext/nexabout.html#Objectives).

### *Maxima*

Maxima (Dissemination of external costs of electricity supply – making electricity external costs known to policy makers) focused on the development of tools, indicators and operational parameters for assessing sustainable transport and energy systems performance (economic, environmental and social) and effective methods for communicating these beyond the scientific community. See the Maxima website for further detail (http://maxima.ier.uni-stuttgart.de/).

### *EXTERNE-POL*

EXTERNE-POL (Extension of Accounting Framework and Policy Applications) was the continuation of the ExternE project series for the analysis of the external costs of energy. Its objectives were:

- improving, validating and extending the methodology of ExternE;
- providing an assessment of new technologies for energy systems;
- implementing the methodology in the accession countries of Eastern Europe;
- creating a permanent internet site for ExternE.

Further information on EXTERNE-POL can be found at www.externe.info/exterpol.html.

## Subsidies and feed-in tariffs

Transferring the considerable subsidy currently enjoyed by the heavily polluting industries to clean renewable technologies would remove barriers for innovation and provide greater motivation for the uptake of the renewable energy sector. The feed-in tariff system (FIT)[1] is a policy instrument that encourages and promotes the use of renewable energy sources for electricity (RES-E) (Rio and Gual, 2007). For example, subsidies for wind technology and FITs in Spain have been attributed to increased efficiency of use and rapid development of technological innovation in many regions across Spain. These policies appear to have succeeded in connecting formerly isolated rural areas to energy supplies, stimulating further economic growth and security of supply. Investment has resulted in the impetus for technological innovation[2] in line with economic development, triggering rural regeneration, as witnessed in areas such as the Navarre region where 51.7 per cent of total energy consumption is provided by wind power alone (*Nature*, 2007a).

The success of the Navarre Region in north-east Spain led to its commendation in 2005 for having the best regional set of policies in Europe at the Conference for Renewable Energy in Berlin (*Nature*, 2007a). Characteristic of these policies is that adoption of sustainable energy systems is facilitated by engaging individuals from a bottom-up/grass roots level, as empowerment to key stakeholders enables more equitable decision making pertinent to those involved with locally agreed, implemented and managed energy systems.

## Renewable energy supply: electricity, transport and heat sectors

The value given to facilitating the development of renewable energy technologies to supply future energy needs is considerable. Yet, as can be seen in Figure 2.1, the growth rate and consumption of the renewable sector across member states of the EU is highly variable. The absence of coordinated national policies across member states, weak regulatory frameworks and the absence of mandatory frameworks for renewables in the transport sector and heating/cooling

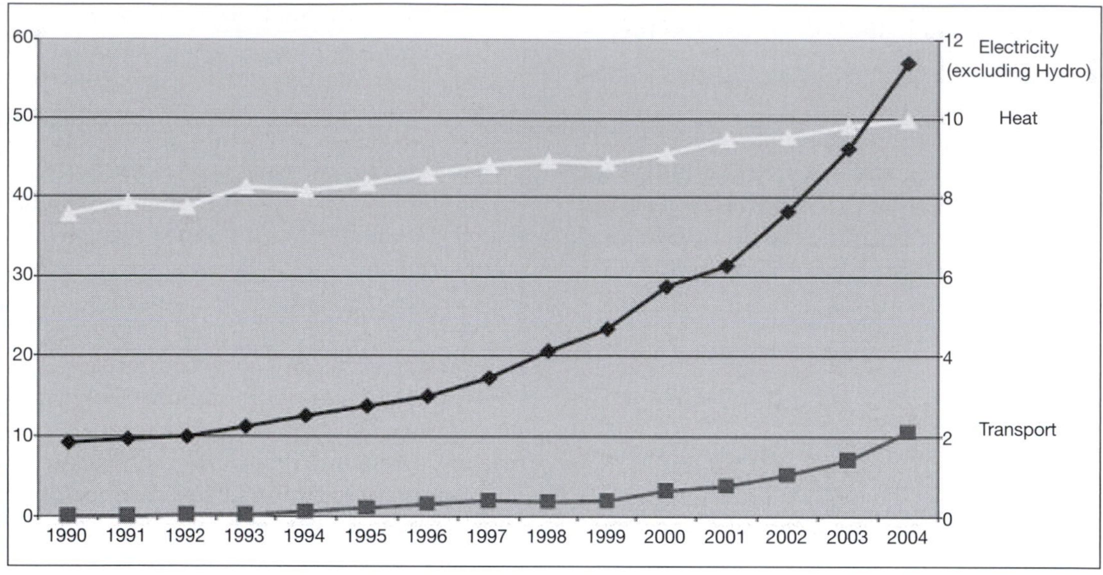


*Source:* COM, 2006

**Figure 2.1** The contribution of renewable energy (electricity, transport and heat) 1990–2004 (mtoe) in EU countries

sectors has resulted in weak growth for both sectors. Conversely, Directive 2001/77/EC, (an EU directive, mandatory from 2001), appears to have driven a steady growth pattern in renewables for the electricity sector.

A viable energy future of low carbon energy economies and security of supply is possible by shifting from conventional fossil fuels to renewables through policy measures. This includes regulatory frameworks requiring compliance to stipulated targets, FITs to encourage efficiency, stimulation of market competitiveness via a range of subsidies to the renewable energy sector and the decentralization of energy systems. The EU recognizes that roadmapping a range of tangible renewable energy futures is essential to achieving both clean energy supply and security, particularly as renewables are quoted as being advantageous because:

> *They are largely indigenous, they do not rely on uncertain projections on the future availability of fuels, and their predominantly decentralised nature makes our societies less vulnerable. It is thus undisputed that renewable energies constitute a key element of a sustainable future.* (COM, 2006, p1)

Policies for renewable energy supply, however, require uniform application and national commitment across all member states to ensure set targets are achieved or even surpassed, to encourage competition and growth in the renewable energy sector, to facilitate a definite departure from the traditional hydrocarbon-based economies of the past. Established on past and present growth rates, there is a clear growth rate in three sectors, suggesting that these three exhibit the greatest potential for transfer to renewable sources: electricity, transportation and space heating (COM, 2006).

## Renewable energy: integrative policies

While societies in the developed world continue with high energy demand, the likelihood of demand-reduction is lowered, as is the likelihood of achieving the net rate of decarbonization. Low carbon energy supply, although less flexible than demand-reduction and requiring initial capital-intensive investment (typically by governments and the private sector), has been established as technically and economically viable through the renewable energy sector (Anderson, 2005). Integrative policies applied (across the electricity, heat and transport sectors) facilitate technological innovation, emission reduction and energy efficiency (WWF, 2008). The example given in Figure 2.2 is for the automotive transport sector. Integrated policies intend to assign appropriate responsibility in achieving the combined objectives of raising energy efficiency and reducing $CO_2$ emissions.

### Economies of scale

Economies of scale also have the potential to reduce the cost of renewable technologies so that they become economically viable, enabling a large scale deployment of low carbon energy supply, but they will require large scale infrastructural changes to accommodate low carbon supply due to the intermittent nature and low-density of renewable energy systems (Muradov and Veziroglu, 2005). These approaches could potentially present a challenge to small and densely populated countries such as the UK (Anderson, 2005).

The increasing need to address the future of energy supply, particularly for electricity, heat and transport, has triggered a series of large scale renewable projects across the UK and within other member states of the EU. This is reflected in the development of large scale renewable energy infrastructures including offshore and onshore wind farms being developed in the UK and EU, echoing the scale and size of previous hydrocarbon infrastructures (O'Brien and O'Keefe, 2006). Regardless of their intermittent nature, low-density and need for large infrastructural investment, where the supply of renewable energy systems has been applied through incentivized schemes with clear, transparent policies that are integrative and horizontal in nature, renewables have transformed not only the visual landscape but they have enhanced many areas economically, particularly those areas formally isolated from the supply of energy (*Nature*, 2007a).

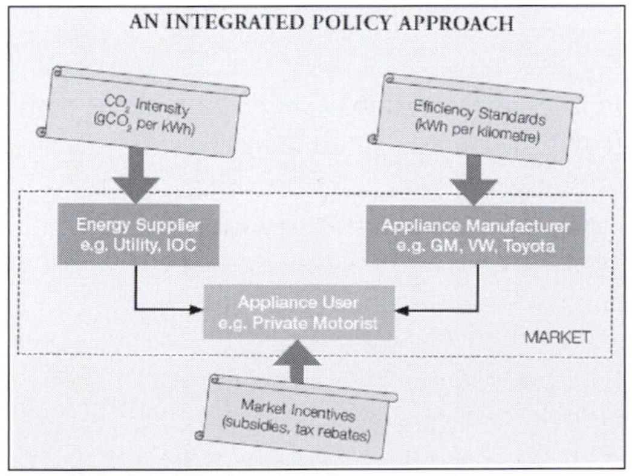


*Source:* WWF, 2008

**Figure 2.2** An integrated approach to policy formation

## The impact of future renewable energy alternatives

The renewable energy sector has the capacity to provide carbon emission reduction and pollution, exploit local and decentralized energy sources such as wind, power, marine and solar while providing the supply of energy to meet demand (Moller et al, 2004; COM, 2006). Equally, where bottom-up approaches have been undertaken, renewables have stimulated technological innovation, and paved the path for rapid growth in profitable renewable energy and economic development (*Nature*, 2007b).

Figures 2.3 and 2.4 illustrate the historical and projected futures for different renewable energy

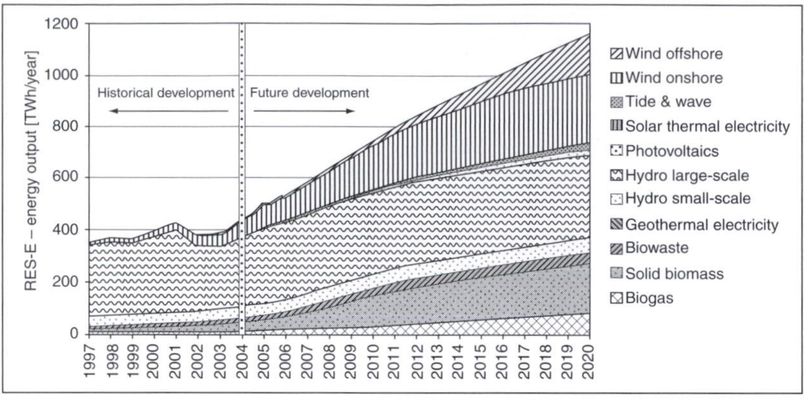


*Source:* COM, 2006

**Figure 2.3** Renewables growth: Electricity projections to 2020

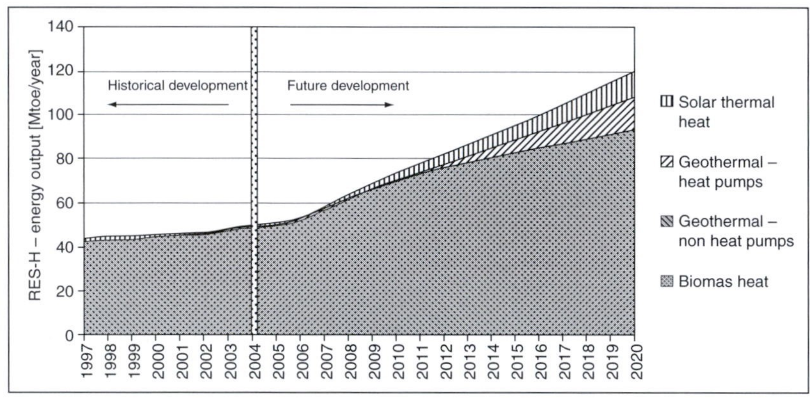


*Source:* COM, 2006

**Figure 2.4** Renewables growth: Heating and cooling projections to 2020

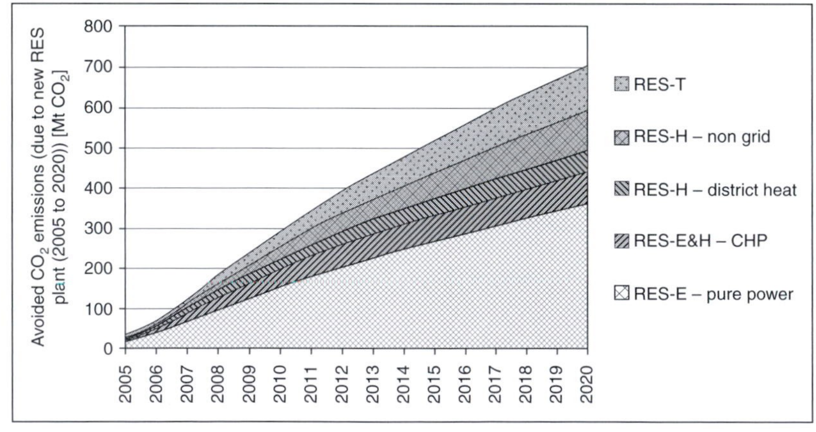


*Source:* COM, 2006

**Figure 2.5** Avoided $CO_2$ emissions due to new RES deployment up to 2020 in the EU

technologies generating electricity and heat (renewable energy source – electricity = RES-E and renewable energy source – heat = RES-H, transport = T), and their output capacities, projected until 2020 (COM, 2006). Figure 2.5 outlines projected gains up until 2020, in terms of carbon reduction via the adoption of a range of renewable energy alternatives. Each scenario indicates a positive exponential growth pattern in all sustainable renewable energy technologies, for heat, electricity production and transportation sectors and correspondingly links renewable trends with rising $CO_2$ reductions.

## Commercial energy supply strategies

Diminishing domestic resources of finite fuels are driving exploration, threatening to encroach on the world's last remaining protected areas such as the Arctic National Wildlife Refuge (ANWR), extraction, and use of dirtier, hydrocarbons, including heavy oils, shale and tar-sands (WWF, 2008). These issues raise questions as to: (i) the adequacy of existing environmental policies concerning GHGs, environmental protection and sustainable energy; and (ii) the level of commitment of nation states within the developed world to lead and move away from the traditional primary energy mix in favour of sustainable energy futures. Contrary to the opening sentence of this section, an added dimension to the hydrocarbon debate is the fact that 'Peak Oil' theory is not universally accepted. For instance, Odell, (2004) critiques the theory, arguing the case for technological development in extraction. This highlights the idea that oil scarcity and 'Peak Oil' theory do not necessarily translate to the non-existence of oil, although extraction of finite resources is increasingly expensive and unsustainable (*The Economist*, 2008a). Conflicting critiques illustrate the complexity of the energy debate.

Energy is a fundamental necessity and crucial element for living within the contemporary world. It is vital for economic development and an equally integral factor for the sustainability of societies' needs, despite the known fact that many energy systems are not neutral in terms of environmental consequences (Dincer, 2002). It has been established in the previous section that energy lies at the crux of the sustainable development paradigm as few activities adversely affect the environment as much as those linked to energy use. More than three-quarters of the world's finite resources (including oil, coal, gas and uranium) used for energy are consumed by one-quarter of the world's population. This lopsided global energy balance is a dynamic that forces inequitable energy consumption and scale of use in favour of the developed world, dominated by the supply of finite resources, the nature of which will, despite polemical debate, undoubtedly lead to an overall shortage of supply (Stanford, 1997).

Crude oil, by far the most widely supplied of hydrocarbons is characterized by vertically integrated systems of corporate control over almost the entire supply chain, from upstream (exploration to distribution via pipeline or tanker) to downstream (refining, blending, storage, distribution and sale of completed products). When considering energy futures, O'Brien and O'Keefe (2006) emphasize the importance of energy security and price stability, particularly as end-users are concerned primarily with the service provided, not the nature of the supply.

The three primary sectors, transportation, electricity production and space heating remain heavily reliant on the supply and use of fossil fuels. Despite polarization of the climate change debate towards GHG emissions, in the context of the global energy balance, energy futures are currently steered towards and appear likely to continue dependency on conventional hydrocarbons (COM, 2000). Energy policy formation in its present form, does not address the possibility that future shortages of the conventional hydrocarbon primary mix could lead to the use of coal and heavy oils, shale and tar-sands with a propensity to emit even higher GHG emissions.

Changes in the ownership of energy systems, primarily, from state control through privatization have resulted in the commodification of energy (O'Brien and O'Keefe, 2006). Although climate change issues and the twinned paradigm of sustainable development are increasingly driving international policy formation in the context of energy systems, commercial supply of energy in market-based economies remains dependent on conventional hydrocarbons, indicating that the Common Market is a significant factor also influencing policy formation (COM, 2000).

Debate in the EU is increasingly mirroring the wider international arena as the future of energy supplies and international commitments to reduce GHG emissions tend to be viewed as interrelated issues concurrently driving energy policy. Concerns of high-demand energy societies are integrally linked with supply. Box 2.5 later on in this chapter outlines some of the implications of the supply and demand of hydrocarbons that are currently shaping the international arena.

## Primary energy mix: Hydrocarbons

Supply of a primary energy mix of finite fuels including coal, oil and natural gas currently drives the existing energy systems of the developed world, having shaped and continually influenced economic development (*Nature*, 2007c; UN, 2007). Yet, these fuels are often supplied via systems that are linear, top-down and inflexible. Additionally, fossil fuels are synonymous with high GHG emissions, are unsustainable and in the absence of meaningful regulation, global demand is likely to increase in the future, as illustrated in Figure 2.6.

In the European Union, as domestic resources decline imports increase and security of supply will become dependent upon unpredictable and changeable geo-political structures (COM, 2002). The EU's international commitment to reducing GHG emissions is unlikely to be achieved unless additional measures are introduced. Despite its firm commitment to sustainable development, the EU has witnessed gradual increases in GHGs since 2000, although these have been attributed (by the EU) to economic development (EEA, 2007) raising concern as to the ability to meet internationally agreed commitments for the reduction of GHGs. Additionally, according to scenario projections from 2000 to 2030, a demand increase of 0.4 per cent per annum is expected in the EU (WETO, 2003).

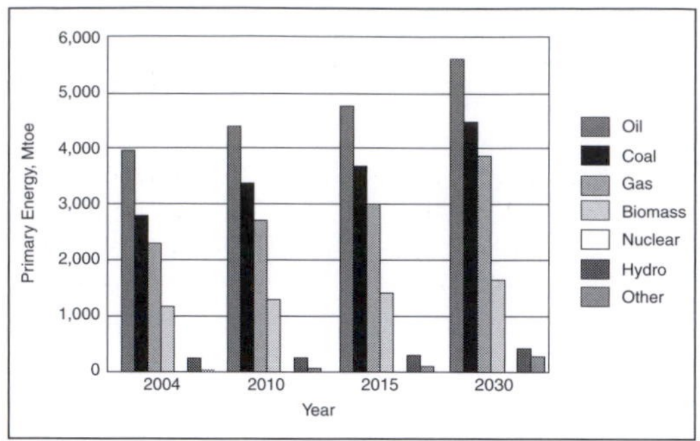


*Source:* WWF, 2008

**Figure 2.6** Projected evolution of the world's primary energy demand by fuel according to the International Energy Agency, 2006 using a 'reference scenario'. Data for 2004 are actual

## Is the nuclear route a viable 'green' alternative?

The shift from fossil fuels to nuclear power is not considered a sustainable option for the future in this report due, principally, to complexities associated with waste and a reliance on uranium (a finite resource). The case against a path towards the nuclear route is outlined in Box 2.3.

## The challenge of global biomass

Total production of wood in 2000 reached approximately 3,900 million cubic metres (CUM), of which 2,300 million CUM was used for wood fuels. At first sight, this would mean that approximately 60 per cent of the world's total wood removals from forests and trees outside forests are used for energy purposes; energy is the main application of woody biomass from forests

## Box 2.3 Is the nuclear route a viable 'green' alternative?

The projected rise in demand for intensive energy is often met with an argument for a shift of supply from fossil fuels to nuclear and renewables. Superficially, the nuclear route has the capacity to twin energy supply with low carbon emissions (IAEA, 2006). However, contradictory to Lovelock's opinion that '*nuclear is the only practical answer to the challenges of global warming*' (Lovelock, 2006), nuclear is no panacea to the problem of securing clean energy supply (O'Brien et al, 2007; Stanford, 1997), as the push for nuclear is fundamentally flawed with issues encompassing uranium availability, capital-intensive public sector investment, proliferation risks, operating safety, toxicity, storage, disposal and security (Toth and Rogner, 2006). Additionally, nuclear energy programmes are essentially polemical in approach. Nuclear is large scale. Nuclear reactors in the UK have never been delivered on time or on budget, or succeeded in delivering expected levels of performance (O'Brien and O'Keefe, 2006; BBC, 2005).

The nuclear option requires expensive set-up and decommission costs, has long lead times (10–20 years[3]), is inflexible (requiring a period of 25 years before cost–benefits are realized) and will probably 'lock-in' future generations' public money to long-term contracts, exhaust funding and resources that could be used for the development of alternative fuels (O'Brien and O'Keefe, 2006). In addition, the nuclear option does not solve the problem of carbon emissions from the transport and domestic sectors and is not directly competitive with oil (Toth and Rogner, 2006).

The issues outlined previously are not exhaustive but do underpin the principal dilemmas of nuclear supply over renewables. In terms of cost, due to the high initial upfront costs of nuclear and equally high amortization periods (Toth and Rogner, 2006), the cost–benefit calculation for nuclear is much poorer than renewables that exhibit more immediate, shorter lead-times (1–2 years) and has more quantifiable risks. Nuclear is ultimately not renewable, being reliant on finite sources of uranium; promises of nuclear fusion are distant;[4] and amortization costs detract from the possibility of developing more sustainable, viable and cheaper alternatives.

and trees outside forests (FAO, 2008). However, this is not the whole story. Wood is only energy in its final phase of consumption; before that final consumption biomass resources offer a range of multiple and simultaneous end uses such as fodder provision, medicines, foliage for roofing, and so forth. It is difficult to plan wood for energy as a single end use except in developed countries where the system of mono-cropped plantations on marginal land is the dominant forestry paradigm. The situation is reversed in developing countries where forest resources are frequently the limitation on agricultural colonization; certainly there is evidence from parts of Africa that it is difficult to plant wood commercially because its relative abundance means that it is still regarded as 'rubbish' (Van Gelder and O'Keefe, 1995).

There has been significant debate about the production of biomass based commercial fuels. FAO have recently completed a global review of biofuel development; the key points are summarized in Box 2.4. In a separate scientific study, FAO looked at traditional biomass consumption at the household level in the developing world. It concluded that biomass resources could be used efficiently at small scales, especially if attention was paid to appropriate end use technologies. It noted that it was possible to develop virtuous cycles of local energy use rather than vicious cycles of deforestation. In general, virtuosity could be sourced by emphasizing improved insulation in traditional end use technologies. This applied to both household and small scale production activities such as charcoal making, fish processing and tobacco curing. Increasingly sophisticated market chains were found to be linking rural and coastal producers to centres of urban demand, where prices continued to rise,

## Box 2.4 Challenge of global biomass

- Demand for agricultural feedstocks for liquid biofuels will be a significant factor for agricultural markets over the next decade and perhaps beyond. It may help reverse the long- term decline in real agricultural commodity prices. All countries and all agricultural markets will face the impact of liquid biofuel development – whether or not they participate directly in the sector.
- Rapidly growing demand for biofuel feedstocks has contributed to higher food prices, threatening the food security of the poor.
- In the longer term, expanded demand and increased prices for agricultural commodities may represent an opportunity for agricultural and rural development. However, higher commodity prices alone are not enough; investments in productivity- and sustainability-enhancing research, enabling institutions, infrastructure and sound policies are also urgently needed. A strong focus on the needs of the poorest and least resource-endowed population groups is crucial.
- The impact of biofuels on greenhouse gas emissions differs according to feedstock, location, agricultural practices and conversion technology. In some cases, the net effect is not favourable. The largest impact is determined by land-use change – for example, through deforestation – as the agricultural area is expanded. Other possible negative environmental effects – on land and water resources, as well as on biodiversity – also depend to a large extent on land-use changes.
- Harmonized approaches for assessing greenhouse balances and other environmental impacts of biofuel production are needed. Criteria for sustainable production can contribute to improving the environmental footprint of biofuels, but they must focus on global public goods, be based on internationally agreed standards and must not put developing countries at a competitive disadvantage.
- Liquid biofuels are likely to replace only a small share of global energy supplies. Land requirements would be too large to allow displacement of fossil fuels on a larger scale. The possible future introduction of second-generation biofuels based on lignocellulosic feedstocks would greatly expand potential.
- Given existing technologies, production of liquid biofuels in many countries is not currently economically viable without subsidies. However, the competitiveness of biofuels varies widely, according to the specific biofuel, feedstock and location. Also, economic viability can change as a result of changing market prices for inputs and oil and as a result of technological advances in the biofuel industry. Investment in research and development is critical for the future of biofuels.
- Policy interventions, especially in the form of subsidies and mandated blending of biofuels with fossil fuels, are driving the rush to liquid biofuels. However, many of the measures being implemented by both developed and developing countries have high economic, social and environmental costs.

*Source:* Adapted from FAO, 2009

suggesting that returns to producers will increase. There was little evidence that local biomass activity was in direct competition with local food production but, rather, that they could be complementary. This shows that building on traditional biomass practice could produce local community business generation (FAO, 2009).

The first generation of biofuels – largely produced from food crops such as grain, sugar beet and cane as well as oil seeds – are limited in their ability to achieve targets for oil substitution. Sugar cane is perhaps the exception but, in all cases, there is competition for land and water that could be used for food and fibre production. A second generation of biofuels from lingocellulosic materials, such as cereal straw, bagasse, forest residues and vegetation grasses holds more promise. The International Energy Agency (IEA)

(2008) estimate such fuel could be delivered for US$0.80–1.00 per litre but there is no clean technology best pathway between biochemical and thermo-chemical routes. Without government subsidy, again with the exception of sugar cane, such crops for oil are a long way from the market.

## International commitments

By referring to the previous paragraphs, it quickly becomes evident that consideration of existing and future energy supply is of significant importance for both developed and developing countries. Moreover, an additional aspect of this challenge is the fact that the global energy context is rapidly being shaped by the escalation in climate change policy formation. Therefore, it is crucial to consider energy supply in the context of international climate change policy.

Since its inception in 1997, the Kyoto Protocol has triggered a series of mega-formulaic international conventions, endless meetings and debates, which have, in themselves, resulted in intense widespread international dialogue regarding the future of energy. Ongoing debate regarding the shortage of conventional hydrocarbons has also raised awareness of the significance of adopting the Precautionary Principle in the context of climate and energy futures. Climate change issues have consequently acted as a backdrop to and provided the impetus for energy policy formation (O'Brien and O'Keefe, 2006).

Unless GHG emissions are mitigated, EU international commitments may not be met. In the UK, research by the Tyndall Centre acknowledges that reducing emission levels by 60 per cent by 2050 will be challenging but could be achieved through additional measures such as energy efficiency and renewable energy technologies (Anderson, 2005; O'Brien and O'Keefe, 2006).

The UK's Kyoto target of a 20 per cent domestic reduction of $CO_2$ below 1990 levels by 2010, however, appears overly ambitious at the very least, as approximately only a 4.2 per cent reduction in $CO_2$ levels had been achieved by 2004 (FOE, 2005). Presently, renewables contribute 2 per cent of the overall energy supply of the electricity, heat and transport sectors as illustrated in Figure 2.7 (BERR, 2008; Stanford, 1997).

Conversely, the EU's drive to become a world leader in GHG reduction and provider of clean energy alternatives has seen the setting for an

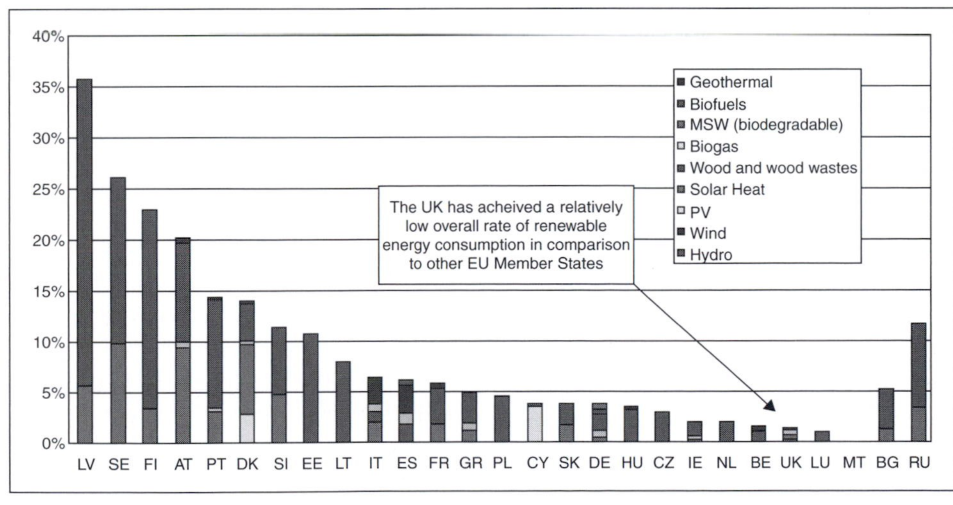


*Source:* COM, 2006

**Figure 2.7** Renewable consumption in EU countries, 2004

ambitious target of 25 per cent of overall energy consumption to be comprised of renewable energies by 2020. A mandatory target of 20 per cent has been set for the same year (EIA, 2008a; COM, 2006).

## Policies for the internalization of externalities

Policies enforcing the internalization of externalities associated with the production and use of energy have the potential to drive the competitiveness of cleaner energy alternatives (Toth and Rogner, 2006). Such policies can take various forms. One could be a cost–benefit analysis approach where estimates of external costs are factored. Another form could be the introduction of an eco-tax, which would tax those fuels and technologies that cause adverse external costs (the level of which would have to be determined) by adding an additional charge to fuel bills (EC, 2003).

## The rapidly changing world of commercial energy supply

Energy security and energy supply are becoming increasingly important to those developed countries whose domestic crude oil is rapidly declining in the face of increasing demand. Energy supplies are prone to volatile market distribution and geopolitical forces leading to market uncertainty and price volatility (WWF, 2008). Currently, more than 75 per cent of proven petroleum reserves are located in OPEC countries[5] with a further 7 per cent in the Russian Federation (WWF, 2008). It is also no coincidence that those countries with the remaining reserves are the most politically unstable where political restructuring in order to exert and enforce control over supply is likely. Examples are outlined in Box 2.5 which summarizes recent activity on Russia's energy markets.

### An incentive for change?

Gazprom is also negotiating with E.On, a German utility company, to increase access to gas via asset swaps, although talks are currently hampered by high oil prices (*European Weekly*, 2008). Amidst a plethora of economic negotiations, most recently, a signed agreement between Gazprom and a German gas company, Verbundnetz Gas AG (VNG), was confirmed on 30 April, 2008, to build €350 million worth of natural gas storage facilities in eastern Germany (Sharewatch, 2008).

In Brazil a recent discovery of oil, billed as the largest world discovery since 2000 by state-owned Petrobas and another discovery nearby of what is estimated as a possible 33 billion barrels, potentially, as *The Economist* describes, '*the third-largest field ever found*' (*Economist*, 2008b, p81) raise questions as to whether these finite, non-sustainable and GHG emitting energy systems will ever foster frugality of energy use, or encourage energy efficiency and support for alternative fuels.

## Energy planning scenarios

Scenarios are a powerful strategic tool used by a diverse multitude of organizations to systematically ascertain the most likely future outcome for any given situation in order to assist decision making and planning by identifying problems, threats and/or opportunities (Bradfield et al, 2005) Essentially, scenario planning is the amalgamation of several rational futures selected from a number of probabilistic outcomes that are developed in enough detail to be highly plausible, using extrapolative, prospective and/or reductionist approaches (Kelly et al, 2004).

Generally, scenarios fall into two broad categories, exploratory or normative. Explorative or descriptive scenarios typically centre on the outcome of decision making with conceivable end-points (O'Brien et al, 2007; Ratcliffe, 2003, cited by Kelly et al, 2004). These scenarios can be defined as a narrative examination of a consistent chain of factors including trends, events or probable outcomes most likely to follow from existing

## Box 2.5 Russia's energy markets

Energy in the form of hydrocarbons such as oil, gas and coal is subject to a diverse range of variables. A significant variable is that of geo-political forces. Oil rich countries from Ecuador to Kazakhstan, although keen to boost profits from natural resources, are often either reluctant in accepting foreign private investment or are entirely exclusionary (*The Economist*, 2008a, p10). For example, the infamous switching off of the gas supply by state owned Gazprom from Russia to the Ukraine in January 2006 not only resulted in the loss of energy supply to the Ukraine, but also affected supply to much of the EU (Parfitt, 2006) – an example of the power of exerted state control over other nation states via energy supply. In the current global climate, nationalism increasingly shapes the supply of primary energy such as oil and gas. For example, Royal Dutch Shell would have greater capacity to supply the developed world if it had freer access to Russia (*Economist*, 2008a, p10).

In the context of international power via nationalization, the Kremlin's tightening control over its energy industries has been marked over the last decade. Acquisition of Yukos in just 4 minutes (*New York Times*, 2008), through State owned Rosneft, an oil and gas company with a revenue of US$33 billion, has been dogged by political controversy but has not halted increased share acquisition (state control is 50.01 per cent of shares) of Gazprom, which currently supplies the EU with 25 per cent of its entire gas needs. The dynamic and unpredictable nature of such swift political and economic manoeuvring is illustrated in Russia's pivotal shift from 1990s liberalism to a more authoritarian corporatism. A case in point is the dismantling of energy giants such as Yukos (formerly Russia's largest oil company) with the arrest in October, 2003 of the owner, a Russian Oligarch, Mikhail Khordokosvky (who was given an eight year jail term, imprisoned in Siberia but now undergoing a second trial in Moscow which opened on 31 March 2009) that paved the way for continuing greater state control of its energy resources (*The Economist*, 2009). Russia's mounting leverage in economic control is steadily extending to international acquisitions. On 29 April 2008, Italian multinational energy company, ENI SpA, was reported to be in talks with Gazprom to sell a share of its Libyan oil field (Dempsey, 2008).

actions (forecasting) including policies, attitudes and behaviour (Schwartz, 1991; Huss, 1988). Normative or strategic scenarios encompass the rationale of starting with a desired outcome and working backwards, termed 'backcasting' or as a reference case, to arrive at the most sustainable scenario (IEA/OECD, 2003; O'Brien et al, 2007) or desirable future that also accounts for a consistent suite of trends and/or patterns (Ratcliffe, 2003, cited by Kelly et al, 2004).

However, a successful range of scenarios does not necessarily mean a wholly accurate depiction of events is ultimately achieved (Bradfield et al, 2005; Shwartz, 1991) Rather, the establishment of scenarios provides a vehicle to enhance decision making by extrapolating probabilistic factors to shape a range of alternative future conditions, therein, equipping planners and policy makers with the knowledge to mitigate or adapt to future conditions. Schwartz (1991) defines scenarios in the following way:

> *Scenarios are not predictions. It is simply not possible to predict the future with certainty... Rather, scenarios are vehicles for helping people learn.* (Schwartz, 1991, p6)

## Parameters shaping scenarios

Energy supply and demand and, increasingly, GHG emissions as well as price and cost parameters, play a determinant role within the suite of factors shaping scenario formation. This range of parameters can be complex, as an emphasis on scenario planning is to significantly challenge the current 'business as usual' approach (Ravetz, 2000, cited by Kelly et al, 2004), incorporating:

1 a synthesis of pertinent information;
2 development of a consistent descriptive and plausible set of explanations of possible future scenarios via a structured methodological application; and
3 an evaluation of the implications of the scenarios in question for the said context (O'Brien, 2004).

Scenarios can be qualitative or quantitative and increasingly, have the propensity to be a synthesis of both (Sluijs et al, 2003). In the context of energy scenarios, technological, quantitative scenario planning has increased over the last decade as demonstrated by the development of energy and climate model futures, readily utilized by the IEA on the international scale (OECD/IEA, 2008).

## Scenarios as a Planning tool

In the face of escalating concern over energy supply and demand and the need to ensure sustainable energy futures, scenario planning is becoming a useful tool in achieving sustainable development in the UK[6] at local and regional levels (Kelly et al, 2004). In the multi-faceted and multi-dimensional paradigm of sustainable development, the aspiration of sustainable development is the need to drive forward to reduce the schism between environmental protection and economic development. Scenario planning lubricates the transition from technologies, practices, regulations of energy systems traditionally reliant on hydrocarbons, towards a more sustainable suite of energy systems capable of reconciling economic growth with carbon emission reductions whilst fulfilling the equivalent social functions and needs of society (Anderson, 2005; Kelly et al, 2004). In the international context, scenario planning is also widely used in determining sustainable energy futures (IEA/OECD, 2003).

Assessments of the future are complex and subject to many uncertainties. A key aspect of energy scenario planning is the role of energy efficiency: from the supply side through to transformation to end-user, encompassing transfer of energy, stability and energy security. However, in order to drive change towards low carbon energy futures and a reduction of emissions via the application of scenarios, Anderson (2005) recognizes the importance of collaboration and the inclusion of all sectors, including both private and public, as necessary to ultimately achieving sustainable energy futures and lower carbon emissions.

## Shell's use of scenarios

Scenario planning energy systems, although a relatively contemporary approach, is not new. An early indicator of the success of modelling scenarios of energy supply and demand was demonstrated by the ability of Royal Dutch Shell to rise to economic prominence despite a series of oil crises during the 1970s. Research by Huss and Honton show that the success of Royal Dutch Shell in averting the oil crises of 1973–1974 and 1979 (and in doing so, leveraging itself into a position of economic prominence) was largely attributable to successful anticipation via scenario planning (Huss and Honton, 1987).

Shell's holistic and qualitative approach in scenario planning stemmed from accounting for uncertainty by recognizing the significance of externalities influencing future outcomes, including change in economic markets, the competitive arena, technology and demographic change (Kelly et al, 2004). As such, during the 1980s, the methodology of scenario planning was influenced principally by Shell's range of qualitative approaches to energy scenarios.

## Scenario planning for future sustainability

The uncertain nature of the future enables, via the instrument of scenario planning, the adoption of approaches such as Sustainable Development (SD) and the Precautionary Principle, both widely promulgated by the EU, to prepare a series of narratives in order to pursue the most desirable

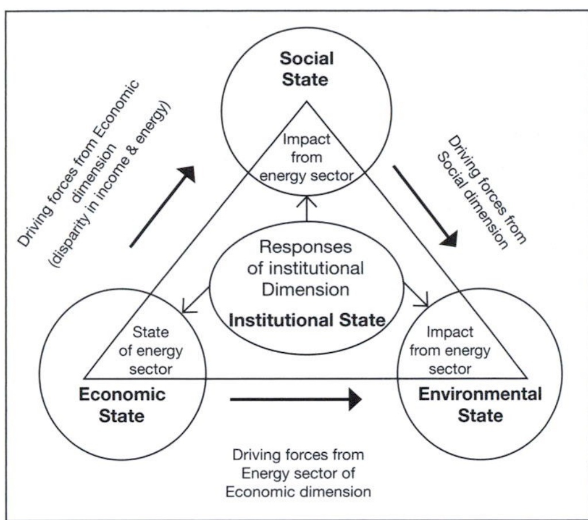


*Source:* IEA, 2008

**Figure 2.8** Interrelations between sustainability dimensions of the energy sector

'path'. The scenario approach is thus particularly suitable for the analysis of possible environmental impacts of energy trends and facilitates analysis of alternative policies as it recognizes the fundamental interrelationship between social, economic and environmental factors as shown in Figure 2.8.

## Scenarios used for energy and emission reduction

Mounting concern about the adverse effects of climate change linked with GHGs has predominantly resulted from increasing climatic changes reported in the media, government reports and peer-review literature. For instance, evidence of anthropogenic climate change has been witnessed in far-flung areas such as the Arctic, where near-surface warming is reported in a recent publication of *Nature* (2008) as being twice the global average over the past two decades (Graversen et al, 2008). Thus, the flexibility of scenario planning has led to a range of scenarios using epistemological theories as indicators to map energy and climate change futures, demonstrated via the application of scenario planning by many governments and international agencies.

In response to the fragmented nature of international debate,[7] scenarios also provide a means by which a range of energy futures can be proffered, to enhance and provide greater clarity as to the best practicable way forward. The value of scenario planning by the Intergovernmental Panel on Climate Change (IPCC) has been widely acknowledged in facilitating future impact analyses of various climate change scenarios, including GHG projections incorporating energy, supply and demand, demographic change, policies and socio-economic factors to create scenarios (narratives or storylines) describing futures that could be rather than futures that will be. Projections in this instance are long-term. For example, the Special Report on Emissions Scenarios (SRES) prepared for the IPCC Third Assessment Report has scenarios projecting to 2100 (IPCC, 2008). However, a major constraint of projecting too far ahead, resulting in inaccuracies the further ahead a scenario is planned, is illustrated by the fact that IPCC predictions of GHG emissions have increased more rapidly than projected (EA, 2008b).

Nevertheless, if constraints are recognized, the efficacy of scenarios in energy futures enable projections to be made and policy decisions undertaken in the face of uncertainty (Mietzner and Reger, 2004). Multifarious groups including military analysts, governments, global institutions and the private sector, such as the EU, the IPCC, the International Institute for Applied Systems Analysis (IIASA) and Royal Dutch Shell (and other energy suppliers) have used scenarios as a tool to initiate or conduct alternative energy policies (EA, 2008b).

## Scenarios for different energy and climate futures

In the UK the Climate Impacts Programme (UKCIP) has produced a wide range of technological climate change scenarios and predictions of future GHG emissions, the latest of which, published in November 2008, are known as UKCIP08. These modelled scenarios detail probabilistic scenario futures for climate change over 30 year intervals from 2010 to 2099 on a 25 × 25km scale within the UK (UKCIP, 2008). According to the latest UKCIP information booklet:

> *UKCIP08 will be the fifth generation of UK climate change scenarios, describing how the climate of the UK might change during the 21st century.* (UKCIP, 2008)

Even staunch critics of the climate change phenomenon recognize the usefulness of scenario planning. For instance, the US Climate Change Technology Program (2005) recognizes that, despite the inherent uncertainties of climate science (in terms of ascertaining the accuracy of levels of greenhouse gas emissions), a global perspective reflecting shifting trends of demographic dynamics and future economic activities as well as the efficacy and financial cost of technological innovation in the renewable energy sector, although characteristically variable, need to be accounted for in scenario planning to mitigate greenhouse gases, climate change, future problems of energy supply and demand, reliability and security (CCTP, 2006).

Advancement in various 'clean' technologies is increasing, one – fuel-cell technology, particularly as hydrogen fuel technology – offers enormous potential for the automotive sector, as reflected in the investment of US$1.2 billion by the US Government and recent talks between the US Senate, the US National Renewable Laboratory, General Motors, the California Fuel Cell Partnership and other research and design (R&D) organizations, on 16 April, 2008 (Adfero, 2008). These advancements in technological innovation enable greater efficacy in the projection of different futures via scenario analysis. Azar et al (2003) similarly advocate the future of hydrogen to fuel transportation through a range of different transport scenarios based on global energy scenarios that meet stringent $CO_2$ emissions constraints.

Although scenario planning is typically centred on developing alternative futures under a range of policy alternatives, many of the existing models are founded on long-term projections based on current, and expected future trends to determine any given future scenario, often termed as a 'reference case'. This approach incorporates top-down and macro-models, such as the IEA models and other models analysed under the Energy Modelling Forum as well as bottom-up approaches such as the technology specific, Western European Markal model under the Energy Technology Systems Analysis Programme (ETSAP) (a generic model tailored by input data over a 40–50 year period for a specific energy system) (Koen and Morales, 2004).

The most socially conscious scenario formulation covering all resources is that undertaken by the Great Transition Initiative (GTI) using Polestar, which is the core programme of the Tellus Institute. It explicitly focuses on research and action for a global civilization of sustainability, equity and well-being (www.gtinitiative.org). More clearly than other scenario building it shows how political economy dictates resource outcomes including energy and emission loadings.

## Limitations

Conflictive indicators are intrinsically difficult to incorporate within energy scenario planning, although various phenomena allow different narratives to be created enabling more comprehensive analysis of policy futures. The efficiency of scenario planning is difficult to judge when scenario planning itself can be built upon other scenarios from different sources as well as a range of indicators. Clarification of the level of impact and the degree of uncertainty is of paramount importance. As such, the complexity of global systems such as climate change and energy, demographic transition and economies of scale can often be reduced too simplistically in order to provide an answer most desired by the government, company and/or international body/agency in question (Kelly et al, 2004). This approach to scenario planning can essentially overlook the importance and influence of complexity and uncertainty.

In its reductionist form, based on past and current trends, scenario planning via forecasting can unintentionally be inherently erroneous, too narrow in focus and too conservative in approach (MacKay and McKiernan, 2003). Conversely, scenario planning can be applied with an approach that is too wide and liberal. Gaining a middle ground is thus problematic given the complexity of energy systems within the socio-economic and geopolitical dynamics that ultimately shape the futures of energy supply, demand and end use. Moreover, the further forward the scenario planning is, the more inaccurate the predictions become (Kelly et al, 2004).

## Conclusion

The most successful approaches to achieving renewable energy capacity and reducing GHGs are witnessed in EU states such as Denmark which have national legislative policies providing viable alternatives to conventional hydrocarbon reliance. An example of an effective policy instrument is the use of FITs, which ensure a profit margin to anyone selling excess electricity back to the national grid as well as requiring utility companies to purchase electricity from renewables (O'Brien and O'Keefe, 2006). Denmark set the pace from the 1980s with a suite of integrative environmental policies and a series of targeted action plans transcending a broad spectrum of Danish industry, simultaneously stimulating economic growth, stabilizing energy consumption, and reduction of GHGs. Rapid development in the renewable energy sector has enabled Denmark to establish itself as a world leader in renewable technologies, particularly wind turbine technology (Moller et al, 2004).

In the case of the EU, it is necessary for the uptake of integrative and transparent policies including the internalization of externalities within national policy regimes across all member states, to ensure the shift from conventional hydrocarbons to sustainable renewable energy sources becomes the norm rather than the privilege and speciality of a few nation states. Integrative policy approaches encourage a broad-based, diverse range of small-scale, local, regional and bottom-up renewable energy systems with grid connection that have the capacity to provide effective, flexible, resilient and secure energy supplies. Renewable technology uses are dependent on indigenous resources, are close to the point of source, and usually managed by the communities in which they serve, ensuring meaningful ownership, local management and security of supply (O'Brien and O'Keefe, 2006; *Nature*, 2007a). Unlike the highly structured, top-down and centralized system of the UK, other member states of the EU, including Denmark and Germany have shown that market intervention, deregulation and a move away from centralized policies (typically small-scale, more interconnected with more interactive users) are demonstrably and arguably more effective in decoupling energy from economic growth (COM, 2006).

Economies of scale have the potential to emulate the large scale of former hydrocarbon infrastructures using indigenous resources for renewable energy, such as on- or offshore wind farms by subsidizing new technologies in order to remove trade and innovation barriers and encourage diversification of RESs. German and Spanish legal regulation, for instance, requires utility companies to purchase electricity from RESs, generating rapid growth in the renewable energy sector, both on the large and small scale (Rio and Gual, 2007).

Ultimately, against conventional hydrocarbon-based energy sources, the costs of renewable technologies are falling, becoming more commercially viable and feasible as alternatives for energy futures. Renewables have exhibited a steady reduction in cost and efficiency in design over the last 20 years. Wind energy cost per kWh has fallen by 50 per cent over the last 15 years whilst the performance of turbines has increased by a factor of 10. Technological advances in the efficiency of solar photovoltaic systems are continuing while the cost of systems is 60 per cent cheaper than in 1990 (COM, 2006).

Additionally, lead times are short, at around 1–2 years and amortization costs are considerably lower than traditional alternatives such as nuclear.

Increasingly, those member states in favour of devolved energy system approaches have engaged wide public support for both the adoption of renewable energy systems capable of reducing the impact of GHGs and a commitment to investment in a diverse range of alternatives to the coal and nuclear options (O'Brien and O'Keefe, 2006). This seems particularly appropriate to many developing countries where power capacities per individual plant are lower than those in developed countries and where there is no opportunity to access cheaper large scale transmission systems.

## Notes

1 A feed in tariff system is an RES-E (renewable energy source – electricity) promotion scheme via a price-based policy which sets the price to be paid for renewable energy per kWh generated, combined with a purchase obligation by utilities (supply systems or grid systems). Costs are borne by consumers or the public budget (Rio and Gual, 2007).

2 For example, the capacity of renewables to supply substantial energy is often flawed by its intermittent nature. To combat the intermittent nature of renewables in the Navarre Region, Spain, 'combined cycle' plants that harness their own waste heat have been developed to meet demand (*Nature*, 2007a).

3 The nature of nuclear energy means that 'lead-times' are long – that is the time from point of order to the delivery of the product (O'Brien and O'Keefe, 2006).

4 Proponents of nuclear often point to the prospect of nuclear fusion which offers the potential for almost limitless, clean power but nuclear fusion research is expensive, remains in the embryonic stages of research and design (R&D) and is not guaranteed of success (*New Scientist*, 2006).

5 Members of OPEC are as follows: Saudi Arabia, Iran, Iraq, Kuwait, Qatar, United Arab Emirates, Algeria, Libya, Nigeria, Indonesia and Venezuela (WWF, 2008).

6 The UK Local Government Act (2000) resulted in a modernization programme for local government incorporating strategic, integrated and futures-orientated approaches as conditions for local governance (Kelly et al, 2004).

7 International debate on climate change has traditionally hinged on various factors such as the US's reluctance to comply with the Kyoto Protocol despite the US being the world's largest GHG emitter. Debate has re-aligned itself towards the emerging economies of India and China as the US has recently been overtaken by China in terms of emissions, due to its burning of fossil fuels and manufacture of cement, although in per capita terms, the US continues to be the most profligate carbon emitting country (*Nature*, 2007b). Additionally, despite international consensus on the dangers of climate change (aside from a recalcitrant few nations, namely the US, no absolute consensus has been agreed as to the 'tipping point' value of GHGs (O'Brien and O'Keefe, 2006).

## References


Adfero (2008) Hydrogen and Fuel Cell Talks Planned by US Senate. Available at: www.fuelcelltoday.com/online/news/articles/2008-04/Hydrogen-and-fuel-cell-talks-pla

Anderson, K. (2005) Decarbonising the UK – Energy for a Climate Conscious Future, Tyndall Centre for Climate Change Research. Available at: www.tyndall.ac.uk/media/news/tyndall_decarbonising_the_uk.pdf

Azar, C., Lindgren, K. and Andersoon, B. A. (2003) 'Global energy scenarios meeting stringent $CO_2$ constraints – cost-effective fuel choices in the transportation sector', *Energy Policy*, vol. 31, pp961–976

BERR (2008) UK Renewable Energy Strategy. Available at: www.berr.gov.uk/energy/sources/renewables/strategy/page43356.html

Bradfield, R., Wright, G., Burt, G., Cairns, G. and Heijdun, V. D. (2005) 'The origins and evolution of scenario techniques in long range business planning', *Futures*, vol 37 pp795–812

Burt, G. and Van der Heijden, K. (2003) 'First steps: towards purposeful activities in scenario thinking and futures studies', *Futures*, vol. 36, no. 10, pp1011–1026

Castles, I. and Henderson, D. (2003) 'Economics, emissions scenarios and the work of the IPCC', *Energy & Environment*, vol. 14, no. 4

COM (2000) 769: Green Paper: Towards a European strategy for the security of energy supply, European Union, Directorate of Energy and Transport, Brussels

COM (2002) Communication from the Commission to the Council and the European Parliament, Final report on the Green Paper 'Towards a European strategy for the security of energy supply', 321 final. Available at: http://ec.europa.eu/energy/green-paper-energy-supply/doc/green_paper_energy_supply_en.pdf

COM (2006) Renewable Energy Road Map Renewable energies in the 21st century: Building a more sustainable future. Available at: http://eur-lex.europa.eu/LexUriServ/LexUriServ.do?uri=COM:2006:0848:FIN:EN:PDF

Dempsey, J. (2008) 'Gazprom and Eni Plan Gas Pipeline in Libya', *New York Times* Online. 9 April. Available at: www.nytimes.com/2008/04/09/business/worldbusiness/09pipeline-web.html?_r=1&scp=2&sq=turkey&st=nyt

Dincer, I. (2002) 'The role of Exergy in energy policy making', *Energy Policy*, vol. 30, pp137–149

EA (2008a) EU Renewable Energy Policy. Available at: www.euractiv.com/en/energy/eu-renewable-energy-policy/article-117536

EA (2008b), Shell backs carbon pricing in 2050 energy scenario. Available at: www.euractiv.com/en/energy/shell-backs-carbon-pricing-2050-energy-scenario/article-171435

EC (2003) External Costs: Research Results on Socio-environmental Damages due to Electricity and Transport. Available at: http://ec.europa.eu/research/energy/pdf/externe_en.pdf

ECB (2002) Macroeconomic stability and growth in the European Monetary Union. Available at: www.ecb.int/press/key/date/2002/html/sp021216.en.html

*Economist* (2008a) 'Peak nationalism, oil keeps getting more expensive – but not because its running out', *TheEconomist*, 5 January, vol. 386, no. 8561, p10

*Economist* (2008b) 'More bounty, could Brazil become as big an oil power as it is an agricultural one?' *The Economist*, 19 April, vol. 387, no. 8576, p81

*Economist* (2009) 'The Khodorkovsky case, A new Moscow show trial', *The Economist*, vol. 391, no. 8625, pp35–36

EEA (2007) Greenhouse gas emission trends and projections in Europe 2007. Available at: http://reports.eea.europa.eu/eea_report_2007_5/en/Greenhouse_gas_emission_trends_and_projections_in_Europe_2007.pdf

EIA (2008a) International Energy Outlook, EIA/DOE, Washington DC. Available at: www.eia.doe.gov/oiaf/ieo/pdf/0484(2008).pdf

EIA (2008b) World Energy and Economic Outlook, Paris: OECD. Available at: www.eia.doe.gov/oiaf/ieo/world.html

ESMAP (2007) Technical and Economic Assessment of Off-grid, Mini-grid and Grid Electrification Technologies. Available at: www.esmap.org/filez/pubs/4172008104859_Mini_Grid_Electrification.pdf

ETSAP (2008) Energy Technology Systems Analysis Programme, Annex VIII: Exploring Energy Technology Perspectives. Available at: www.etsap.org/markal/main.html

European Commission (2003) External Costs: Research results on socio-environmental damages due to electricity and transport, Luxembourg. ISBN 92-894-3353-1. Available at: http://ec.europa.eu/research/energy/pdf/externe_en.pdf

*European Weekly* (2008) Gazprom, EON ink Yuzhno-Russkoye development deal, 6 October, Issue 802. Available at: www.neurope.eu/articles/90100.php

FAO (2008) The State of Food and Agriculture, 2008, Biofuels: Prospects, Risks and Opportunities. Available at: www.fao.org/docrep/011/i0100e/i0100e00.htm

FAO (2009) Small-Scale Bioenergy Initiatives: Brief description and preliminary lessons on livelihood impacts from case studies in Asia, Latin America and Africa. Available at: ftp://ftp.fao.org/docrep/fao/011/aj991e/aj991e.pdf

FOE (2005) Emissions Breach Kyoto Target, 5 September, Press Release: Friends of the Earth (FoE). Available at: www.foe.co.uk/resource/press_releases/emissions_breach_kyoto_tar_02092005.html

Graversen, R. G., Maurisen, T., Tjernstrom, M., Kallen, E. and Svensson, G. (2008) 'Vertical structure of recent Arctic warming', *Nature*, 541, 3 January, pp53–55

Huss, W. R. (1988) 'A move toward scenario analysis', *International Journal of Forecasting*, no.4, pp377–388

Huss, W. R. and Honton, J. E. (1987) 'Scenario planning what style should you use?', *Long Range Planning*, vol. 20, no. 4, pp21–29

IAEA (2006) Nuclear Power and Sustainable Development, IAEA, Vienna. Available at: www.iaea.org/OurWork/ST/NE/Pess/assets/06-13891_NP&SDbrochure.pdf

IEA (2005) World Energy Outlook, IEA/OECD, Paris. Available at: www.iea.org/Textbase/npsum/WEO2005SUM.pdf

IEA (2008) Interrelations between sustainability dimensions of the energy sector. Available at: www.iea.org/textbase/papers/2001/csd-9.pdf

IEA/OECD (2003) Energy to 2050: Scenarios for Sustainable Futures IEA/OECD, Paris. Available at: http://iea.org/textbase/nppdf/free/2000/2050_2003.pdf

IPCC (2000) IPCC Special Report: Summary for Policy Makers. Available at: www.ipcc.ch/pdf/special-reports/spm/sres-en.pdf

IPCC (2005a) Expert meeting on Emission Scenarios, 12–14 January, Washington, DC, Meeting report, p4. Available at: www.mnp.nl/ipcc/docs/IPPC_Expert_meeting_emissions_Washington_2005.pdf

IPCC (2005b) Workshop on New Emission Scenarios 29 June–1 July, Laxenburg, Austria, Meeting Report. Available at: www.ipcc.ch/meet/othercorres/ESWmeetingreport.pdf

IPCC (2008) Socio-Economic Data and Scenarios. Available at: http://sedac.ciesin.columbia.edu/ddc/index.html?method=all&format=long&sort=score&config=htdig&restrict=&exclude=&words=scenario+projection&goButton=go#

Jacobson, M. Z. (2006) Addressing Global Warming, Air Pollution Health Damage, and Long-Term Energy Needs Simultaneously, Dept. of Civil and Environmental Engineering, Stanford University, CA. Available at: www.stanford.edu/group/efmh/jacobson/ClimateHealth4.pdf

Kahane, A. (1992) 'Scenarios for energy: sustainable world vs global mercantilism', *Long Range Planning*, vol. 25, no. 4, pp38–46

Kelly, R., Sirr, L. and Ratcliffe, J. (2004) 'Futures thinking to achieve sustainable development at a local level in Ireland', *Foresight*, vol. 6, no. 2, pp80–90

Koen, E. L. and Morales, S. R. (2004) 'Response from a MARKAL technology model to the EMF scenario assumptions', *Energy Economics*, vol. 26, pp655–674

Lovelock, J. (2006) *The Revenge of Gaia: Earth's Climate Crisis and the Fate of Humanity*, Allen Lane, London

MacKay, R. B. and McKiernan, P. (*2003*) 'The role of hindsight in foresight: refining strategic reasoning', *Futures*, vol 36, no 2, pp161–179

Mietzner, D. and Reger, D. (2004) Paper 3: Scenario Approaches – History, Differences, Advantages and Disadvantages. Available online at: http://forera.jrc.ec.europa.eu/fta/papers/Session%201%20Methodological%20Selection/Scenario%20Approaches.pdf

Moller, H. B., Sommera, S. G. and Ahringb, B. K. (2004) 'Methane productivity of manure, straw and solid fractions of manure', *Biomass and Bioenergy*, vol. 26, pp485–495

Muradov, N. Z. and Veziroglu, T. N. (2005) 'From hydrocarbon to hydrogen carbon to hydrogen economy', *International Journal of Hydrogen Energy*, vol. 30, pp225–237

*Nature* (2007a) 'Graphic detail, gas exchange: $CO_2$ emissions 1990–2006, 28 June, vol. 447, pp1038

*Nature* (2007b) 'Energy-go-round', *Nature*, 28 June, vol. 447, pp1047–1048

*Nature* (2007c) 'Market watch', *Nature*, 28 June, vol. 447, pp1045

*Nature* (2008) 'Vertical structure of recent Arctic warming', *Nature*, 3 January, vol. 541, pp53–55

*New Scientist* (2006) No future for fusion power says top scientist. Available at: www.newscientist.com/channel/fundamentals/dn8827-no-future-for-fusion-power-says-top-scientist.html

*New York Times* (2008) Rosneft Wins Yukos Stake in 4 Minute Sale. NYTimes Online, 28 March 2007. Available at: www.nytimes.com/2007/03/28/business/worldbusiness/28yukos.html

O'Brien, F. A. (2004) 'Scenario planning: Lessons for practice from teaching and learning,' *European Journal of Operation Research*, vol. 152, no. 3, pp709–22

O'Brien, G. and O'Keefe, P. (2006) 'The future of nuclear power in Europe: A response', *International Journal of Environmental Studies*, vol. 3, no. 2, pp121–130

O'Brien, G., O'Keefe, P. and Rose, J. (2007) 'Energy, poverty and governance', *International Journal of Environmental Studies*, vol. 64., no. 5, pp607–618

Odell, P. R. (2004) *Why Carbon Fuels Will Dominate the 21st Century's Global Energy Economy*, Brentwood, UK: Multi-Science Publishing

O'Riordan, T. (2001) *Environmental and Ecological Economics in Environmental Science for Environmental Management*, Upper Saddle River, NJ: Prentice Hall Publishers

OECD/IEA (2008) *From 1st to 2nd Generation Biofuel Technologies*, Paris: OECD

Parfitt, T. (2006) 'Russia turns off supplies to Ukraine in payment row, and EU feels the chill', *Guardian* Online, 2 January. Available: : www.guardian.co.uk/world/2006/jan/02/russia.ukraine; http://themes.eea.europa.eu/Sectors_and_activities/energy/indicators/EN35%2C2007.04/EN35_EU25_External_costs_2006.pdf

Peterson, G. D., Cumming, G. S. and Carpenter, S. R. (2003) 'Scenario planning: a tool for conservation in an uncertain world', *Conservation Biology*, vol. 17, no. 2, pp358–366

Rio, del P. and Gual, M. A. (2007) 'An integrated assessment of the feed-in tariff system in Spain', *Energy Policy*, vol. 35, pp994–1012

Schwartz, P. (1991) *The Art of the Long View, Currency*, Doubleday

Sharewatch (2008) Gazprom, VNG to invest 350 million euros in German gas storage facilities. Available at: www.sharewatch.com/story.php?storynumber=302915

Sluijs Van der. J., Kloprogge P., Risbey, J. and Ravetz, J. (2003) A Synthesis of Qualitative and Quantitative Uncertainty Assessment: Applications of the Numeral, Unit, Spread, Assessment, Pedigree (NUSAP) System, Copernicus Institute for Sustainable Development and Innovation, Department of Science Technology and Society, Utrecht University, The Netherlands. Available at: www.nusap.net/downloads/JvdS_NUSAP_abstract.pdf

Stanford, A. (1997) 'A vision of a sustainable energy future', *Renewable Energy*, vol. 10, no. 2/3, pp417–422

Toth, F. L. and Rogner, H.-H. (2006) 'Oil and nuclear power: Past, present, and future', *Energy Economics,* 28, pp1–25. Available at: www.iaea.org/OurWork/ST/NE/Pess/assets/oil+np_toth+rogner0106.pdf

UKCIP (2008) Combining climate & socio-economic scenarios. Available at: www.ukcip.org.uk/index.php?option=com_content&task=view&id=204&Itemid=9

UN (2007) The International Development Agenda and the Climate Change Challenge. Available at: www.un.org/esa/policy/devplan/2007%20docs/climate.pdf

US CCTP (2006) U.S. Climate Change Technology Program: Strategic Plan, U.S. Lead-Agency Department of Energy. Washington DC. Available at: www.climatetechnology.gov/stratplan/final/CCTP-StratPlan-Sep-2006.pdf

Van Gelder, B. and O'Keefe, P. (1995) *The New Forester*, London: IT Publications

Vicini, G., Gracceva, F., Anil, M. and Constantini, V. (2005) Security of Energy Supply: Comparing Scenarios from a European Perspective FEEM Working Paper No. 89.05. Available at: http://ssrn.com/abstract=758225

WEC (2007) World Energy Policy Scenarios to 2050, London: World Energy Council. Available at: www.worldenergy.org/documents/scenarios_study_online_1.pdf

WETO (2003) World energy, technology and climate policy outlook 2030. Available at: http://ec.europa.eu/research/energy/gp/gp_pu/article_1257_en.htm

WETO-H2 (2006) World Energy and Technology Outlook 2050 – WETO-H2, Brussels: European Commission Directorate-General for Research. Available at: ftp://ftp.cordis.europa.eu/pub/fp7/energy/docs/weto-h2_en.pdf

WHO (2005) WHO Air quality guidelines for particulate matter, ozone, nitrogen dioxide and sulfur dioxide: Global update 2005: Summary of risk assessment, WHO. Available at: http://whqlibdoc.who.int/hq/2006/WHO_SDE_PHE_OEH_06.02_eng.pdf

WWF (2008) Plugged In: The end of the oil age. Available at: http://assets.panda.org/downloads/plugged_in_full_report___final.pdf

# 3

# Energy and Development Planning

This chapter is about the development of principles that should underlie energy policy on a global basis. They are commitments to a sustainable energy economy and to energy security. Given these concerns, a commitment to a model of 'good practice' must be constructed and implemented in both developed and developing countries. This inevitably demands a reflexive policy that takes into direct consideration the heterogeneous nature of energy supply technologies, and local, regional and national cultural traditions. The following discussion suggests a number of socio-economic, technical and environmental considerations for designing a sustainable energy economy for implementation in developing countries. Their implementation is illustrated by a case study of wood energy in Africa.

## Re-thinking energy systems

The United Nations Commission on Sustainable Development (UNCSD) argues that access to affordable energy services is a starting point for thinking about energy systems (CSD, 2002). An energy system must deliver the appropriate service, at the point of need with the following characteristics:

- It should enhance livelihood strategies. Women in India regularly spend between two to seven hours each day collecting fuel for cooking and, in rural sub-Saharan Africa, many women carry 20kg of fuelwood an average of 5km every day. This time could be spent on childcare, education, socializing and income generation (Wakeford, 2004).
- It should not contribute to climate change but should use renewable sources or, at a minimum, be carbon neutral.
- It should be democratic. Ownership and management should be local.

Thinking about energy technologies usually focuses only on the supply side. System thinking recognizes the demand side, that users are an essential component. But users are people and they have many roles in developing and managing the energy system. One simple example demonstrating the lack of energy services is that more than a third of the world's population (around 2 billion people) rely on traditional fuels (biomass, wood, charcoal and dung) as a principal source of fuel for cooking and space heating. This figure is set to rise by a further 200 million by 2030 should current trends continue. Tragically, these fuels increase morbidity and mortality among more than 1.6 million women and children a year from the pollution of indoor cooking fires (ITDG, 2002).

Adding to the problem is that environmental degradation from commercial logging and agricultural colonization accelerates rural to urban migration. This accelerated rate of urbanization leads to a corresponding, and therefore growing, demand for charcoal and increases pressure on the local biomass resource (Munslow et al, 1988; Leach and Mearns, 1998). With increasing pressure on the biomass resource due to charcoal use, induced by human settlement together with ecologically insensitive and destructive land use practice, accelerated environmental degradation becomes an inevitable and unavoidable consequence of energy poverty (Hardoy et al, 2001). Such practices not only raise the number of hazards and disasters dramatically at the local level but contribute significantly to heightened levels of risk and reducing resilience (Abramovitz, 2001).

Technology transfer as a single component of an energy approach which aims for sustainable, long-term energy provision is insufficient (Wakeford, 2004). Reliance on single or 'one-off' technological 'fixes' without consideration of the need for adaptive capacity to build capability and robustness is erroneous, as such solutions cannot be maintained, adapted or used for future and naturally evolving energy service requirements. At best, this approach works in the short term. At worst, such interventionist solutions are expensive, inappropriate, ill-thought out and ultimately reduce peoples' resilience and therefore capacity and capability to spread risk in terms of energy sources and service provision. Resilience is therefore key to any energy development process. Resilience is not a science, but a process, using human capacity and ingenuity (both of which are socially constructed) to mitigate vulnerabilities and reduce risks. It is a measure of adaptive capacity. Adaptation options or strategies can be identified using participatory techniques based upon indigenous knowledge and local understanding to develop solutions. This is a social learning process, a collective action and reflection that occurs among individuals and groups as they work to improve a situation (Keen et al, 2005).

The concept of resilience has become an increasingly important factor within the context of sustainable development as it is precisely the element of resilience that needs to be considered within development to avoid disastrous consequences and increased risk to both human populations and the environment (Abramovitz, 2001). Resilience is not only a central facet of the sustainable development arena (Batabyal, 1998; Van der Leeuw and Leygonie, 2000; Waller, 2001; Johnson and Wielchelt, 2004), but it is core to the framework for disaster risk reduction as the global issue of climate change which affects people at the local level, often overlapping and converging with broader international agendas. For instance, the Millennium Development Goals (MDGs) and the Hyogo Declaration of the United Nations International Strategy for Disaster Reduction (UN/ISDR) 2005 World Conference declare:

> *We recognize the intrinsic relationship between disaster reduction, sustainable development and poverty eradication, among others.* (UN/ISDR, 2005)

The focus of actions within the international development arena are framed by ideas of governance, risk identification and reduction and preparedness (UN/ISDR, 2005). The Fourth Assessment Report of the IPCC also goes some way to emphasize that the climate change issue is not just about increased warming but that it is about climatic variability. Those most vulnerable to such variability and uncertainty live in developing countries which will be increasingly subject to droughts, floods and storm surges. Therefore, a greater emphasis on pre-disaster planning for natural hazards with an emphasis on resilience and recovery of livelihood systems is paramount in planning energy system futures (IPCC, 2007). Energy sub-systems at the livelihood level necessarily require a coherent governance response. The response needs to drive towards a resilience focus within development to allow space for local coping mechanisms and to ensure these resilience mechanisms are inbuilt within appropriate technologies (O'Keefe et al, 2003; O'Brien et al, 2006).

## Approaches to energy in addressing energy poverty: From theory to practice – the case of electricity

The process of energy technology development in developing countries is generally approached with a thinking that embodies a series of progressive steps in the form of an energy ladder (O'Keefe, 1993; Barnes and Floor, 1996). The progression of this series of steps within households is typified in conventional thinking with a baseline from biomass to kerosene, then gas and ultimately, electricity, as illustrated in Figure 3.1. Unfortunately in developing countries, as incomes increase, there is no automatic step up the energy ladder; as people access improved fuels and technologies, they tend to retain their older carbon intensive technologies – such as wood and charcoal – because of insecurity of supply and availability of technology.

In the new millennium, where there is considerable doubt about the long-term future of carbon-based fuels such as kerosene, both in terms of their availability and contribution to anthropogenic acceleration of climate change, it seems nonsensical to assume the same progression for the energy poor. It is not simply an issue of fuels but also of transmission systems. The frequent assumption has been that the desired energy routing is from local acquisition of biomass through to electricity delivered by a national transmission system. But these transmission systems themselves provide significant problems, not least because they lead to assumptions that large scale energy technologies must feed them, working against efficiency and renewables.

Ladder 1 in Figure 3.2 illustrates the conventional model that restricts efficiency and renewable strategies. What is needed is a move to electrification that is not sourced to a single large supplier. Persuading the middle classes in developing countries that their access to audio visual and IT services does not require large scale generation in some distant city is a challenge which is directly linked to trying to increase energy access for the poor.

For those currently at the biomass level of Ladder 1 (Figure 3.2) a different approach is needed. The step change shown by the shaded area in Figure 3.2 represents a shift from biomass to the first step on Ladder 2. This is the area

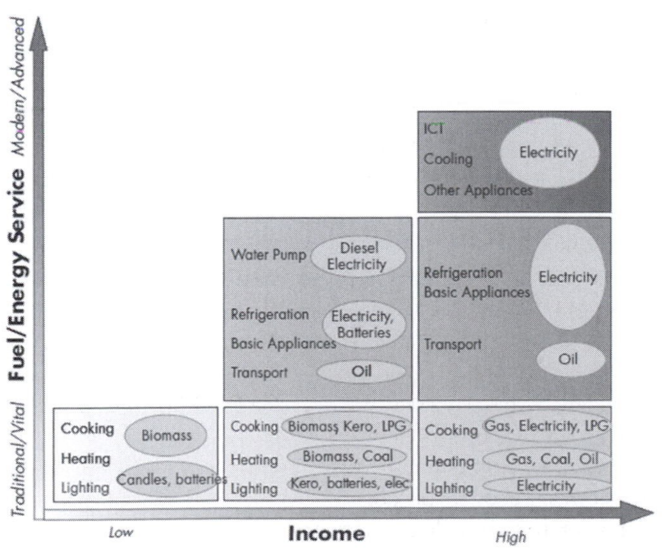


*Source:* IEA, 2002

**Figure 3.1** Conventional energy ladder approaches

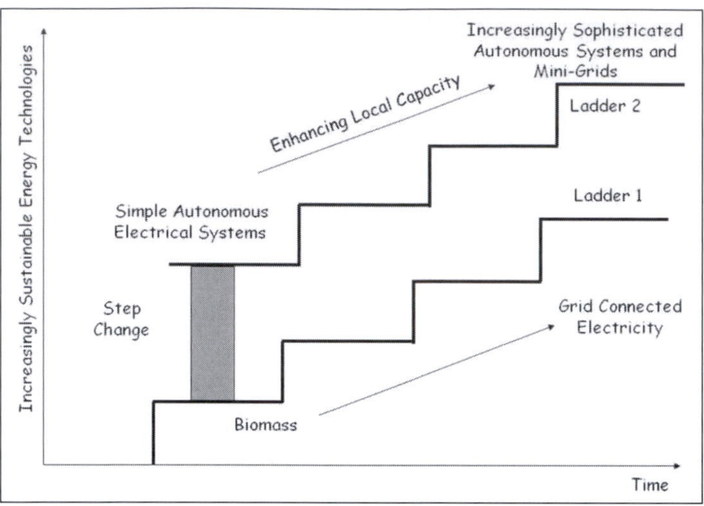


*Source:* O'Brien et al, 2006

**Figure 3.2** Alternative energy ladder approaches

where interventions are needed by developmental programmes to accelerate the change from traditional fuel sources represented in Ladder 2 (EASE, undated). Ladder 2 then represents a new development pattern – different from that of Ladder 1. It assumes that the technologies used throughout the ladder will use renewable resources. Second, the end point assumes the development of autonomous small scale grid systems. The overall goal of this approach is poverty alleviation but in conjunction with the development of governance expressed through the ownership, management and development of the energy system (O'Brien et al, 2006).

There are three reasons behind this necessary logic. First, many renewables are scattered resources, scattered through space and time. They are ill suited to large scale interconnected grid systems even when the potential surplus power can be fed back into the transmission system. Despite this caveat, renewable energy can be built as a clean and continuous system.

Second, there are significant issues about the longevity of large scale interconnected approaches. Mycle Schneider, director of the World Information Service on Energy (WISE-Paris), writing for the World Wide Fund for Nature (WWF), clearly thinks that the days of large grid connected systems are over. This argument, applicable to much of the developing world does not, however, contradict developments in Western Europe where the UK national grid and the French electricity system are the most appropriate technology to supply part of a pan-EU supranational grid. Schneider lists the benefits of small scale systems as low capital and maintenance costs, high investment flexibility and low grid losses (Schneider, 2000). David Appleyard claims that in Latin America, where around a third of all people do not access mains electricity, distribution companies are estimated to be losing around 40 per cent of generated electricity due to theft, poor maintenance and inefficiency. Similar losses can also be found in the distribution networks of Africa, Southeast Asia, the former Soviet Union and large swathes of eastern Europe (Appleyard, 1999). Dunn makes a different point that 24/7 electricity supply at high voltages are a specific requirement to industrialized societies, not the developing world (Dunn, 2000). Decentralized micro-system approaches act to reduce vulnerability through increasing diversity in power supply options and thus increase resilience.

Third, autonomous systems, operated and owned locally, can act as a lever for developing governance. Essentially, they are democratic and reflect the history of the development of energy resources in developed countries where in the first instance most systems were developed by local authorities. Such systems are still operational in developed countries; for example, on the Hebridean Isle of Unst unused power from a wind turbine is used to generate hydrogen that powers fuel cells during periods of low or no wind (PURE, 2000).

In order to progress such system development, there are four levels of planning that are required, namely:

- *Needs assessment*: ensuring that a clear understanding of energy needs is generated.
- *Energy mapping*: knowing what local energy resources exist.
- *Support systems* (technical, human and financial) that are clearly identified.
- *Appropriate level*: defining the entry level (ITDG, 2002; Brunt et al, 2004).

All case studies show the importance of micro-finance schemes owned by the community in ensuring the long-term success of small scale renewable energy projects. It is not a question of technological capacity but one of political will creating the correct regulatory environment to make markets work. Table 3.1 summarizes appropriate dimensions and characteristics for sustainable projects.

**Table 3.1** Features of sustainable energy systems

| Dimension | Characteristics |
|---|---|
| Appropriate | Matched to the needs of the community and to cultural norms renewable resource |
| Exploits indigenous renewable resource | Exploits local renewable resources such as water, solar, wind, geothermal, etc. |
| Capacity enhancing | It should enhance local capacity and time to devote to other productive endeavours such as income generation, education, socializing, etc. |
| Adaptable | It should be capable of expanding and developing along with the capacity of the community |
| Easy to repair and maintain | Ease of use and repair freeing the local community from dependence on outside expertise and distant supply lines |
| Upgradeable | They should be able to integrate technological improvements in a seamless manner appropriate with the development of capacity of the users |

*Source:* Adapted from O'Brien et al, 2006

## Bridging the energy gap: What do developed countries offer through technology transfer?

The main way of bridging the energy gap between rich and poor is through technology transfer. Technological developments are rooted in a system which sees the poor paying significant monies to the developed world in patent rights, capital repayments and insurance that substantially works against the development of local systems. In terms of scale, the technologies are large and complex. This leads to problems of fragility and vulnerability.

The issue of vulnerability of energy systems, however, is very real in the developed world. Transferring such vulnerability to the developing world merely leads to a build up of problems. Lovins and Lovins describe the vulnerability of the US energy systems as: the 'unintended side effect of the nature and organization of highly centralized technologies' (Lovins and Lovins, 1982, p2).

Energy systems in the developed world depend upon the availability of energy supplies, which are vulnerable to global geopolitical forces.

Interruptions to supply can lead to economic shocks and concerns have been raised within the European Union (EU) regarding the security of energy supply (O'Brien et al, 2006). The EU Green Paper on Energy Security points out that 70 per cent of EU energy supplies will be imported by 2030 (COM, 2000). The Commission's response to the Green Paper acknowledged that it drew attention to the structural weaknesses and geopolitical, social and environmental shortcomings of the EU's energy supply, notably as regards European commitments in the Kyoto Protocol (COM, 2002). Energy systems in the developed world are not democratic. Ownership is generally concentrated either through the state or, more recently, the market. Both approaches are hierarchical. Privatization did not bring a move to a horizontal industrial structure or a more democratic shareholding base. As Thomas argues: 'there is a serious risk that the electricity industry will become a weakly regulated oligopoly with a veneer of competition' (Thomas, 2004, p3).

Technology transfer debates to date have not addressed these issues of scale. In particular, technology transfers do not deal directly with the issue of the access of the poor to the services generated, for example, the Clean Development Mechanism (CDM) of the Kyoto Protocol. A recent Organization for Economic Cooperation and Development (OECD) survey summarized the emerging trend: 'a large and rapidly growing portion of the CDM project portfolio has few direct environmental, economic or social effects other than GHG mitigation, and produces few outputs other than emissions credit' (Ellis et al, 2004, p32).

CDM is simply shifting the location of where the greenhouse gas emission reductions are generated and is failing to make a contribution to the sustainable development of the host.

## Addressing the obstacles

The strongest obstacle to the delivery of a small scale renewable future is the objections of existing producers (Dunn, 2000). The conventional energy sector receives some US$200 billion in subsidies, making it difficult for emerging technologies such as renewables to compete. These subsidies, coupled with a range of institutional and policy barriers inhibit the development of the renewable energy sector (Sawin, 2004). The International Renewable Energy Conference in 2004 concluded that removing subsidies and internalizing external costs would be necessary to establish a level playing field (O'Brien et al, 2006).

Despite the obstacles, the costs of renewable energy technologies continue to fall; the renewable energy market reached a high of US$30 billion in 2004. Grid-connected Solar PV outstripped all other energy technologies in the world, by growing in existing capacity by 60 per cent each year from 2000 to 2004. Harkins (2000) provides an excellent case study of the growth in Kenya. In second place was wind power, which grew by 28 per cent in 2004 (Martinot et al, 2005). A number of commentators believe that falling costs and increasing demand for new energy technologies suggests renewables could become a significant part of the mainstream energy economy within a decade (Flavin, 2003; Sawin 2004).

## Facing the energy poverty issue

There are still some 1.4 billion people who will be without grid connection by 2030 (IEA, 2002). The IEA argues that alternative pathways are under development; but a comprehensive approach would exercise the technology and develop capacity in transferring it to the beneficiary. This process would enhance local capacity and governance whilst taking a step towards breaking the energy poverty cycle (IEA, 2005). Markets respond to signals. But industry often misses signals when governments do not legislate and regulate clearly, creating disruptive technologies (Christensen, 1997; Christensen and Raynor, 2003). In the energy sector the emergence of small scale renewable technologies offers the scope for a new approach to energy production and management but only if there is a clear framework of legislation and regulation and no subsidies to existing supply (O'Brien et al, 2006).

There are examples of how incentives can be generated. For example, the Conference of Parties (COP) has a mechanism in place through the CDM for promoting small scale renewable energy and efficiency projects. But as Brunt et al (2004) show, despite changes to reduce costs and approval time there remain considerable barriers to a greater take-up of small scale projects, particularly in rural areas where many are excluded from the grid (Brunt et al, 2004). One possibility would be to assign a greater value to the carbon emission reductions (CERs) generated by such projects. Similar actions were adopted by the World Bank Community Development Carbon Fund (CDCF) that places a 15–20 per cent premium on carbon (Carbon Market Update, 2005). Helme argues that promoting a sectoral, bottom-up approach, based on technological feasibility and cost-effectiveness, could prove to be administratively and politically feasible (Helme, 2005). To leave the poor without power militates against international commitments set out in the MDGs. The private sector is unlikely to act without an indication from the international community that it must lead the fight against energy poverty (O'Brien et al, 2006).

## Learning the lessons?

Three simple lessons emerge. The energy problem cannot be solved without solving the poverty problem but the poverty problem cannot be solved without solving the energy problem. Top-down systems have not delivered beneficial results to date and tend to generate vulnerability while simultaneously precluding the development of decentralized renewable systems. Markets must be steered by strong international legislation and regulation that together promote sustainable energy futures with an emphasis on renewables and energy efficiency that can be accessed by the poor.

## Africa's energy challenge

Africa, and to a lesser extent, Asia and Latin America, find that a dominant energy source is biomass. It is essentially the forced energy choice of the poor, largely used on three stone fires for cooking and space heating. It remains a key energy challenge but one that has to be addressed in a different way because, particularly in rural areas, most of the energy sourcing and use is beyond the market (O'Keefe and Munslow, 1989; Kirkby et al, 2001; Mahiri and Howorth, 2001).

It is a mistake to attribute large scale deforestation to wood cutting for energy; more important processes are land clearance for new agriculture, grazing and bushfires. However, in local instances, fuelwood consumption can be a major contributory process, particularly near settlements and in sensitive eco-systems. For example, in the arid zones of developing countries, like the Sahel in Africa, stands of trees are being destroyed or threatened by fuelwood cutting. Either there is no replanting or it is insufficient to maintain existing stocks and yields: natural recovery is also hampered by livestock grazing.

The most serious problems arise not from the rural population, but from the urban population which demands a steady supply of fuelwood and charcoal. Loggers have a wide variety of resources from which to cut and will simply look elsewhere if any village or rural region tries to collect a selling price (in stumpage fees). In short, local people have no bargaining power with which to enhance their management of local resources. Experience shows, however, that the only people who can protect and replant trees are local villagers. Planting by authorities is difficult to manage on the required scale and it is expensive. But as long as prices to producers remain low and levels of responsibility remain unclear, local people often have little interest in planting or protecting trees and, consequently, sustainable wood production is virtually impossible (Soussan et al, 1992).

### Charcoal

In rural areas cutting fuelwood for conversion to charcoal (essentially an urban fuel) is a major source of income and non-agricultural employment. End-users prefer charcoal to woodfuels for a number of reasons. It is cleaner, produces less smoke, is easier to handle, easier to light, involves shorter cooking times, and is free from insect attack and wetting. It is usually made by stacking wood, covering it with a layer of earth and letting it burn with a limited supply of air. The efficiency of these simple earthen kilns is low, typically ranging from 40 to 60 per cent. One bag of charcoal is equivalent in volume to ten bags of wood. Charcoal is a 'priced' fuel and because it is popular, and its method of production is poor, a number of ways to promote efficiency in both production and use now exist. The introduction of more efficient kilns is one obvious move. If capital investment is made, ranging from a few hundred dollars for simple modifications to traditional kilns to $100,000 or more for a modern continuous retort, higher energy efficiencies can be achieved (Khristoferson and Bokalders, 1987). The source of wood is also a matter for concern. Indiscriminate felling for charcoal has had detrimental effects, especially when it happens close to urban areas, where, among other things, woodland as an amenity is destroyed. One option is to identify charcoal production areas which can then be institutionalized and improved. It is a mistake to ban charcoal production altogether, as in The Gambia, because, where there is sufficient demand, banning simply leads to cross-frontier transfers. This, in turn, can impoverish other areas (in the case of The Gambia it is the Cassamance region of Senegal), which have no forms of redress. As even the best traditional rural technologies lose 60 per cent of the energy content of wood in converting it into charcoal, and as the more customary loss is around 75 per cent, urban household demand exacerbates deforestation. Even though charcoal stoves use slightly less energy than the corresponding wood stoves, in the end changing over from wood to charcoal doubles, if not triples, wood consumption.

A global problem that parallels both energy use and changes in land use is the conservation of tropical rainforests and their biological diversity. They account, very roughly, for 50 per cent of all species of flora and fauna. Since 1950, over 10,000 species have been lost, largely through the destruction of tropical rain forests. However, although it does not have such an immediate impact on biodiversity, deforestation is larger in scale in semi-arid and arid regions than in tropical rainforests, where logging and land clearance are for agriculture, not for fuelwood. More emphasis has to be given, in energy policies, to the sustainable management of local land use systems in which woody biomass for energy is just one component. Energy efficiency technologies that reduce the cost and secure the supplies of energy for rural and low income urban households must also be emphasized (IEA, 2002).

## Energy and equity

The challenge of maintaining and enhancing the biosphere is not simply a matter of polluting less and moving from non-renewable to renewable resources, but also of equity. Solving one environmental problem can frequently lead to others, resulting in a position where this is no solution that will work without greater equity in access to the biosphere. The operation of an expanding, open economic system in a finite closed biosphere poses several policy issues. It is important to screen technologies to minimize biosphere disruption, to use environmental impact assessment (EIA) and cost–benefit analysis (CBA), and environmental profiles to outline the impact of investment on the biosphere. But the major problem remains. It is that current energy production technologies are largely hydrocarbon based and cause significant environmental damage. Yet it is precisely these technologies that developing countries need most for their national economic development.

Rural women and children pay an increasing price in time in gathering biomass fuels to secure their energy subsistence. It is commonly two to three hours per day and in some cases

## Box 3.1 Women and energy

Women not only provide the backbone of rural Third World agriculture, undertaking some 70 per cent of all farming activities, but they also dominate household energy collection and utilization. Added to the time spent collecting wood (bundles may weigh up to 5kilos) is an hour and a half pounding and grinding foods, and anything from one to six hours fetching water. As land degradation spreads, especially in sensitive environments, women have to spend more hours walking greater distances to collect fuelwood. In this situation there may also be a corresponding shortage of water. This dual increase in the burden on women creates serious problems for other tasks such as planting, weeding and harvesting, and this accelerates the impoverishment of women. So far, it has proved difficult to design specific energy projects for women which can successfully break this vicious circle. Attaching an energy component to an ongoing women's project offers better development opportunities. The burden on women is caused by the gender division of labour; men are largely the income generators and women largely do the unpaid, unrecognized work. Women frequently do not have access to opportunities which generate income and, as a consequence, have few opportunities for purchasing energy or labour saving devices. They are not only constrained by societal traps but also by finance. In a systematic end use approach, the views of women will be central to the design of programmes and projects, in which energy is included. Training and extension programmes in agroforestry specifically designed for women must be strengthened to include income generation and security of energy supply. To this end, more support should be given to 'training of trainers' programmes in local biomass production, linked to the continuing training programmes for low investment agriculture.

*Source:* Energia, 2009

can be up to five hours. Poorer urban households pay an increasing cash price because for most of them, collecting fuel is not possible as supplies around towns become more and more depleted. A proportion of income in urban areas is spent on switching to more advanced fuels. In Addis Ababa, Ethiopia, for example, by the late 1980s, 70 per cent of households had switched to kerosene because wood and charcoal prices had risen. Paradoxically, as the price of energy rises, the demand does not fall. Therefore in developing countries, over the last 20 years, oil consumption continued to grow steadily, reflecting the original low consumption base, while in developed countries it varied, following the movement in oil prices. This indicated both a higher volume of consumption and the ability to lower consumption through investment in energy efficiency. Developing countries must have a right to the economic opportunities enjoyed by developed countries. For that purpose, more investment in state-of-the-art technologies that maximize efficiency, minimize environmental damage and are budget neutral, especially for poor people, is needed. Furthermore, the concentration of energy investment in urban economic centres does not encourage significant rural development without which rural resource depletion and migration to urban centres will grow (Energia, 2009). See Box 3.1.

There is little doubt that a key issue in equity is the access of poorer households to fuel supplies. Their need for energy is irreducible but the cost in collection time or in cash has been rising over the last 20 years. Institutional support for poorer households must concentrate on minimizing the risk of fuel shortage. In rural areas, especially in the case of refugees (the most vulnerable of the vulnerable), where fuel scarcity is compounded by large local increases in population, policy must provide greater access to and control of productive resources by local people. In the case of refugee settlements in Sudan, government policy puts a ceiling on their access to resources by insisting that it should be no better than that of other local people. At the same time, the activities of commercial farmers and commercial charcoal producers have led to a great reduction in the amount of fuelwood available. Initiatives aimed at improving local fuel supplies through farm woodlands, social forestry and community forestry, however, have been successful in some cases. Another way of supporting the more disadvantaged in both rural and urban areas is through income support schemes that allow for the purchase of energy. In rural areas such income may be achieved by smallholder cash cropping, service industry and domestic based manufacturing industries. Experience of refugee schemes in Tanzania and Sudan shows that, even in the case of refugees, it is possible to achieve sustainable energy development within a project (DANIDA, 1999).

## The fuelwood problem in Africa

Fuelwood is the issue at the centre of energy planning in Africa. That wood energy cannot be provided by planting fuelwood trees is, at last, being recognized. Agroforestry, in which trees are a multi-purpose resource in a land use system, seems to provide an opportunity for addressing the problem. However, women who are usually the wood gatherers and farmers of Africa, need to be key participants in any solution. More importantly, energy systems in urban areas need to be addressed. Many of them still rely heavily on charcoal. By encouraging a change from charcoal to fuels which are both preferable and more easily obtained, the pressure on rural wood resources could be reduced. As we have already remarked, this transition will ease the problem of wood depletion and rural environmental degradation. Given Africa's heavy dependence on wood for energy the call is now for a form of forestry which will contribute to the process of sustainable development. Implicit in this is a need for new forestry initiatives which contribute to a participatory, equitable, decentralized and self-sustaining process of rural development throughout Africa.

Fuelwood stress is also conditioned by social and demographic trends within African countries; both overall population growth and rates of urbanization are exacerbating the problems. In particular, fuelwood policies do not fully consider the significance of urbanization processes. Urban growth rates of 10 per cent are the norm in Africa and what were rural societies are becoming increasingly urban. Population growth in rural areas affects fuelwood use in much the same way as it does other forms of resource exploitation. However, the varying circumstances of different people and places make generalization about fuelwood stress problematic and the problems can rarely be summarized. Fuelwood use and scarcity reflect complex and variable interactions between local production systems and the environmental resource on which they are based. The significance and origins of fuelwood problems vary as much as do local environments and societies. Given this, a number of criteria must be satisfied before a sustainable energy policy can reflect the heterogeneous nature of fuelwood supply and end use:-

- Biomass resources in different areas must be measured and this will give an indication of the maximum available as potential fuel. The areas taken will equate to agroclimatic zones, which are a broad indication of land productivity.
- The characteristic rural economy, including population densities, forms of settlement and dominant types of agriculture must be identified. This will indicate the level of demand for energy and the characteristic patterns of land management.
- The socio-economic condition of the area must be described because it determines access to the resource base for fuel by different sections of the local community.
- Factors which produce significant exports of woodfuel resources, such as commercial logging, and the influence of urban woodfuel and charcoal markets, must be incorporated.
- Major forms of structural change which seriously affect the fuelwood situation in a locality must be analysed. They will include

land colonization, demographic change and urbanization, major developments such as roads and hydro-electric power schemes and catastrophic drought or conflict.

This is a complex list of criteria, but their consideration is essential if fuelwood problems and solutions are to relate to the condition of the people experiencing fuelwood stress.

## The African fuelwood experience

Most forestry plans in the past 25 years have treated the biomass problem simply as one of supply and demand. It was argued that people were extracting more biomass than the environment could sustainably produce and the solution was self-evident – if projected demand exceeded supply, either plant more trees or devise policies to reduce demand. As a result, foresters tried to increase tree supplies by various large scale means like monocultural plantations, peri-urban woodlots, community woodlots, and increased policing of forests and woodlands. Their object was to plant as many trees as quickly as possible. Unfortunately, and all too often, decisions to spend large sums of money planting trees have been taken without considering other options or the consequences of existing market and policy failures. Foresters have only themselves to blame for excluding options and courting failure (Van Gelder and O'Keefe, 1995).

### The Kenyan experience

From 1980, there was an attempt, in Kenya, to address the broad energy and development problem, particularly that of fuelwood, in a systematic manner. Fuel switching was considered but, because of the comparative expense of oil, it seemed insufficiently attractive. Energy conservation was mooted but for fuelwood in particular there were substantial limits to investment in this area. Finally, the analysts looked at the possibility of expanding wood energy supplies and this led to the formulation of the Kenyan Woodfuel Development Programme (KWDP). This programme was run from the Ministry of Energy but had a level of contact with the District Forestry administration. The KWDP was focused on the Kakemega and Kisii districts of western Kenya. They are densely populated and undergoing rapid land consolidation, and were thought to be the best areas in which to explore potential models of agroforestry. Early surveys came to the striking conclusion that deforestation does not necessarily occur in densely populated areas. Quite the reverse – there is a great deal of evidence to suggest that farmers, given the necessary inputs, will increase the amount of woody biomass on their farms (Bradley, 1991; O'Keefe and Raskin, 1985; Hosier and O'Keefe, 1983).

## Towards solutions

Although there has been much work in recent years on local demand for trees and tree products, it has largely been led from social science and has not addressed the issues of local production. What is urgently needed is a new form of social forestry which provides wood near where people live. This requires integrating wood into existing land use patterns in the farming system; it requires production design for a new agroforestry. It is important, even at the risk of repetition, to correct the three most popular misconceptions about the problem of declining wood resources. First, it is frequently assumed that deforestation is caused by commercial logging and cutting for fuelwood; this is simply not true – agricultural colonization is the major cause. Second, it is frequently assumed that forests are a primary source of woodfuel for rural people; this is wrong – in Africa over 90 per cent of biomass fuel comes from agroforestry systems. Third, it is assumed that rural people fell trees for domestic energy use – this happens very infrequently because woodfuel is a residue from other uses of wood in the rural economy. Quite simply, woodfuel is what is left over. Any new agroforestry project must recognize that:

- trees can be combined with crops and/or animals in many different systems of land use;
- a range of goods would be produced, not one single product;
- indigenous, multipurpose trees and shrubs are the core of intervention;
- such complex production systems are probably better suited to fertile environments rather than to fragile conditions;
- and use systems will actively reflect sociocultural values.

The integration of a project which takes these things into account is a step-by-step process which allows the people themselves to control it, to follow its progress and to adapt the programme in a gradual way without deviating from its objectives.

At the turn of the century, implementing this kind of social forestry has largely fallen to policy advocacy by development non-governmental organizations with implementation by local populations. There is a very uneven national policy framework towards agroforestry and more broadly towards biomass energy. While some training exists in university forestry departments, progress has been piece-meal and very patchy indeed. Yet again, the conclusion seems to be that we know enough, but the political will to address the problem of biomass energy for poor people is not in place.

## References


Abramovitz, J. N. (2001) Unnatural Disasters. Worldwatch Paper 158. Washington, DC: Worldwatch Institute

Appleyard, D. (1999) 'Power theft: an insidious menace', *Power Economics*, July

Barnes, D. and Floor, W. M. (1996) 'Rural energy in developing countries: A challenge for economic development', *Annual Review of Energy and Environment*, vol. 21, pp497–530

Batabyal, A. A. (1998) 'The concept of resilience: Retrospect and prospect', *Environment and Development Economics*, vol. 3, no. 2, pp235–239

Bradley, P. N. (1991) *Woodfuel, Women and Woodlots*, London: Macmillan Educational

Brunt, C., Luce, P. and Peters, R. (2004) The Clean Development Mechanism (CDM) an International Perspective and Implications for the LAC Regions (Pembina Institute for Appropriate Development). Available online at: www.pembina.org/pdf/publications/Review_of_Current_Status_of_CDM_and_LAC_Implications_para_web.pdf.

Carbon Market Update (2005) Development Carbon Fund. Issue 1, UNEP, IETA. Available online at: www.ieta.org/ieta/www/pages/getfile.php?docID=901.

Christensen, C. M. (1997) *The Innovator's Dilemma*, Cambridge, MA: Harvard Business School Press

Christensen, C. M. and Raynor, M. E. (2003) *The Innovator's Solution*, Cambridge, MA: Harvard Business School Press

COM (2000) European Commission, Green Paper COM 769 final, 'Towards a European strategy for the security of energy supply'. Available online at: http://europa.eu.int/comm/energy_transport/doc-principal/pubfinal_en. pdf.

COM (2002) Communication from the Commission to the Council and the European Parliament, Final report on the Green Paper 'Towards a European strategy for the security of energy supply', 321 final. Available online at: http://europa.eu.int/comm/energy_transport/livrevert/final/report_en.pdf.

CSD (2002) Ninth session, Agenda Item 4. Decision. Energy for Sustainable Development, Section 6.22, New Yor: United Nations

DANIDA (1999) Available online at: www.um.dk/en/menu/DevelopmentPolicy/Evaluations/Publications/ReportsByYear/1999/DanEval09Synthesis.htm.

Dunn, S. (2000) Micropower: The Next Electrical Era. Worldwatch Institute Paper 151. Washington, DC: Worldwatch Institute

EASE (Undated). Available online at: www.ease-web.org/html/why_energy_poverty.html.

Ellis, J., Winkler, H., Corfee-Morlot, J. and Winkler, H. (2004) Taking Stock of Progress under the CDM. OECD and IEA Information Paper, COM/ENV/EPOC/IEA/SLT 2004, 4. Paris: OECD/IEA. Also available at: www.erc.uct.ac.za/Research/publications/04Ellis%20etal%20-%20CDM%20stock%20taking.pdf

Energia (2009) Available online at: www.energia.org/.

Flavin, C. (2003) Renewable Energy Enters Boom Period, Washington, DC: Worldwatch Institute.

Available online at: www.worldwatch.org/press/news/2003/07/10/.

Hardoy, J. E., Mitlin, D. and Satterthwaite, D. (2001) *Environmental Problems in an Urbanising World*, London: Earthscan

Harkins, M. (2000) *A Case Study on Private Provision of Photovoltaic Systems in Kenya in ESMAP Energy Services to the World's Poor*, Washington, DC: World Bank

Helme, N. (2005) Center for Clean Air Policy, Sector-Based Approach: Overview and Possible 'Straw' Proposal. Presentation to the Economic Commission for Latin America and the Caribbean, 13–14 September, Santiago, Chile. Available online at: www.ccap.org/Presenations/ECLAC%20Sector-Based%20Options%20Sept05NH.pdf

Hosier, R. and O'Keefe, P. (1983) 'Planning to meet Kenya's household energy needs: An initial appraisal', *GeoJournal*, vol. 7, no. 1, pp29–34

IEA (2002) *World Energy Outlook*, Paris: OECD/IEA. Available online at: www.iea.org/textbase/nppdf/free/2000/weo2002.pdf.

IEA (2005) *World Energy Outlook*, Paris: OECD/IEA

IPCC (2007) 'Summary for Policymakers', in: M. L. Parry, O. F. Canziani, J. P. Palutikof, P. J. van der Linden and C. E. Hanson, eds., *Climate Change 2007: Impacts, Adaptation and Vulnerability. Contribution of Working Group II to the Fourth Assessment Report of the Intergovernmental Panel on Climate Change*, Cambridge: Cambridge University Press, pp7–22. Available online at: www.ipcc.ch/pdf/assessment-report/ar4/syr/ar4_syr_spm.pdf

ITDG (2002) Sustainable Energy for Poverty Reduction: An Action Plan. IT Consultants IT Power ITDG Latin America. Available online at: www.itdg.org/html/advocacy/docs/itdg-greenpeace-study.pdf.

Johnson, J. L. and Wielchelt, S. A. (2004) 'Introduction to the special issue on resilience', *Substance Use & Misuse*, vol. 39, no. 5, pp657–670

Keen, M., Brown, V. A. and Dyball, R. (2005) 'Social learning: A new approach to environmental management', in: M. Keen, V. A. Brown and R. Dyball, eds., *Social Learning in Environmental Management: Towards a Sustainable Future*, London: Earthscan

Kirkby, J., O'Keefe, P. and Howorth, C. (2001) 'Introduction: Rethinking environment and development in Africa and Asia', *Land Degradation & Development*, vol. 12, pp195–203

Kristoferson, L. A. and Bokalders, V. (1987) *Renewable Energy Technologies. Their Applications in Developing Countries*, Oxford: Pergamon Press

Leach, G. and Mearns, R. (1998) *Beyond the Woodfuel Crisis: People, Land and Trees in Africa*, London: Earthscan

Lovins, A. B. and Lovins, L. H. (1982) *Brittle Power: Energy Strategy for National Security*, Amherst, NH: Brick House. Available online at: www.rmi.org/images/other/EnergySecurity/S82-03_BrPwrParts123 .pdf.

Mahiri, I. and Howorth, C. (2001) 'Twenty years of resolving the irresolvable: Approaches to the fuelwood problem in Kenya', *Land Degradation & Development*, vol. 12, pp205–215

Martinot, E. et al (2005) *Renewables (2005) Global Status Report*, Washington, DC: Worldwatch Institute and GTZ GmbH

Meyer, N. I. (2004) 'Renewable energy policy in Denmark', *Energy for Sustainable Development*, vol. VIII, no. 1, pp25–35. Available online at: www.ieiglobal.org/ESDVol8No1/05denmark.pdf.

Munslow, B., Ferf, A., Katerere, Y. and O'Keefe, P. (1988) *The Fuelwood Trap: A Study of the SADCC Region*, London: Earthscan

O'Brien, G. and O'Keefe, P. (2006) 'The future of nuclear energy in Europe: A response', *International Journal of Environmental Studies*, vol 63, no 2, pp121–130

O'Brien, G. O'Keefe, P. Rose, J. and Wisner, B. (2006) 'Climate change and disaster management', *Disasters*, vol. 30, no. 1, pp64–80

O'Keefe, P. (1993) 'The energy ladder', *Boiling Point*, no 31, p46

O'Keefe, P, and Munslow, B. (1989) 'Understanding fuelwood I: A critique of existing interventions in southern Africa', *Natural Resources Forum*, vol. 13, pp2–10

O'Keefe, P. and Raskin, P. (1985) 'Fuelwood in Kenya: Crisis and opportunity', *AMBIO*, vol. 14, pp220–224

O'Keefe, P., Wilson, L. and Cheetham, K. (2003) *From Poverty to Climate Change. A Note for CoP 9*, London: ETC UK Ltd

PURE (2000) Promoting Unst Renewable Energy Project. Available online at: www.pure.shetland.co.uk/index.html.

Sawin, J. L. (2004) Mainstreaming Renewable Energy in the 21st Century, Worldwatch Paper 169, Worldwatch Institute, Washington DC

Schneider, M. (2000) *Climate Change and Nuclear Power*, Gland: World Wide Fund for Nature

Soussan, J., O'Keefe, P. and Mercer, D. E. (1992) 'Finding local answers to fuelwood problems: A typological approach', *Natural Resources Forum*, vol. 16, pp91–101

Swain, J. (2004a) *Mainstreaming Renewable Energy in the 21st Century*, Washington, DC: Worldwatch Institute

Swain, J. L. (2004b) National Policy Instruments, Thematic Background Paper. Bonn: International Conference for Renewable Energies

Thomas, S. (2004) The British Model in Britain: Failing slowly, Paper presented at International Workshop on: 'Thirty Years of World Energy Policy – cum – Editorial Board Meeting of Energy Policy', Hong Kong Energy Studies Centre and Department of Geography, Hong Kong Baptist University, 23–25 March

UN/ISDR (2005) World Conference on Disaster Reduction, 2005, Hyogo Declaration, paragraph 2, 18–22 January, Kobe, Hyogo, Japan. Available online at: www.unisdr.org/wcdr/intergover/official-doc/L-docs/Hyogodeclaration-english.pdf.

Van der Leeuw, S. E. and Leygonie, C. A. (2000) 'A long-term perspective on resilience in socio-natural systems'. Paper presented at the workshop on system shocks–system resilience held in Abisko, Sweden, 22–26 May

Van Gelder, B. and O'Keefe, P. (1995) *The New Forester*, Intermediate Technology Publications, London

Wakeford, T. (2004) Democratising technology: Reclaiming science for sustainable development, ITDG discussion paper. Available online at: www.itdg.org/docs/advocacy/democratising_technology_itdg.pdf.

Waller, M. W. (2001) 'Resilience in ecosystemic context: Evolution of the concept', *American Journal of Orthopsychiatry*, vol. 71, no. 3, pp1–8

# 4

# Efficiency of End Use

This chapter looks at the issues of end use efficiency in the developed world.[1] End use efficiency for the developing world is dealt with in Chapter 3.

## Introduction

Efficiency determines the quantity of useful energy services that can be obtained from a given energy resource. Both the transformation process of a primary resource – such as natural gas into electricity – (covered in Chapter 5) and the conversion of an energy source – for example, electricity – into a useful service or end use have associated losses. Less than 10 per cent of the power supplied to an incandescent light bulb is converted into useful light. The rest is lost as heat and in parts of the spectrum that are not visible to the human eye. Improving end use efficiency is important for a number of reasons:

- *Climate goals*: the EU estimates that at least 20 per cent of energy is wasted due to inefficiencies (EU, 2006). Using resources efficiently can reduce greenhouse emissions, hence slowing climate change and minimizing the risk of adverse climatic shifts.
- *Energy security*: many OECD countries are increasingly reliant on imported supplies. Geopolitical uncertainties could threaten supply lines. Improved efficiency would reduce the demand for imported fuels.
- *Price volatility*: energy prices rose steeply in 2008 and are likely to remain high, driven by supply constraints and increasing global demand. Rising prices against a background of rising demand for increasingly scarce resources could threaten global economic stability, or even trigger recession.

But there are other equally important reasons for improving end use efficiency. Conventional or fossil fuels are highly polluting. Reducing use has both environmental and health benefits. Those on fixed incomes such as the elderly are vulnerable to fuel poverty and higher efficiencies can reduce the amount spent on fuel. Improving efficiency is both a technological and social challenge. Policy makers are now focusing on ways in which different policy instruments can influence technological developments and behaviour with respect to energy efficiency. This chapter explores developments in the transport sector, the built environment and in goods, appliances and electrical motor systems.

## Efficiency

Generally speaking, efficiency is the ratio of total energy output to total energy input and is usually expressed as a percentage. In the energy system there are broadly two areas where efficiency is important. The first is the transformation of a primary energy resource into a secondary energy

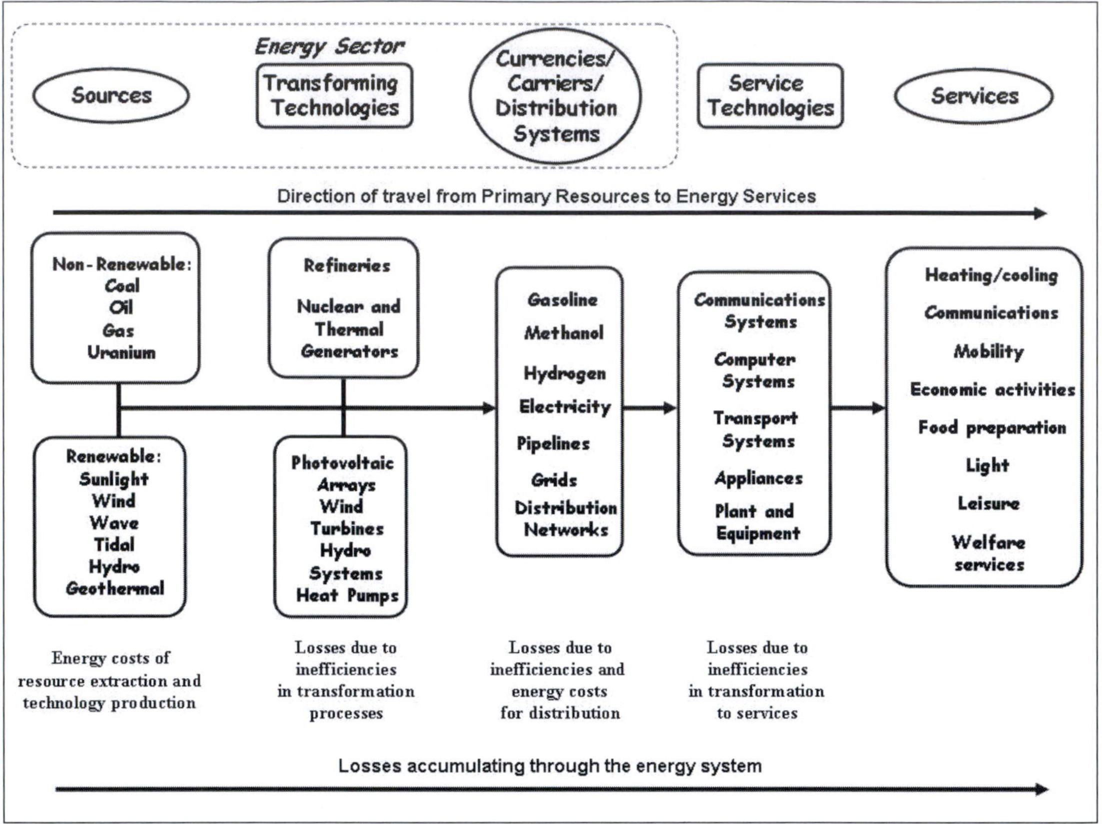


*Source:* Adapted from Scott, 1995

**Figure 4.1** The energy system

resource or energy carriers, such as coal into electricity. The second is the transformation of secondary energy sources into energy services. Figure 4.1 gives an overview of the energy system.

The left hand side of Figure 4.1 lists energy resources that through a transformation process can eventually be realized as energy services. The transformation process for a non-renewable resource, such as coal into electricity, is very different to that of wind or photovoltaics. Different processes are governed by different laws or rules. The efficiency of the transformation process is the first step in determining the efficiency of a particular transformation/energy service pathway. The majority of power is produced by processes that are governed by the laws of thermodynamics. The laws of thermodynamics arose from early experiments that linked heat and mechanical work and are shown in Box 4.1.

The laws of thermodynamics set the limitations of the heat engine. The heat engine is widely used in transformation processes, for example, in thermal power stations, the internal combustion engine (ICE) and jet engines. In fact any engine that uses heat to produce mechanical work is a heat engine. The heat engine works by using temperature difference to produce mechanical work (Figure 4.2).

Energy stored, for example, in steam, is then passed through a turbine to produce mechanical work. If the turbine is connected to an electrical generator then electrical power will be produced. The efficiency of the heat engine is:

## Box 4.1 The laws of thermodynamics

**1st Law:** Energy cannot be created or destroyed.

This is also known as the Law of Conservation of Energy. When heat energy is added to a system, the energy appears either as increased internal energy or as external work done by the system.

**2nd Law:** Energy and Materials tend to be transformed in one direction

This is also known as the Entropy Law, where entropy is a measure of disorderliness. The direction of transformation can be seen in everyday objects, for example, objects will go from hot to cold and ice will turn to water if there is no external intervention such as addition or withdrawal of heat.

**3rd Law:** As a system approaches a temperature of absolute zero (degrees Kelvin (0°K) or minus 273 degrees Celsius (–273°C)) all processes cease and the entropy of the system approaches a minimum value.

In reality absolute zero has never been reached, but it is possible to reach a temperature that is very close to absolute zero.

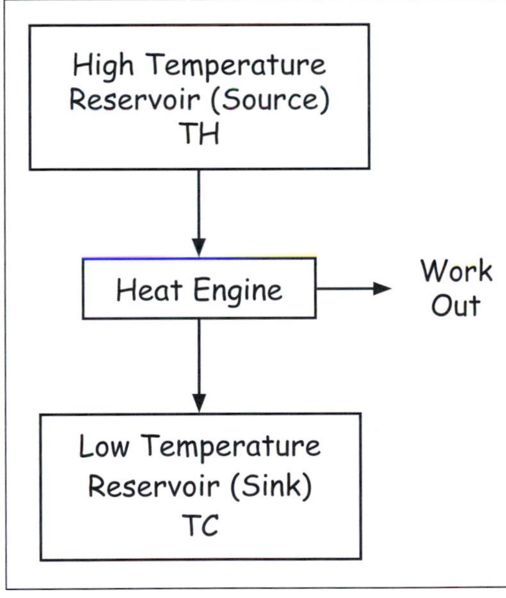


**Figure 4.2** The heat engine

Efficiency (%) = (1 – TC/TH) × 100

Where TH is the temperature of the heat source and TC is the temperature of the heat sink.

In short, to attain 100 per cent efficiency then TC must be absolute zero. In reality this is unattainable and typically TC would be the temperature, in the case of a steam engine, of the steam condenser, typically about 13°C (286°K). TH, the temperature of steam entering the turbine, is limited by mechanical factors. There is a limit to the temperature of the system as the metals used in turbine construction would need to remain sufficiently cool for them to retain mechanical rigidity. Steel begins to creep at 265°C. If TH, the temperature of superheated steam entering the system was, for example, 250°C (523°K) then the efficiency of the system would be:

Efficiency (%) = (1 – 286/523) × 100 = (1 – 0.55) × 100 = 45%

This is a very rough estimate and ignores losses in the turbine and power needed for ancillary equipment. In addition, the boiler that supplies the steam has inefficiencies associated with the combustion process and the transfer of heat. In reality the efficiency of the conversion of heat to mechanical power will always be less than 100 per cent. Typically the efficiency of a thermal power station is less than 40 per cent. The waste heat can be captured and used to provide heating. This is known as combined heat and power (CHP). Although this does significantly improve the efficiency of the overall system, it does require that the power station is near to the point of use. Typically large coal fired power stations are in remote locations, making the use of CHP impractical because of the losses associated with transporting low-grade heat over long distances. For smaller plants located near to urban areas, CHP is viable. And CHP can be realized at the household level through micro-generators that typically use gas as the fuel to produce electricity and heat.

From Figure 4.1 we can see that there are losses (or energy costs) associated with each step in the energy system:

- *Resources*: gathering resources can require an energy input, for example, into mining of coal, drilling and transporting oil or gas and in the construction of technologies to capture renewables.
- *Transformation*: losses associated with the transformation of primary resources into secondary resources and in refining primary resources.
- *Carriers and distribution*: losses associated with electrical distribution systems and energy costs associated with pumping systems for pipelines and freight distribution of fuels.
- *Service technologies*: the losses associated with the range of technologies that utilizes transformation energy for useful services, such as the light bulb providing light and the vehicle providing mobility.

This chapter has its focus on service technologies and deals with three sectors; the built environment, appliances and equipment, and transport. The efficiency issues associated with other parts of the energy system are discussed in Chapters 5 and 7.

## Energy trends

Within OECD countries the decline in energy intensive industries, growth of knowledge-based industries and the service sector, changing demographics and increased personal mobility are changing the end use energy profile. Figure 4.3 shows the evolution of the energy economy within the EU from 1990 and projects forward to 2030, typifying these changes within the OECD.

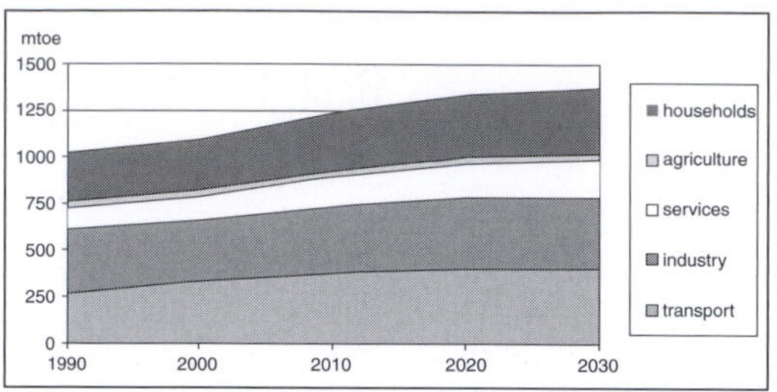


*Source:* EU Commission, Directorate-General for Energy and Transport, 2006

**Figure 4.3** Final energy consumption by sector

Final energy demand could rise by 25 per cent from 2000 to 2030 with household growth rising by 29 per cent, services by 49 per cent, transport by 21 per cent, industry by 19 per cent and agriculture by 10 per cent. Although all sectors increase, households and services dominate reflecting demographic and lifestyle changes and changes to the economic mix.

To influence the future use of energy will require intervention. Supply-side interventions can influence the energy mix and promote the use of more renewable resources. But of equal importance are interventions on the demand-side to influence the ways in which we use energy and the efficiencies of end use technologies.

## Energy and the built environment

The built environment refers to all buildings, whether for domestic, public or private use. The built environment is the physical expression of societal development and reflects the social, economic, cultural, spiritual and political evolution of society. The developed world, or OECD countries, are highly urbanized, but the shift to urbanization is a global phenomenon. In 2005 global urbanization reached 49 per cent of the global population and is predicted to reach 60 per cent of the global population by 2030 (UN, 2005).

Within the EU, and typically within OECD countries, about 40 per cent of overall energy demand is used in buildings (EU, 2006). Demand is likely to grow, fuelled by the construction boom in rapidly industrializing nations such as India and China (WBCSD, 2007). Energy is used in the built environment to provide heating and cooling, lighting and to power a range of equipment and appliances. There are two broad approaches to improving the efficiency of the built environment:

- Improve the thermal efficiency of buildings to reduce demand for heating or cooling services.
- Improve the end use efficiency of the equipment and appliances that are used within buildings.

The next section deals with the thermal efficiency of buildings. Equipment and appliances are dealt with later in this chapter.

## Thermal efficiency of buildings

Buildings use 40 per cent of energy output within the EU (EU Commission, 2005, p45). Improving the thermal efficiency of buildings can reduce the amount of energy used either for heating in cooler climates or cooling in warmer climates. The thermal efficiency of a building is a function of a number of factors such as:

- siting and organizing the building configuration and massing to reduce loads;
- reducing cooling loads by eliminating undesirable solar heat gain;
- reducing heating loads by using desirable solar heat gain;
- using natural light as a substitute for (or complement to) electrical lighting;
- using natural ventilation whenever possible;
- using more efficient heating and cooling equipment to satisfy reduced loads;
- using computerized building control systems.

*Source:* Federal Energy Management Programme, 2001

Building form and energy management strategies are not the only determinants of efficiency. The materials used and the construction standards are also important. The following sections will look at how building efficiency is specified and how differing construction techniques can increase the thermal efficiency of a building.

## Material efficiency

The thermal efficiency of materials is known as the U-value. The U-value is the measure of heat loss through a material and represents the amount of heat lost in watts per square metre of material (for example, a wall, roof, glazing and so on) when the temperature is one degree lower outside. Simply put, the more thermally efficient the material used, the lower the U-value and subsequent heat loss. For example, single glazed windows have a typical U-value of 5.6 while double glazed windows have a typical U-value of 2.8. Table 4.1 lists typical U-values of materials used in building construction.

The methods for determining U-values of building elements are based on standards that were developed in the European Committee for Standardisation (CEN) and the International Organisation for Standardisation (ISO) and published as British Standards. New materials and building design make the determination of U-values complex and procedures are regularly updated to ensure compliance with regulations (Anderson, 2006).

## Thermal efficiency

The efficiency standard or energy rating of a building is influenced by climatic factors and existing regulations at the national level. Comparison between different nations is problematic. Within the EU, for example, there is a range of methods for determining the energy rating of a building (Míguez et al, 2006). In the UK the method for rating a building is known as the Standard Assessment Procedure (SAP). In force since 1995, SAP rates the energy efficiency of

**Table 4.1** Typical U-value of construction

| Cavity wall insulation | U-value | Roof insulation | U-value |
|---|---|---|---|
| 100mm blown polystyrene | 0.3 | 150mm glass wool | 0.25 |
| 100mm blown mineral wool | 0.3 | 150mm rock wool | 0.23 |
| 100mm blown cellulose fibre | 0.3 | 150mm sheep's wool | 0.23 |
| 60mm extruded polystyrene insulation | 0.4 | 200mm glass wool | 0.19 |
| Timber frame 150mm, mineral quilt | 0.25 | 200mm cellulose fibre | 0.16 |
| Timber frame 140mm, cellulose fibre | 0.19 | | |
| External wall insulation | | Windows | |
| 60mm moulded polystyrene | 0.44 | Single glazing | 5.6 |
| Internal wall insulation | | Double glazing | 2.8 |
| 50mm expanded polystyrene | 0.48 | Double glazing, with argon | 2.6 |
| 38mm polyurethane | 0.45 | Double glazing, low-e | 1.8 |
| | | Double glazing, low-e with argon | 1.5 |

*Source:* Adapted from the Irish Energy Centre, undated

buildings on a scale of 0–120 based on thermal performance, heating system and energy prices, as shown in Box 4.2. This lack of standardization has made international comparisons difficult. The average SAP for domestic dwellings in England is about 48 points (DEFRA, 2007b). This compares to around 90 points for a well-insulated dwelling in Scandinavia. Lack of standardization can act as a barrier to the transfer of good practice between EU member states. As part of the EU Kyoto commitment, the Energy Performance of Buildings Directive (EPBD), which came into force in 2003, is designed to standardize the method for assessing the energy performance of buildings (OJ, 2003). It also sets minimum standards for new buildings, requires the production of Energy Performance Certificates and inspection regimes for heating and cooling systems. The EU estimates that the savings potential in the building sector is around 28 per cent, which

## Box 4.2 Definition of SAP rating

The Standard Assessment Procedure (SAP) is a government-specified energy rating for a dwelling. It is based on the calculated annual energy cost for space and water heating. The calculation assumes a standard occupancy pattern, derived from the measured floor area so that the size of the dwelling does not strongly affect the result, which is expressed on a scale of 1–120. A higher number means a more efficient building. The individual energy efficiency or SAP rating of a dwelling depends upon a range of factors that contribute to energy efficiency, namely:

- thermal insulation of the building fabric;
- efficiency and control of the heating system;
- ventilation characteristics of the dwelling;
- solar gain characteristics of the dwelling;
- the price of fuels used for space and water heating;
- renewable energy technologies.

*Source:* BRE, 2005

in turn can reduce the total EU final energy use by around 11 per cent. The deadline for adoption of the directive was 2006, with measures such as Energy Performance Certificates expected to be in place across the EU by 2008 (EU Commission, undated-b).

## Improving building thermal efficiency

Given the high energy use of buildings and the fact that the majority of that use is either for heating or cooling, it is not surprising that there is considerable worldwide interest in improving the thermal performance of buildings. Improvements in the construction standards of buildings are enabled through regulations, standards or codes that mandate minimum energy efficiency standards. These can be set at the national level, at the level of a group of countries, for example, in the EU through Directives, or at the level of a sub-national region inside a federal country, for example, at the state level in the US. Almost all OECD countries in Europe and about half of the remaining OECD countries have mandatory efficiency standards (WEC, 2008). A strong trend in the building sector is a move towards zero energy buildings (ZEBs). These are based on two principles. First the thermal efficiency of the building itself must be very high. Second the energy requirements of the building should be met from renewable or non-polluting resources. The following section looks at the principles underpinning ZEBs.

## Zero energy buildings (ZEBs)

ZEB is a general term used to describe a building that has a net energy consumption of zero over a specified period of time, for example, a year. There is no agreed definition of what is meant by a zero emission or zero energy building. A ZEB can be measured in a number of ways relating to either energy, emissions or cost. Setting the boundaries, that is whether or not to consider the performance of the building itself, or the whole lifecycle (including the energy used in construction and embodied within the construction materials) is also problematic. The majority of studies have been focused on the building during its operational phase or lifetime, as this could undermine reduction claims during operation.

In general, a zero emissions/energy building is one that has greatly reduced energy needs through efficiency measures and the energy needs it does have are met through renewable resources. There are two ways of defining a ZEB: (i) the building should meet its energy needs from embedded capacity in the building or from capacity within its boundary (also known as an Off-Grid ZEB (OGZEB)); (ii) the building is connected to external sources. Table 4.2 sets out the options for both of these definitions.

## Low energy buildings

The starting point for a ZEB is to ensure that the building energy requirements are as low as possible through low energy building technologies and techniques such as high insulation standards, daylighting, high efficiency HVAC (heating, ventilation and air conditioning), natural ventilation and evaporative cooling. The design of low energy buildings maximizes the use of such technologies and techniques and eliminates aspects of construction that can lower efficiency, such as thermal bridges. Thermal bridges can be created when materials that have poor insulation characteristics come into contact, allowing heat to flow through the path created. The Passive House, shown in Figure 4.4, illustrates how these techniques and technologies can be incorporated into the building design and can significantly reduce demand for space heating, as shown in Figure 4.5.

A Passive House is a building in which a comfortable interior climate can be maintained without active heating and cooling systems. The concept was developed by Professor Bo Adamson and Dr Wolfgang Feist, who founded the Passive

**Table 4.2** ZEB supply options

| ZEB on-site supply options | |
|---|---|
| Use renewable energy sources available within the building footprint | PV, solar hot water and wind located on the building |
| Use renewable resources available at the site | PV, solar hot water, low impact hydro and wind located at the site but not building mounted |
| ZEB off-site supply options | |
| Use renewable energy sources off site to generate energy for use on site | Biomass, wood pellets, ethanol or biodiesel that can be imported for off-site or waste streams form on-site processes that can be used to generate electricity and heat |
| Purchase off-site renewable energy sources | Utility based wind, PV, emission credits or other 'green' purchasing options. |

*Source:* Adapted from Torcellini et al, 2006

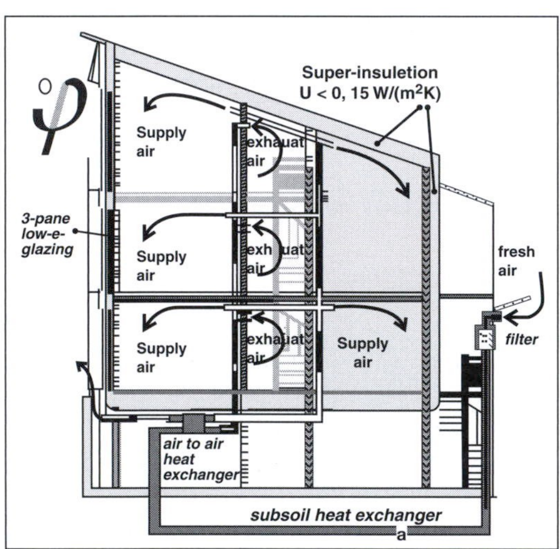


*Source:* Passive House Solutions, undated

**Figure 4.4** Passive House concept

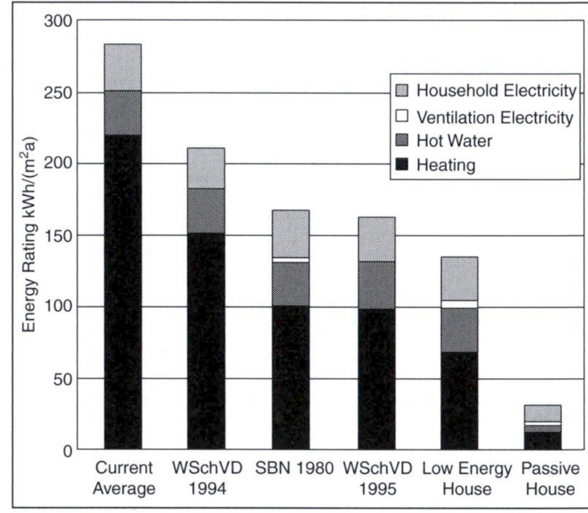


Key:
WSchVO = German Heat Protection Regulation
SBN = Swedish Construction Standard

*Source:* Passive House Institute, undated-a

**Figure 4.5** Comparison of energy ratings of homes

House Institute in 1996 (Passive House Institute, undated-a). A Passive House is defined as follows:

> *A Passive House is a building, for which thermal comfort (ISO 7730) can be achieved solely by post heating or post cooling of the fresh air mass, which is required to fulfil sufficient indoor air quality conditions (DIN 1946) – without a need for recirculated air.* (Passive House Institute, 2006-b)

The heating requirement for a Passive House is less than 15kWh/(m$^2$a) (4755Btu/ft$^2$/yr) and overall combined primary energy consumption for heat, hot water and household electricity should not exceed 120kWh/(m$^2$a) (38039Btu/ft$^2$/yr). Table 4.3 lists some of the basic features of Passive House construction.

Although building design has a significant impact on thermal efficiency, other factors such

**Table 4.3** Basic features of Passive House construction

| Compact form and good insulation | All components of the exterior shell of the house are insulated to achieve a U-factor that does not exceed 0.15W/(m²K) (0.026Btu/h/ft²/°F) |
|---|---|
| Southern orientation and shade considerations | Passive use of solar energy is a significant factor in passive house design |
| Energy-efficient window glazing and frames | Windows (glazing and frames, combined) should have U-factors not exceeding 0.80W/(m²K) (0.14Btu/h/ft²/°F), with solar heat-gain coefficients around 50% |
| Building envelope air-tightness | Air leakage through unsealed joints must be less than 0.6 times the house volume per hour |
| Passive preheating of fresh air | Fresh air may be brought into the house through underground ducts that exchange heat with the soil. This preheats fresh air to a temperature above 5°C (41°F), even on cold winter days |
| Highly efficient heat recovery from exhaust air using an air-to-air heat exchanger | Most of the perceptible heat in the exhaust air is transferred to the incoming fresh air (heat recovery rate over 80%) |
| Hot water supply using regenerative energy sources | Solar collectors or heat pumps provide energy for hot water |
| Energy-saving household appliances | Low energy refrigerators, stoves, freezers, lamps, washers, dryers, etc. are indispensable in a passive house |

*Source:* Passive House Institute, undated-a

as building orientation can play an important role in both heating and cooling by using natural sources. For example, orientation can maximize solar gain and reduce heating requirements. Shading can attenuate heat gain in warmer climates and warm seasons through the use of shutters, screens or shade trees that will reduce the need for additional cooling. Massing in the building, for example concrete walls and floors can be used as heat stores to retain heat during the warm periods of the day that can be released during cooler evening periods.

## Defining a zero energy building

As stated earlier there is no single agreed definition for a ZEB. Torcellini et al (2006) identify four commonly used definitions that take into consideration differences between uses of the term in North America and Europe.

1. *Net zero site energy*: in this ZEB the amount of renewable energy produced on the site (including on the building) is equal to that used by the building. This definition applies, in general, to zero energy buildings in North America.
2. *Net zero source energy*: in this ZEB the amount of renewable energy produced is equal to the amount of energy it uses over a year when accounted for at the source. Source energy refers to the primary energy used to generate and deliver energy to the site. This definition covers sites that typically need to import energy at certain times but can also export energy. Care is needed in calculating the buildings' total source energy as off-site generators and transmission systems are inefficient. These must be accounted for.
3. *Net zero energy costs*: in this ZEB the amount of money paid to the owner of the building for energy exported to a utility is equal to the amount paid for imported energy.
4. *Net zero energy emissions*: this is also known as zero carbon or zero emissions, terms typically used outside North America, and in this definition a ZEB produces as much

emission-free renewable energy on site as its uses from emissions producing sources.

The different definitions of a ZEB can influence the design of buildings. For example, a housing development with BIPV (building integrated photovoltaics) will produce electricity during the day, but not at night when the demand for electricity is likely to be high. However, for office buildings demand is likely to be high during the day and low during the night. Different buildings have different energy use profiles and the design and purpose of the building will influence which supply-side strategies and consequently which ZEB definition is the most appropriate. Table 4.4 lists some of advantages and disadvantages of the various approaches to ZEBs.

One further issue that is not addressed within these definitions is embedded or embodied energy. The majority of buildings will consist of materials that require energy for their manufacture. For example, concrete is a common building material. Cement, a component of concrete is manufactured and accounts for some 6 per cent of global industrial energy use. Many of the materials used in buildings such as steel, glass and bricks require substantial amounts of energy to produce (IEA, 2007a). The building process also carries an energy cost.

Conceptually a ZEB implies that some sort of energy technology will be needed to meet its energy requirements. To avoid pollution and greenhouse gas production requires renewable technologies. This is a different conceptual approach, as typically the energy density and reliability of renewable resources is much lower than conventional fossil fuels, meaning efficiencies need to be as high as possible to minimize the scale of renewable deployment.

## Implications of ZEBS

If all buildings reached the efficiency levels of a ZEB this would reduce energy use, theoretically,

**Table 4.4** Advantages and disadvantages of ZEBs

| Type | Advantages | Disadvantages |
|---|---|---|
| Net zero site energy | Conceptually easy to understand.<br>Verifiable by on-site measurements.<br>Encourages energy-efficient building designs. | Reliant on specific skills to be realized that may not be readily available.<br>Cost of renewable technologies, such as PV, are high. |
| Net zero source energy | Good model for understanding impact on national energy system.<br>Easier ZEB to realize. | Does not account for non-energy difference between fuel types (supply availability, pollution).<br>Difficulty of calculating source energy because of generation and transmission inefficiencies |
| Net zero energy costs | Easy to implement and measure.<br>Verifiable from utility bills. | Highly volatile energy prices make it difficult to track over time.<br>Requires net-metering which is not yet fully established |
| Net zero energy emissions | Accounts for non-energy differences between fuel types (pollution, greenhouse gases).<br>Easier ZEB to realize. | Requires clear knowledge of emission factors for off-site generators |

*Source:* Adapted from Torcellenni et al, 2006

in the OECD countries by 40 per cent. As the turnover of building stock is slow, the impact of ZEBs will not be felt for some time. But, as noted earlier, the starting point for a ZEB is ensuring that the thermal efficiency of buildings is as high as possible. This is more easily achieved for new build than for refurbishment. For new build, standards for thermal efficiency can be mandated. For existing stock, this will require some form of intervention, for example, financial support for a refurbishment programme. Research conducted in Europe has identified that there is considerable potential for Passive House standards to contribute to climate, security and energy poverty reduction goals (PEP, 2006). The drive behind the Passive House concept is to reduce heating requirements to a level of 15kWh/m$^2$/year. Figure 4.6 shows the range of existing standards in a number of EU countries and how these compare with new and Passive House standards.

The study assumes levels of market penetration for both new build and refurbishment as shown in Table 4.5. These figures are based on an analysis of national trends in housing markets up to 2020. Note that Germany has the highest projected market penetration reflecting its higher national aspirations and leading status in Passive House development.

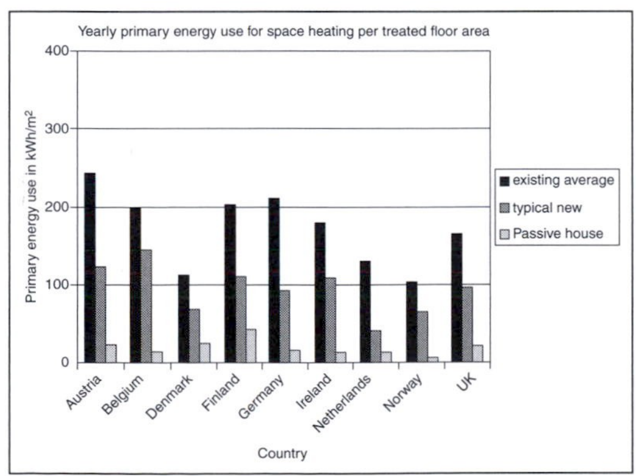


Source: PEP, 2006, p12

**Figure 4.6** Yearly primary space heating energy uses per dwelling, per existing, typical new and passive house

**Table 4.5** Projected market penetration of new build and refurbished dwellings

| Country | New build (per cent) | Refurbishment (per cent) |
|---|---|---|
| Austria | 20 | 15 |
| Belgium | 20 | 15 |
| Denmark | 20 | 15 |
| Finland | 20 | 15 |
| Germany | 50 | 30 |
| Ireland | 20 | 15 |
| Netherlands | 20 | 15 |
| Norway | 20 | 15 |
| UK | 20 | 15 |

*Source:* Adapted from PEP, 2006

The study finds that over the surveyed period emissions fall by 0.03MtC in 2006 to 1.09MtC in 2020 giving a cumulative reduction in carbon emissions of 4.65Mt. The study finds that the energy saving potential for a single residence represents a carbon reduction of about 50–65 per cent. The energy saving per country is very much dependent on energy sources used. Table 4.6 shows the range of energy sources in the study countries.

For those countries using imported fossil fuel supplies, such as natural gas or oil, there are additional benefits in terms of energy security. Overall the project demonstrates the contribution that would be made to climate goals. For example, the EU Kyoto commitments are for a 0.4 per cent per year reduction below 1990 levels and this programme realizes a figure of 0.46 after the first 2 years, which exceeds the Kyoto target for that sector (PEP, 2006).

Although the participants in the study and the Commission are committed to Passive House principles for new build and refurbished dwellings, there are considerable obstacles to be overcome. The most frequently encountered barriers in partner countries were: limited know-how; limited contractor skills; and limited acceptance of passive houses in the market. Elswiijk and Kaan (2008) identify a number of issues such as translating the Passive House principles into different

**Table 4.6** Considered energy sources for space heating in the study countries

| Energy sources for space heating | Austria | Belgium | Denmark | Finland | Germany | Ireland | Netherlands | Norway | UK |
|---|---|---|---|---|---|---|---|---|---|
| Electricity | x | | x | | x | x | | x | |
| Gas | | x | x | | x | x | x | | x |
| Oil | x | | x | | x | x | | x | |
| Wood pellets | x | | | | | | | x | |
| District heating | x | | x | x | x | | | x | |
| Renewables | | | x | | | | | | |
| Coal | | | | | x | x | | | |

*Source:* Adapted from PEP, 2006, p10

building traditions and into acceptable building codes. Supply-side issues were also problematic, with a lack of high efficiency components such as glazing. Overcoming these barriers requires effort, through information and awareness raising, to influence the market so that the new designs and refurbishment to higher thermal standards become embedded as the norm.

There are cost implications for Passive House features, which are typically higher than those of a standard, or low energy house, but this difference is strongly related to energy prices. The higher the price of energy the more attractive the Passive House concept becomes (Audenaerta et al, 2007). The large increase in 2008 of the cost of fossil fuel is likely to shift the balance towards the development of more efficient buildings as well as increase their attractiveness to the public. Recently both the UK and Germany have announced programmes that include support for refurbishing existing stock to Passive House standards. The interest in zero energy developments is not limited to individual buildings. A number of proposals seek to develop new cities based on the principles of very thermally efficienct buildings powered by renewable technologies. Masdar City in Abu Dhabi and Dongtan Eco-City in China are examples of cities that are being planned along ZEB principles (BBC, 2008a; Langellier and Pedroletti, 2006). The internet-based resources listed in Box 4.3 provide further detail on ZEB and Passive House principles.

## Box 4.3 Internet resources

EcoSmart Show Village:
www.barratt-investor-relations.co.uk/media/releases/Content.aspx?id=1318
Greenspace Research:
www.greenspaceresearch.com/
Leonardo-energy:
www.leonardo-energy.org/drupal/3dforum
EON.UK:
www.eon-uk.com/about/2016house.aspx
The Green Home Guide:
www.greenhomeguide.org
Low Carbon Cities:
www.lowcarboncities.co.uk
Passive House Platform:
www.passiefhuisplatform.be/multimedia/001/
Passive House:
www.passivhaustagung.de/Kran/First_Passive_House_Kranichstein_en.html

## UK perspective

The UK has some of the least efficient building stock in Europe, particularly in terms of housing. The UK housing sector comprises some 25 million buildings, of which approximately 21 million are in England. In 2001 the English Housing Condition survey found that about one third, roughly 7 million, were found to be non-decent when compared to the criteria established for the survey, one of which is providing a reasonable degree of thermal comfort (UK Government, 2001). Much of the UK's housing is inefficient and typical energy use for space heating is twice that for equivalent buildings in the Nordic region (Lapillonne and Pollier, 2007; Olivier, 2001). Almost 90 per cent of current buildings will still be in use in 2050 (Sustainable Development Commission, 2006). The majority of UK building stock, almost 80 per cent, was built prior to the 1980s, before the energy efficiency standards were introduced into the building regulations. Box 4.4 lists some of the key policies and measures related to building thermal efficiency.

Improvement in the thermal efficiency of existing stock is needed to make progress in meeting climate goals, improving energy security and reducing fuel poverty. Although new build can achieve high thermal efficiencies, the low rate of demolition and new build means that considerable effort will be needed to refurbish existing stock. The major energy use in dwellings is for space heating and hot water, as shown in Figure 4.7. The highest use, around 84 per cent, is for space and water heating and 83 per cent of this demand is currently met by gas.

In 2006 the UK government published an update to the 2000 building regulations that set higher standards for energy efficiency for new and existing buildings (UK Government, 2006). These regulations raise the energy efficiency of new buildings by 40 per cent, compared with the 2002 requirements. Building standards in the UK are increasingly influenced by the EU. In response to the EU Directive European Energy Performance of Buildings: Directive 2002/91/EC (OJ, 2002) the UK government introduced

## Box 4.4 Key energy efficiency policies and measure

**Home-focused energy saving policies**

- *Energy Efficiency Commitment (EEC)*: energy suppliers must achieve targets for promoting improvements in domestic energy efficiency by helping householders make energy savings through installing cavity wall and loft insulation, energy efficient boilers and so on. Phase 1 (EEC1) is complete with all targets met. EEC2 runs from 2005 to 2011.
- *Home Energy Conservation Act (HECA) (1995)*: local authority 'HECA' officers to deliver by 2010 a 30 per cent improvement in energy efficiency over 1996 levels. Although not mandatory, a 12 per cent improvement has been reported so far.
- *Part L Building Regulations*: set out the legal requirements for energy use in buildings and have recently been updated. Changes applied from April 2006 should see extra carbon savings of ~1Mt/yr by 2010.
- *Directive on Energy Performance of Buildings (2002/91/EC)*: applies minimum standards for energy performance in new buildings and refurbishments of existing larger buildings. Implemented in the UK in 2007 partly through Home Information Packs, these require energy efficiency information to be presented to home buyers.
- *Decent Homes*: alongside other goals, requires local authority housing to provide a 'reasonable degree of thermal comfort'.

**Fuel poverty policies**

- *Warm Front*: the main programme for tackling fuel poverty in England. It aims to bring homes up to a satisfactory SAP rating of 65; the national stock average is currently 51.

*Source:* UK Government, 2005

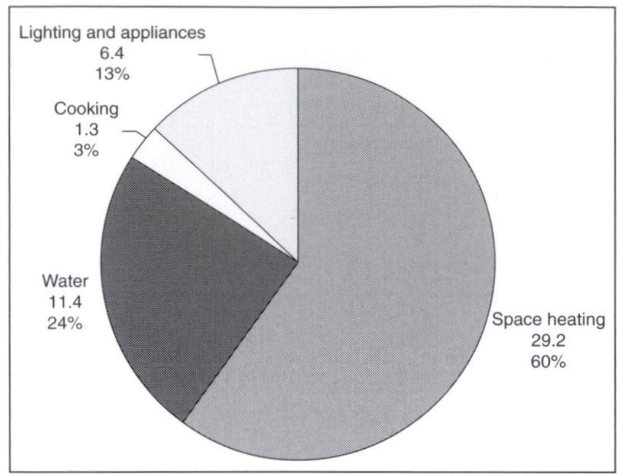


*Source:* ACE, 2005

**Figure 4.7** Domestic final energy by end use (mtoe), 2003

Energy Performance Certificates (EPCs) in 2007 for domestic dwellings that are either sold or let (UK Government, undated).

More recently the UK government has set mandatory standards for new buildings and requires that all new build from 2016 will be zero carbon (UK Government, 2007). As part of this drive the government has introduced a new set of building codes known as the Code for Sustainable Homes (UK Government, 2008). The code is a national standard that defines the sustainability of a home based on energy requirements and on criteria such as waste, ecology, water and materials. Houses are grouped into 6 star categories, with each category representing a percentage improvement on homes that are built to Part L of the Building Regulations in 2006. For example a 3-star rating is 25 per cent better and this is expected to be achieved for all new build by 2009. A 5-star rating represents a 100 per cent improvement. Carbon zero homes are given a 6-star rating. The government has introduced fiscal incentives to encourage the 6-star rating by abolishing stamp duty (a tax paid when houses are sold) on 6-star rated zero carbon homes up to a value to £500,000 and a £15,000 reduction on stamp duty for houses over £500,000. The code only applies to new build and influencing improvements in existing stock will require other interventions.

The EU is committed to the Kyoto Protocol 8 per cent target but has unilaterally set targets to cut its carbon emissions by 20 per cent by 2020 compared with a 1990 benchmark and has offered to deepen this to 30 per cent if other major emitters follow suit (COM, 2007). These commitments do not prevent individual states using a variety of domestic measures to comply with national standards that exceed those of the EU, providing these are implemented in ways that do not conflict with EU policy. In the UK the Climate Change Bill, expected to become law in 2008, sets a target reduction of 29 per cent below 1990 levels by 2020 and a 60 per cent reduction by 2050 (UK Government, undated-a).

Studies do show that considerable improvements in the thermal efficiency of existing buildings are achievable through passive measures such as increased insulation levels (cavity and solid walls and lofts), high efficiency glazing, air tightness and ventilation (Boardman et al, 2005; ESD, 2004; Johnston et al, 2005). It is unlikely that passive measures alone will achieve the UK government target of a 60 per cent reduction in emissions by 2050, but, according to ESD, these measures along with active generation and heat capture technologies can reach that level for the UK housing sector. The recently announced UK Renewable Energy Strategy, which is undergoing consultation at the time of writing, contains a range of measures related to renewable energy technologies and on improving the efficiency of existing housing stock (UK Government, 2008). The findings of the consultation are expected in 2009.

However, to date much of the UK Government's policy towards housing has been very patchy and has had little impact on improving very much of the housing stock. Boardman, in a study conducted into reducing emissions from UK housing by 80 per cent by 2050, points out that UK housing policy is highly fragmented and sets out a number of areas where action is needed to raise the thermal efficiency of UK housing from its current average SAP rating of 48 to an average of 80 by 2050. These include an education and awareness programme that will demonstrate the benefits of moving up the EPC bands, subsidized loans, stamp duty rebates and targeted investment to encourage the majority of householders to move up the EPC ratings, carbon targets for local authorities related to their housing stock and financial incentives for innovation to encourage a learning process around new skills, techniques and products, which combine to exceed minimum standards (Boardman, 2007, p47).

## Non-domestic buildings

This refers to all buildings, either public or private, that are used for non-domestic purposes. This is a very mixed group, ranging from office buildings, factories, hospitals, schools, commercial, retail and leisure. As in the domestic sector non-domestic buildings are subject to standards that regulate their efficiency, but there is a growing recognition that Passive House principles are also relevant to this sector in a drive to produce zero energy or zero emission buildings. Research undertaken by the World Business Council for Sustainable Development (WBCSD) into the perception of energy efficiency issues in buildings identified three key barriers to the implementation of higher efficiency measures:-

- lack of information about building energy use and costs;
- lack of leadership from professionals and business people in the industry;
- lack of know-how and experience as too few professionals have been involved in sustainable building work (WBCSD, 2007).

Although lack of awareness is a continuing issue, there are many examples of buildings that have moved beyond that which is typically required by building regulations. For example, the Council House 2 building in Melbourne Australia, opened in 2006, has an energy demand of around 15 per cent that of typical existing buildings (City of Melbourne, undated). Buildings that have been designed to minimize energy use through passive techniques and capture renewable resources are not new. For example the Solar Office in Doxford Park, Tyne and Wear, UK incorporates many passive design features. The building has been designed to have a considerably lower energy requirement than a typical air conditioned building (85kWh/m$^2$/year, while a conventional air conditioned office uses 200–400kWh/m$^2$/year). The energy strategy for the building is based on minimizing heat losses, using building mass to control temperatures, utilizing natural ventilation and saving energy consumption by maximizing natural daylighting. The building design has active features and produces energy from the integrated PV in the façades (Lloyds Jones et al, 1989).

It is clear that policy is driving towards the ZEB concept, but there is a lack of clarity on which form of ZEB will be pursued. The Code for Sustainable Homes at present defines a ZEB as one that produces all of its energy requirements on site. The UK Green Building Council argues that in many cases this may not be achievable and argues that zero carbon should be allowed to use off-site renewable energy, but only where every effort has been made to first install on-site renewables (UK Green Building Council, 2008). This issue has yet to be resolved.

## Appliances and lighting

Figure 4.8 shows that electricity use for lighting and appliances throughout the OECD area has risen quickly. This is projected to continue to rise if additional measures are not taken to improve end use efficiency. The OECD estimates that about one-third of overall electricity production is used in the domestic sector with overall growth predicted to rise by 25 per cent by 2020 in the OECD. Without intervention to encourage manufacturers to produce more efficient lights and appliances and for consumers to purchase them, the IEA (International Energy Agency) predicts a growth of about 30 per cent in domestic electricity demand between 2008 and 2030 in the OECD (IEA, 2003). One of the main drivers

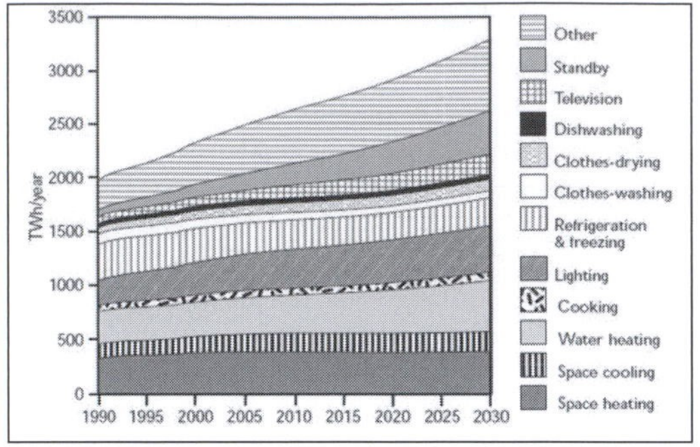


*Source:* IEA, 2003

**Figure 4.8** Projected IEA residential electricity consumption by end use with current policies

behind this increase is the rise in the number of electrical appliances per household. For example, in the UK the number has increased from an average of 17 electrical items in 1970 to around 47 in 2004 (EST, 2006, p9). According to DEFRA, 25 per cent of the UK's total electricity consumption is used to power lighting and appliances in the home. This is predicted to rise by 20 per cent by 2020 (DEFRA, 2007b).

The IEA estimates that considerable reductions in demand could be made, up to 33 per cent by 2030, with what it terms cost-effective energy efficiency policies. These are a series of measures targeted to strengthen residential appliance and equipment policies to target – as a minimum – the least lifecycle cost (LLCC) for each appliance class. The IEA argues that the most effective, reliable and cost-effective are mandatory Minimum Energy Performance Standards (MEPS) and comparative energy labelling. However, it does advocate for initiatives such as information campaigns, certification, voluntary agreements, technology procurement programmes and economic incentives as effective complements to standards and labelling and could encourage producers and consumers to go beyond minimum standards.

Lights and appliances are traded goods and have a significantly shorter lifetime than, for example, a house. As efficiency standards improve and these goods are replaced, over time domestic electricity is argued to fall. However, the IEA solution argued above does not advocate measures above and beyond those likely to be driven by national standards, that is to say, what is acceptable politically. This means different countries within the OECD area are likely to respond at different rates driven by a mixture of domestic political considerations, technological capacity to develop more efficient products, fiscal and other incentives, information campaigns, etc. In essence the mix of push and pull factors will determine the speed of take-up (IEA, 2003).

Policy in its widest sense results from the action of a broad range of institutions. Although the market approach advocated above does deliver efficiency improvements, active intervention can drive the process more rapidly. Within OECD countries, active interventions are the responsibility of either national governments or groups of countries that have a common interest, for example, the EU. Programmes that have been introduced at the national level to influence behaviour have achieved energy savings. Box 4.5 lists a number of examples of active measures and market influences that have improved efficiency.

## The EU policy context – lights and appliances

Although the EU is an economic union, this does not extend to control of, for example, taxation policy. That is the responsibility of member states. However the EU can act to set product standards so that one member does not receive a competitive advantage by operating at a lower, and hence cheaper, environmental standard. This approach has applied to both manufacturing processes and the products themselves, including lights and appliances. Because of climate concern, this has recently broadened to setting standards for items beyond traded goods, such as buildings. Examples of the type of measures developed by the EU are shown in Table 4.7.

The EU Energy Action Plan reports that a 27 per cent reduction in domestic energy use is realizable by 2020 (CEC, 2006). This includes measures for improving the thermal efficiency

## Box 4.5 Influence of push and pull factors

In Japan the Top Runner standard which requires all products to reach the same specification as the most efficient in the market within a set timeframe is expected to bring energy savings of 63 per cent for air-conditioning systems and 83 per cent for computers by 2010. By 2030, Japan is aiming for an increase in energy efficiency of 30 per cent.

Rising energy prices in the UK are influencing consumer behaviour. While household expenditure rose by just 1.4 per cent in 2005, there was an 11 per cent rise in consumer purchases of energy efficient appliances to £1.6 billion.

In the US more than 1400 US manufacturers now use the high-efficiency Energy Star logo across 32,000 product models. In 2005 alone, the use of Energy Star products in the US prevented the release of GHG emissions equivalent to 23 million cars and saved US$12 billion in energy costs. The Energy Star programme has run since 1992. Standards are set either by the Environmental Protection Agency or Department of Energy. Manufacturers meeting those standards can use the Energy Star logo.

The Energy Star label has been adopted for office equipment in European countries and is expected to help reduce consumption by about 10TWh/yr by 2015, equivalent to just over 2 per cent of the UK's electricity consumption in 2005.

*Source:* The Climate Group, 2007

of buildings and lighting and appliances. The EU has set legally binding targets of 20 per cent less $CO_2$ emissions and 20 per cent renewable energy use by 2020 and a non-legally binding target of a 20 per cent increase in energy efficiency by 2020. Although the Action Plan is not a direc-

**Table 4.7** EU Directives targeted at improving end use efficiency

| Directive | Title | Scope |
|---|---|---|
| 92/75/EEC | Energy labelling of domestic appliances | Overarching, with several daughter directives, specific to each appliance group. Being redrafted. |
| 2002/91/EC | Energy Performance of Buildings (EPBD) | Basis for Energy Performance Certificates; cost-effective, realizable saving potential of 22% from existing buildings by 2010. |
| 2005/32/EC | Eco-design – requirements for energy-using products (EuP) | Setting minimum environmental performance criteria for products. |
| 2006/32/EC | Energy end use efficiency and energy services | Requires Energy Efficiency Action Plans from individual member states by June 2007, identifying how to achieve a 9% reduction in delivered energy 2008–2016, in relation to baseline, i.e. 1% pa. Covers metering and billing by utilities. |

*Source:* Adapted from Boardman, 2007, p19

tive, it does extend the framework directive on energy-using products (EuP).

Priority Action 1 of the plan is focused on labelling and minimum performance standards. Minimum performance requirements are planned for 14 priority product groups by the end of 2008 and the Framework Directive 92/75/EC on labelling will be reinforced to enhance its effectiveness. Household appliances that will be impacted by minimum performance requirements are shown in Table 4.8.

**Table 4.8** Residential product policy studies for the Energy-using Products Directive

| Product | Measures adopted by Commission after: |
|---|---|
| Battery chargers, power supplies | Study complete |
| Personal computers and monitors (ICT) | July 2008 |
| Televisions (CE) | Study complete |
| Standby and off-mode losses | Study complete |
| Domestic refrigeration (freezers, fridges, etc.) | November 2008 |
| Washing machines, dishwashers | November 2008 |
| Boilers | January 2009 |
| Water heaters | January 2009 |
| Room air conditioning | February 2009 |
| Domestic lighting | March 2009 |
| Simple converter boxes for digital TV | Date unknown |
| Solid fuel small combustion installations (in particular for heating) | Date unknown |
| Laundry dryers | Date unknown |
| Vacuum cleaners | Date unknown |
| Complex set-top boxes (with conditional access and/or functions that are always on) | Date unknown |

*Source:* Adapted from Boardman, 2007, p20

There 10 priority areas outlined in the Action Plan and these together with the Energy Efficiency Action Plans represent a process of transforming the market with respect to efficiency. As the EU points out, the savings will only be realized if all the actions are fully implemented and ongoing work is needed to ensure that standards are increasingly improved. It is difficult to judge at this point the effectiveness of these measures. For example, all EU members have reported their Energy Efficiency Action Plans and initial screening by the Energy Efficiency Watch (EEW) team from the Wuppertal Institute and Ecofys indicate that plans fall into three broad categories:

- those that have invested strong efforts in developing their plans;
- those that submitted plans that have already been adopted in other contexts at national level;
- those that provided only short or draft plans (EEW, 2007).

Although the plans contain a broad array of measures, only Spain and Denmark propose meeting the 9 per cent reduction by the 2016 timeframe and it is too early at this point to judge how effective this measure will be. But the study does show that new member states, such as the Baltic States, Bulgaria, Romania, Poland and the Czech Republic still require basic infrastructure to be upgraded before effective implementation can begin. Other states – like Finland, the UK and Sweden, for example – have a longer tradition of implementing energy efficiency improvements.

To make further inroads into improving end use efficiency may require further intervention at the EU level or at the national level. For example, lighting systems account for around 6 per cent of UK household domestic electricity consumption. Most lights in service are the inefficient incandescent bulb. There are striking differences in the efficiency of different lighting types as shown in Table 4.9. Note that efficiency is expressed as efficacy, which is the amount of light emitted per unit of power used.

Although LED (light emitting diode) lights are still in development and currently mainly

**Table 4.9** Typical efficacies of different light types

| Type | Efficacy (lumens/Watt) |
|---|---|
| Incandescent 40W | 10 |
| Incandescent 60W | 12 |
| Incandescent 100W | 15 |
| Halogen | 25 |
| Compact fluorescent lamp (CFL) | 40–60 |
| Linear fluorescent | 60–80 |
| Light emitting diode (LED) | 150 |

*Source:* Adapted from Boardman, 2007, p26

used in specialist applications such as flashlights or for decorative purpose, their low energy consumption and reliability could have a significant impact on domestic energy use. Boardman reports that electricity consumption for lighting could be reduced from 16, 000Gwh to 2000Gwh by 2030 if a concerted effort were made. However, there are considerable social, economic and technical barriers to be addressed (Boardman, 2007).

Another area that the EU and the global community are aware of is the growing use of ICT and internet related devices in the household. The proliferation of electronics in the home has increased power usage considerably, as shown in Table 4.10, and without intervention is likely to continue to increase. More and more devices use standby power, where electricity is consumed by end use electrical equipment when it is switched off or not performing its main function. The most common users of standby power are televisions (TVs) and video equipment with remote controls, electrical equipment with external low voltage power supplies (e.g. cordless telephones, mobile phone chargers), office equipment and devices with continuous digital displays (e.g. microwave ovens). The actual power-draw in standby mode is small, typically 0.5–30W. However, standby power is consumed 24 hours per day, and more and more new appliances have features that consume standby power. Although consumption by individual appliances is small, the cumulative total is large and standby power consumption is becoming comparable to refrigeration (IEA, 2003).

## EU research and development for energy efficiency

The EU has made some funding available in the area of eco-innovation, although precise figures for expenditure in support of energy efficiency measures within the EU are not available. Efficiency applies very broadly and there are some programmes that will make a contribution to a wide range of efficiency measures. The Seventh Framework Programme (FP7) for research and technological development, for example, provides monies for select research projects during the period 2007–2013 according to different thematic areas, including 2.3 billion euro for energy. To what extent this will support efficiency research is dependent on the projects supported.

**Table 4.10** Estimation of the total energy consumption for various information and telecommunication technologies in European households

| | Average energy consumption in kWh/year | | |
|---|---|---|---|
| | 1996 | 2000 | 2010 No policy |
| Television | 149 | 155 | 272 |
| Receiver | 18 | 31 | 161 |
| Video appliances | 86 | 87 | 79 |
| Audio appliances | 158 | 167 | 195 |
| Personal computer | 32 | 88 | 243 |
| PC monitor | 28 | 42 | 35 |
| PC network/gateway | 1 | 18 | 64 |
| Other (games, phone, etc.) | 30 | 30 | 30 |
| Total consumer electronics (million) | 502 | 618 | 1079 |
| Households in EU (million) | 147 | 152 | 158 |
| Total consumption in EU (Twh/year) | 74 | 94 | 170 |

*Source:* IEA, 2003

## Box 4.6 EU research and development programmes

Seventh Framework Programme. Available at: http://cordis.europa.eu/fp7/home_en.html
Intelligent Energy Europe (IEE). Available at: http://ec.europa.eu/energy/intelligent/index_en.html
Entrepreneurship and Innovation programme. Available at: http://ec.europa.eu/cip/eip_en.htm

Funding within the Competiveness and Innovation Programme (CIP) supports two programmes that are more focused on efficiency. These are the Intelligent Energy Europe (IEE) programme with funding of €730 million and the Entrepreneurship and Innovation programme with €430 million that is focused on eco-innovation. Internet addresses for these programmes are shown in Box 4.6.

## Future approaches to energy efficiency

Considerable effort is, and continues, to be made to improve the efficiency of many energy using products. Often technological developments have reached a point where further large efficiency gains are unlikely, for example, those processes that are governed by the laws of thermodynamics, for example the ICE. The use of hybrids and fuel cells signal a break with the conventional technological development route. There are other examples. The refrigerator is common to almost all households in the developed world and its use is growing throughout the industrializing and developing worlds. The benefits in terms of food storage are self-evident. However, the technology relies on a motor. Even though considerable improvements have been made in refrigeration technology, a significant shift in efficiency requires a new approach. Conventional refrigeration and air conditioning work by compressing a refrigerant, which grows cold as it is allowed to rapidly expand. The refrigerant is then circulated around to remove heat from fridges or air that is then used for cooling.

Other techniques such as magnetic cooling and the electrocaloric effect do require moving parts. Both techniques use ordering and disordering to remove heat. Magnetic cooling has existed for years, employing certain materials with 'magnetic dipoles' that act like tiny compass needles. These are cycled through a magnetic field to produce the order and disorder effect. That approach is projected to be 40 per cent more efficient than conventional cooling. The electrocaloric effect removes heat by the ordering and disordering of polymers, which are distributed in a thin film just a millionth of a metre thick. In an electric field, the molecules spontaneously line up, creating heat. Removing the field causes the polymers to cool down. This effect is thought to be more efficient than magnetic cooling (Neese et al, 2008; Gschneidner and Gibson, 2001). It is difficult to judge if either of these techniques will replace the existing domestic refrigerator.

Other approaches use an effect known as thermoacoustic refrigeration. This is based on a technology that uses high-amplitude sound waves in a pressurized gas to pump heat from one place to another – or uses a heat temperature difference to induce sound, which can be converted to electricity with a high efficiency piezoelectric loudspeaker. This approach is being employed by the SCORE (Stove for Cooking, Refrigeration and Electricity) project aimed at finding more efficient and safe ways of utilizing biomass fuels such as wood. The University of Nottingham, University of Manchester, Imperial College London and Queen Mary, University of London are the primary partners in the project, along with the Los Alamos Laboratories. The idea is to produce a number of energy services from a single fuel (SCORE, 2007).

Hand-held devices such as phones, entertainment systems and small computers are a fact of modern day life. As more and more functions are built into these devices, their efficiency becomes more important. Portable devices rely on battery power and although battery technology is improving, ensuring a useful service length between charges requires efficient power management and components. Considerable effort is being made by large manufacturers such as Intel and Texas Instruments to make electronic devices more efficient. For example, Intel has set out a research programme that looks at improved power management and component design to reduce power use (Chary et al, 2004). Other manufacturers are seeking to improve efficiency as well as look at other techniques. Texas Instruments are developing a range of devices that do not rely on being connected to a power source but generate power from their environment, for example a sensor that monitors a bridge but uses the vibration of traffic to generate power (BBC, 2008b). Devices such as weather stations powered either from photovoltaic systems or wind turbines and a battery are becoming increasingly common. A drive to more efficient systems is likely to see more and more autonomous systems in everyday use.

ICT (Information and Communications Technology) provides the opportunity to improve energy efficiency, but it is also an energy user. Although ICT is an established technology and used extensively, the micro- and nano-electronics industry is highly innovative. Not only has processor power continued to increase in line with Moore's law (that predicts that this will double every two years), the power requirements of a chip of a given capacity have halved every 18 months. This has been achieved through great miniaturization and innovative architecture. Future developments will use techniques that turn off part of the processor when it is not being used, reducing energy needs further.

In Europe, information technologies have a positive environmental impact, for example, the dematerialization of transport through a switch from air travel to videoconferencing and the digitalization of information, for example, through the switch from catalogues to websites. There are, however, adverse environmental effects associated with the manufacture and disposal of ICT equipment as it contains toxic and hazardous substances (Yi and Thomas, 2007). But broadly the benefits that ICT brings can increasingly outweigh the disadvantages. ICT accounts for around 2 per cent of global energy use and around 2 per cent of global carbon emissions. However, the innovations in the sector can help to promote energy savings in other areas, such as the following:

- *The power grid*: With greater decentralization of grids and the use of renewable energy sources and micro-generation, ICT can play a major role not only in reducing losses and increasing efficiency but also in managing and controlling the distributed power grid to ensure stability and security.
- *Energy smart homes and buildings*: Buildings account for some 40 per cent of energy use in the EU. ICT-based energy management systems for both new and old buildings will help to reduce energy use. Smart metering and advanced visualization that continuously monitors building performance can optimize efficiency. Heightened awareness of energy consumption can stimulate behavioural changes at both the household and enterprise level. Studies in Finland have noted gains of 7 per cent in households that have systems providing consumers real-time feedback on their consumption. It is believed that energy savings in companies could be as high as 10 per cent.
- *Indoor and outdoor lighting*: About one fifth of the world's electricity consumption is for lighting. High efficiency LED technology could save 30 per cent of today's consumption by 2015 and up to 50 per cent by 2025. Adding sensing and actuation capabilities to energy-efficient bulbs, so that they automatically adjust to the environment, can lead to further improvements (Communication from the Commission, 2008).

## Efficiency of electrical motors

This section looks at the efficiency of electric motors and electrical motor systems. Although these systems are usually associated with industry and the commercial sector, there are a considerable number used within the domestic sector. In reality electric motors provide a wide range of services from large scale driving pumps in the water networks through to electric fans found in many households. Within the EU, electrical motors account for about 65–70 per cent of all electricity consumed by industry. In the US it is estimated at 67 per cent. Studies show that switching to energy efficient motor systems could save Europe up to 202 billion kWh in electricity consumption, equivalent to a reduction of €10 billion per year in operating costs for industry. It would also create the following additional benefits:

- a saving of €5–10 billion per year in operating costs for European industry through reduced maintenance and improved operations (EU-25).
- a saving of €6 billion per year for Europe in reduced environmental costs (EU-25, calculated using the EU-15 fuel mix).
- a reduction of 79 million tonnes of $CO_2$ emissions (EU-15), or approximately a quarter of the EU's Kyoto target. This is the annual amount of $CO_2$ that a forest the size of Finland transforms into oxygen. If industry is allowed to trade these emission reductions based on energy saved, this would generate a revenue stream of €2 billion per year. For EU-25, the reduction potential is 100 million tonne.
- a 45GW reduction in the need for new power plant capacity over the next 20 years (EU-25).
- a 6 per cent reduction in Europe's energy imports (EU-25) (De Keulenaer et al, 2004).

De Keulenaer et al claim this is the equivalent of:

- 45 nuclear power units (1000MW)
- 130 fossil fuel power units (350MW)

The 202 billion kWh is equivalent to about five times the electricity production of all wind power units in Europe (EU-25) in 2003 (5 × 40 billion kWh).

One of the main reasons for this level of inefficiency in electrical motor systems is cost. Standard motors are very cheap to buy. Typical practice is to over specify the electrical motor. This means that effectively the system has a less efficient motor that has greater power than that needed. Often systems of this kind incorporate inefficient control mechanisms. In total this lowers the efficiency of the overall system. Figures 4.9 and 4.10 show two configurations for a pumping system. Figure 4.9 shows a conventional system. In this case the flow is regulated using a throttle. In Figure 4.10 the approach is based on the use of a Variable Speed Drive (VSD). This regulates the speed of the motor and hence the flow through the system. A throttle is not required. By using a more efficient pump, coupling and pipe-work, the overall efficiency of the system is more than double the conventional approach. The conventional system has total efficiency of 31 per cent and the VSD system has a total efficiency of 72 per cent.

VSDs are electronic systems that are attached to an induction motor that use external data, such as pressure or flow rate, to regulate motor speed so that either pressure or flow rate is kept within specified levels. They can be incorporated into any motor with a variable load, but the most common applications are pumps and fans. Other applications include air compressors. The system works by using a control signal to regulate the speed of the motor. In many applications energy use can be cut by 87 per cent just by adjusting the motor speed. Despite the scale of the potential savings, less than 10 per cent of motors worldwide are combined with a variable speed drive (Business Europe, 2007).

Energy saving is not confined to the industrial sector. In the domestic sector motors are used in a range of items such as vacuum cleaners, washers, dryers and heating systems. In those cases where motors are in use for extended periods, such as those used to drive circulation pumps for domestic heating systems, the energy saving potential

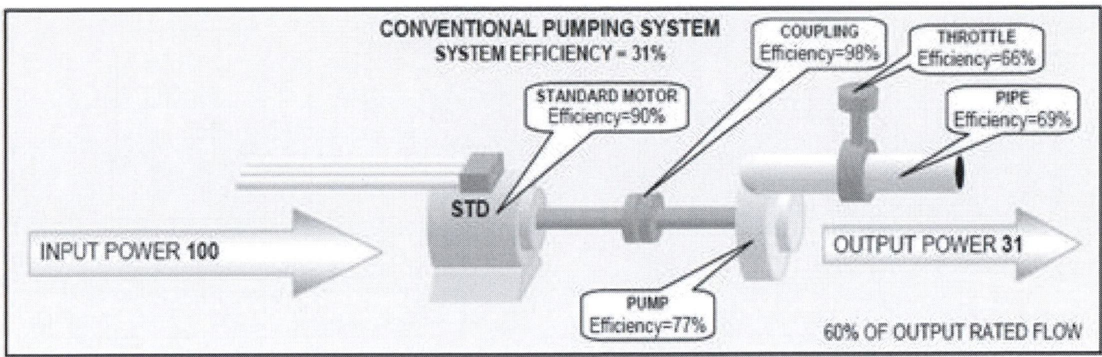


*Source:* De Keulenaer et al, 2004

**Figure 4.9** Conventional system

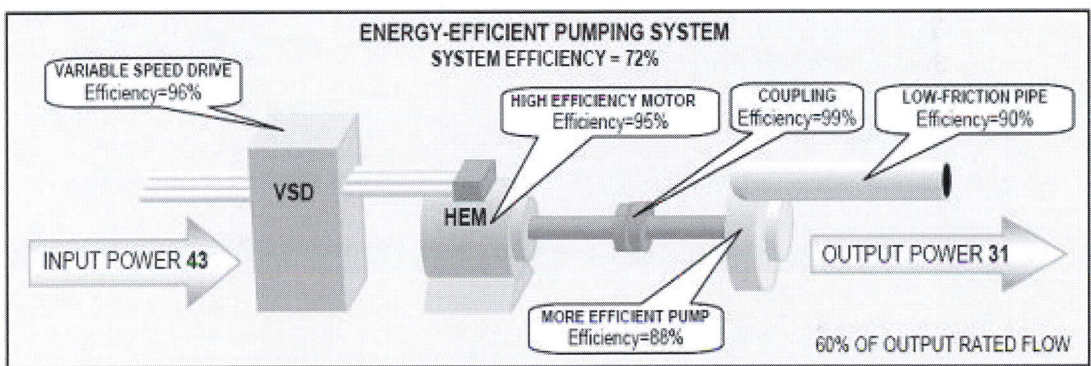


*Source:* De Keulenaer et al, 2004

**Figure 4.10** Efficient system

is significant. There are some 90 million circulation pumps in use in heating systems throughout the EU. Although these are typically less than 250 watts (there are an additional 300,000 circulators above this size – typically used in multi-family dwellings) the overall energy consumption is estimated at 30–40TWh. Studies suggest that this could be reduced by 10–40TWh with existing technology to control motor speed. Electricity is also used in other areas such as thermostats and motorized valves, but the circulatory system offers the greatest scope for savings (Save II, 2001).

## Barriers

The benefits of using energy efficient systems would seem to be self-evident. De Keulenaer et al (2004) report a number of studies that identify the key barriers. Some of them are specific to certain industrial sectors or certain categories of motor systems (e.g. pumps, compressors, fans). Nevertheless, some general observations stand out. The following nine types of market barriers, grouped into categories according to importance, describe the largest part of the problem:

- Major barriers:
  1 pay-back time is too long due to low electricity prices;
  2 reluctance to change a working process;
  3 split budgets;
- Medium barriers:
  4 not all parties in the supply chain are motivated;
  5 lack of correct definitions of motor system efficiency;
  6 oversizing due to lack of knowledge of mechanical characteristics of load;
  7 lack of management time;
- Moderate barriers:
  8 shortage of capital;
  9 other functional specifications conflict with energy efficiency.

With rising electricity prices the major obstacle, pay-back time, may start to decline in importance. The authors of the study do suggest a number of critical success factors:

1 *A legal framework that favours high efficiency motor systems*: Standards are typically voluntary. Although the EU supports the concept of efficient motors through the Motor Challenge Programme, it has yet to propose EU-wide legislation.[2]
2 *Adequate support*: More resources generally means better outcomes.
3 *High quality information, in relevant terms*: Ensuring that the right message about the benefits is effectively communicated.
4 *Streamlining with other programmes*: Ensuring that within the EU a clear and concise message is promoted.
5 *Measuring results and giving feedback*: Clear results presented in a clear way.
6 *Involvement and coordination between different interested parties*: A fragmented approach within the supply and user chains militates against an effective approach.
7 *Differentiating for each separate market*: National markets within the EU do vary and this should be recognized.

## Transport

Today's transport sector, covering systems for road, rail, air and water, is predominantly based on the combustion of fossil fuels, making it one of the largest sources of both urban and regional air pollution and greenhouse gases. It is the cause of other environmental and social ill effects, ranging from loss of land and open space to noise-related nuisance, injuries and deaths arising from accidents. But the movement of goods and people is crucial for social and economic development by enabling trade and providing opportunities for employment, education and leisure. Shifting the transport sector to a more sustainable basis is an urgent challenge. According to the International Transport Forum (ITF) transport emissions are increasing and this is set to continue well into the future. For example, globally, aviation is set to double in less than 20 years as is container traffic by sea and car ownership and use. And this is in stark contrast to the aspirations of many governments to make significant cuts in greenhouse gas emissions (ITF, 2007).

In the transport sector there are three broad areas of action that can both reduce dependence on fossil fuels and improve end use efficiency:

1 *Fuels*: the high dependence of transport on fossil fuels is driving the use of alternatives such as biofuels. Other changes that are likely to occur in the longer term are a move to the use of hydrogen as a new energy carrier.
2 *Technology*: improving the efficiency of existing technologies and introducing new technologies such as fuel cell, electric and hybrid systems.
3 *Behavioural*: this is a broad area covering urban planning, alternative work scheduling, improvements in public transport, changes in driver behaviour and intelligent transport systems (Source: WEC, 2007).

### Fuels

Fuel substitution, that is, the use of non-fossil sources that can either replace exiting supplies or be mixed with them is a mechanism for reducing GHG and improving energy security for those countries that rely on imported primary energy supplies. Alternatives to fossil fuels include:

## *Biofuels*

This category consists of fuels such as biodiesel which is derived from vegetable oil extracted from oilseed crops, such as rapeseed and sunflower, and mixed with a small amount of methanol, and bioethanol which is obtained from the fermentation of sugar-bearing and starch crops such as sugar beet, wheat, maize and potato. Other forms of ethanol production include cellulosic ethanol, and can be produced from a wide variety of cellulosic biomass feedstocks including agricultural plant wastes (corn stover, cereal straws, sugarcane bagasse), plant wastes from industrial processes (sawdust, paper pulp) and energy crops grown specifically for fuel production, such as switchgrass. These are sometimes referred to as first generation biofuels.

## *Synfuels*

This category refers to fuel components that are similar to those of current fossil-derived petrol (gasoline) and diesel fuels and hence can be used in existing fuel distribution systems and with standard ICEs. Synfuels can be derived from biomass (BTL, Biomass to Liquid, applies to synthetic fuels made from biomass through a thermo chemical route. Feedstock biomass that may be used in this process includes wood, straw, corn, garbage and sewage-sludge. Synfuels can also be produced from coal (CTL – coal to liquid) and gas (GTL – gas to liquid). These are sometimes referred to as second generation biofuels.

BTL and cellulosic ethanol have the potential to reduce conventional energy consumption (and GHG emissions) by up to 90 per cent. But this has to be balanced against the energy expended in the production process. BTL is a developing technology and the overall efficiency (or energy cost) of production is dependent upon the biomass inputs and the transformation processes. Other synthetic fuels such as GTL and CTL do increase the diversity of the fuel supply base and, particularly for GTL, are already available and economically viable. On a lifecycle basis, GHG emissions from GTL are comparable to conventional diesel, and for CTL without carbon capture and storage they are approximately double. CTL and GTL also contribute to technological experience and the understanding of synthetic fuels in general, benefiting BTL development (Kavalov and Peteves, 2005; WEC, 2007; World Coal Institute, undated). Synfuels are of particular interest to the aviation sector as they can be readily used as a substitute for aviation kerosene. The best option from a climate perspective is BTL, however, there are implications for land use for the volumes of biomass needed for BTL synfuel (Farmery, 2006).

Although alternative fuels do offer an opportunity to diversify the supply chain, there are some issues that are problematic. In the US biofuel production has displaced existing agricultural production and it is claimed that this change has increased net greenhouse gas (GHG) emissions (Searchinger et al, 2008). The UN advisor on food, Olivier de Schutter, claims that energy crops have driven up food prices and has called for a freeze on biofuel investment (BBC, 2008c). A UK government commissioned review of biofuel policy concluded that although biofuels had a role in a low carbon transport future it was unlikely that the EU target of 10 per cent by 2020 could be met sustainably. A target of 5–8 per cent is claimed to be more viable (Gallagher, 2008).

It is likely that biofuels will continue to play a role in the fuel supply system, although the quantity and types will depend upon market conditions and public acceptability. Other alternative fuels such as hydrogen can be used in the ICE but there are problems related to the onboard storage of hydrogen and the lack of re-fuelling infrastructure. There are examples of efforts to develop new infrastructure to make hydrogen available for vehicles, such as the Hydrogen Highway in California (California Hydrogen Highway, 2008).

Although hydrogen is a clean fuel at the point of use, with water being produced as the byproduct of the combustion process, there are issues around the production of hydrogen. There are two methods:

- The first involves stripping hydrogen from the hydrogen rich natural gas ($CH_4$) methane molecule. This is a proven method for hydrogen production, the byproduct of which is carbon dioxide.
- The second involves using electrolysis utilizing electricity from renewable resources. There are issues around the efficiency of the process, but recent research suggests that production efficiencies can exceed 85 per cent, making the production of hydrogen economically attractive (Dopp, 2007). However, there are still issues about the energy cost of compression and storage to be considered and at present hydrogen offers potential for short predictable journeys because of the problems associated with fuel storage and lack of infrastructure (Bossel, 2003). In the longer term the use of a combination of hydrogen and fuel cell offer the most realistic prospects for vehicle technology.

## Technologies

Alternative strategies to improve efficiency and reduce dependence on fossil fuel are based on re-thinking the role of the ICE as the source of motive power for the vehicle. The hybrid concept has the ICE as the motive power that drives a generator that supplies electrical power to an electric traction system that provides the motive power. There are some variations on the hybrid concept, for example, where the sole role of the ICE is to drive a generator (serial configuration) or where the ICE can drive both a generator and act as a source of motive power (parallel configuration). (See Figure 4.11.) The parallel configuration is the one most used in production motor vehicles. In this configuration the motive power can be produced by the battery or engine and the battery is recharged by the engine. The series configuration is conceptually similar to the battery electric vehicle; that is a vehicle that derives all of its motive power from onboard batteries that are then recharged when the vehicle is at rest. This is often referred to as the Plug-In Hybrid Electric Vehicle (PHEV). In this configuration the engine drives a generator that charges the battery when needed. Motive power is provided by an electric motor.

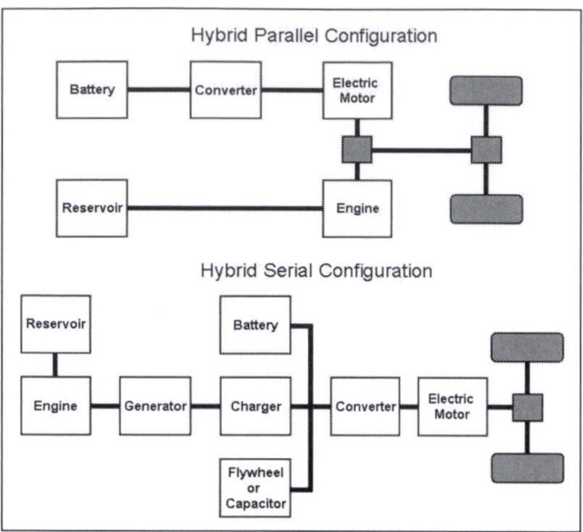


**Figure 4.11** Hybrid vehicle configurations

There is considerable investment in developing hybrid vehicles as they can both improve end use efficiency, that is more distance is travelled per unit of fuel and they can reduce the production of harmful emissions. The most efficient emit between 42 and 47 per cent of the emissions of a conventional car engine. Technological innovation is not sufficient to develop a market for hybrid vehicles. Incentives such as significant tax credits, the waiving of various charges and road regulations applicable to conventional vehicles, government R&D programmes and reduced fuel duty on biofuels are needed. In the UK there are government measures to help support the introduction of hybrids such as low Vehicle Excise Duty and a congestion charge exemption in London.

Since the introduction of hybrid vehicles in 1999–2000 by Honda sales have grown rapidly and are forecast to reach one million in annual unit sales by 2010. The US is the largest hybrid market with Toyota the clear leader there and worldwide. The most popular range is Prius but sales of the Lexus model are increasing. Worldwide sales of hybrids were 750,946 units by August 2006, with just over half sold in the US and 36,470 sold in Europe, an increase of 11 per cent over the previous year (The Climate Group, 2007). Currently most major manufacturers have either introduced hybrid models or are planning to.

Electric vehicles have been in service for some time but have typically been used for specialist or niche areas, for example, the milk float. Battery vehicles rely on an external source of electricity for recharging. There are only a few models available, for example, the electric Smart Car, a development of the Smart Car manufactured by Mercedes Benz, of which 100 units are being trialled in London by local authorities, estate agents, building associations and other fleet users who live inside the capital's congestion charge zone (English, 2008; What Car?, 2007). Both hybrid and electric technologies are also suitable for other types of vehicular transport systems such as buses, truck and light and heavy rail systems and a number of systems are being developed.

## Hydrogen and fuel cells

Hydrogen does not come as a pre-existing source of energy as do fossil fuels, but as an energy carrier. In effect producing hydrogen from the electrolysis of water means that hydrogen stores or carries the energy that was used to power the process. In some respects this is similar to a battery. Hydrogen can be produced from both renewable and non-renewable energy sources. Hydrogen can be used in an ICE, but the technological drive has focused on use in a fuel cell; this is more efficient as the power can be supplied directly to the motors that drive the wheels, eliminating many of the moving parts found in a conventional vehicle. Hydrogen reacts with oxygen inside the fuel cell which produces electricity. Figure 4.12 gives an overview of a PEM (Polymer Electrolyte Membrane) fuel cell.

**Table 4.11** Fuel cell types and applications

| | AFC (Alkaline) | PEM (polymer electrolyte membrane) | DMFC (direct methanol) | PAFC (phosphoric acid) | MCFC (molten carbonate) | SOFC (solid oxide) |
|---|---|---|---|---|---|---|
| Operating temperature (degrees C) | <100 | 60–120 | 60–120 | 6–120 | 600–800 | 800±1000 low temperature (500±600) possible |
| Applications | Transportation<br>Space<br>Military<br>Energy storage systems | | | Combined heat and power for decentralized stationary power systems | Combined heat and power for stationary decentralized systems and for transportation (trains, boats, ...) | |
| Power Output | Small plants 5–150kW modular | Small plants 5–150kW modular | Small plants 5kW | Small - medium sized plants 50kW–11MW | Small power plants 100kW–2MW | Small power plants 100–250kW |

*Source:* Adapted from Carrette et al, 2001

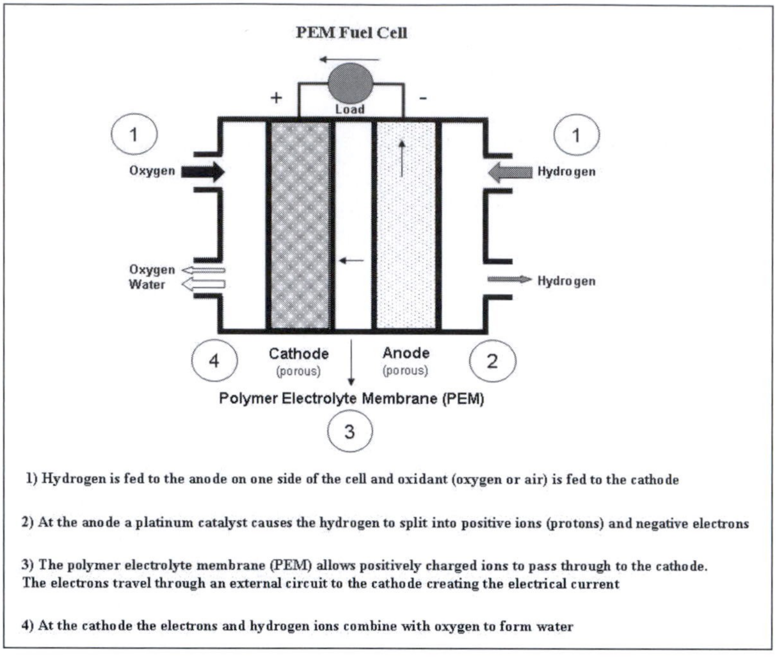


*Source:* Adapted from Carrette et al, 2001

**Figure 4.12** PEM (polymer electrolyte membrane) fuel cell

In order to produce sufficient power to drive a vehicle, fuel cells are connected together in series and parallel in order to produce sufficient voltage and current. These configurations are known as stacks. The PEM fuel cell is the most suitable for vehicles because of its relatively low operating temperature. There are other types of fuel cells that use different materials and can be used for different applications, as shown in Table 4.11.

## Fuel cell vehicles

A primary area of research is hydrogen storage in order to increase the range of hydrogen vehicles, while reducing the weight, energy consumption and complexity of the storage systems. Two primary methods of storage are metal hydrides and compression. Fuel cell technology is still developing and it is unlikely that fuel cell vehicles will gain a significant market share of the road vehicle fleet before 2015 (IEA, 2007b). Although hybrid, electric and fuel cell vehicles are still in development stage, technological problems are not the only barriers to adoption. Others include:

- *Public awareness* – price as opposed to lifecycle cost is a primary public consideration; the range of vehicles is perceived as small – even though it is generally sufficient for most trips – and there is a lack of confidence because of past technical problems such as battery failure.
- *Manufacturers* – as with any new technology there is a lack of standardization for components and test methods.
- *Utilities* – there are safety and demand management issues around connecting large numbers of electric vehicles to the grid for charging purposes (IEA, 2007b).

Despite the development problems associated with hydrogen fuel cell vehicles and the fact that the market is at an early stage, there is considerable interest in fuel cell technology with about 600 units sold in 2006. Many of the top ten car manufacturers – such as GM, Toyota, Ford, DaimlerChrysler and Honda – all plan for commercially viable fuel cell cars to be available by 2015, and DaimlerChrysler, Honda and GM believe there will be a mass market by 2020–2025. DaimlerChrysler and Ford have invested US$100 million into a joint venture with Ballard Power, the largest maker of fuel cells for vehicles. Figure 4.13 shows the NECAR 4, a demonstration vehicle produced by DaimlerChryler, that uses a PEM system fuelled with liquid hydrogen. Toyota plans to lease a new hydrogen-powered fuel cell car in Japan and the US in 2008 and to have 50,000 cars on US roads by 2020. This is very small compared to the current market volume of some 64 million units and the scale of any uptake will be determined by the development of a large scale hydrogen refuelling infrastructure. In the US this is estimated at US$100 billion. In 2005 there were about 140 hydrogen fuel stations globally, with 59 per cent in the US (50 per cent of these in California alone), and 7 per cent in both Germany and Japan. A further 59 were developed in 2006, 29 of them in the US, 16 in California alone (The Climate Group, 2007). The number of stations is likely to continue to grow as the technology develops.

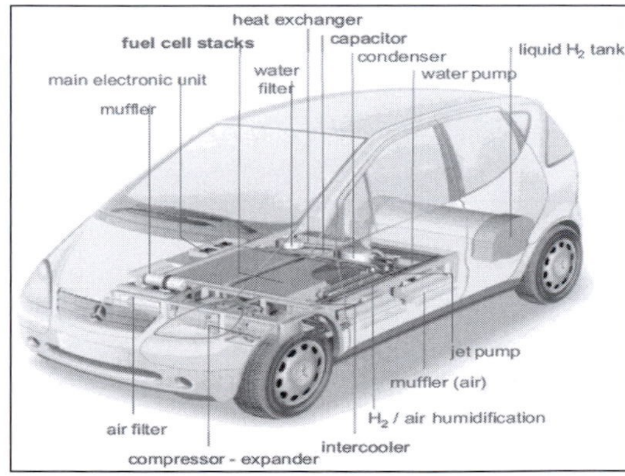


*Source:* Carrette et al, 2001

**Figure 4.13** NECAR PEM vehicle

## Behavioural

Projections of future patterns suggest continued growth in all forms of transport. Since 1990, $CO_2$ emissions from aviation – which are directly related to the amount of fuel consumed – have increased by 87 per cent and now account for around 3.5 per cent of the contribution to climate change of all human activities. The IPCC has estimated that this share will grow to 5 per cent by 2050 (Kahn et al, 2007). With growth in personal transport expected to continue to grow particularly in the industrializing nations of India and China, then alternative measures are needed to curb demand.

Measures to reduce transport energy demand take a number of different forms. The previous section dealt with the technological and fuel issues. Other measures include the policy nexus such as regulations, taxes and pricing measures, and still others that fall into a demand management nexus that are aimed at reducing demand and reducing energy consumption. Broadly these measures are aimed at modal shift – that is trying to encourage transport users to shift from high energy use (and often polluting) vehicles to transport modes that are more efficient in terms of fuel, although in dense urban areas this includes in terms of land use. Such a shift would encompass greater use of walking and public transport as well as working remotely using communication networks.

# EU policy context

Transport was one of the European Community's earliest common policies. Since the Treaty of Rome entered into force in 1958, transport policy has focused on removing obstacles at the borders between member states to facilitate the free movement of persons and goods. The transport industry also occupies an important position in the EU, accounting for 7 per cent of its gross national product (GNP), 7 per cent of all jobs, 40 per cent of member states' investment and 30 per cent of Community energy consumption. Table 4.12 gives more detailed statistics of transport

**Table 4.12** Statistical overview of EU transport (data for 2006, unless otherwise indicated)

| | |
|---|---|
| Employment | The transport services sector employed about 8.8 million persons in the European Union's 27 Sovereign States (EU27). Almost two-thirds (63%) of them worked in land transport (road, rail, inland waterways), 2% in sea transport, 5% in air transport and 30% in supporting and auxiliary transport activities (such as cargo handling, storage and warehousing, travel and transport agencies, tour operators). |
| Household Expenditure | In 2006, private households in the EU27 spent €893 billion or roughly 13.6% of their total consumption on transport.<br>About one-third of this sum (around €297 billion) was used to purchase vehicles, almost half (€440 billion) was spent on the operation of personal transport equipment (e.g. to buy fuel for the car) and the remainder (€155 billion) was spent on transport services. |
| Goods Transport | The demand for the four land transport modes road, rail, inland waterways and pipelines in the EU27 added up to 2595 billion tkm in 2006. Road transport accounted for 72.7% of this total, rail for 16.7%, inland waterways for 5.3% and oil pipelines for the remaining 5.2%.<br>If intra-EU maritime transport (the demand for which is estimated to have been around 1545 billion tkm) and intra-EU air transport (3.0 billion tkm) are added to the land modes, then the share of road transport is reduced to 45.6%, rail accounts for 10.5%, inland waterways contribute 3.3% and oil pipelines add another 3.2%. Maritime transport then accounts for 37.3% and air transport for 0.1% of the total (all referring to the EU27 in 2006). |
| Passenger Transport | Intra-EU27 and domestic transport demand using passenger cars, powered two-wheelers, buses and coaches, railways as well as tram and metro was about 5746 billion pkm or 11,674 km per person in 2006. Passenger cars accounted for 80.1% of this total, powered two-wheelers for 2.7%, buses and coaches for 9.1%, railways for 6.7% and tram and metro for 1.5%.<br>Adding intra-EU air transport (the demand for which is estimated to have been around 547 billion pkm in 2006) and intra-EU sea transport (40 billion pkm) to the land modes reduces the share of passenger cars to 72.7% and the share of powered two-wheelers to 2.4%. Buses and coaches then account for 8.3%, railways for 6.1% and tram and metro for 1.3%. The two additional modes, air and sea, contribute 8.6% and 0.6% respectively (all referring to the EU27 in 2006). |
| Transport Growth | Goods transport: about 2.8% per year (1995–2006).<br>Passenger transport: about 1.7% per year (1995–2006). |
| Transport Safety | Road: 42,953 persons killed in the EU27 (fatalities within 30 days) in 2006, 5.2% fewer than in 2005 (when 45,296 people lost their lives). In comparison with 2000, the number of road fatalities was lower by almost a quarter (23.9%).<br>Rail: 65 fatalities in 2005 (not including suicides).<br>Air: in 2006, 4 air passengers died over EU territory and 1 passenger died on board an EU carrier throughout the world. |

*Source:* Adapted from EU Commission, 2007a, sect. 3.1.1

and Figure 4.14 shows the growth in transport between 1996 and 2006.

Much of the thrust of EU policy has been aimed at ensuring the free movement of goods and products, however, more recently environmental concerns have risen in importance along with concerns about energy security and increasing congestion, especially in urban areas. The White Paper (European transport policy for 2010: time to decide EU Commission, 2001)) points out that the efficiency of the transport system is less than optimal. For example, the trans-European transport network itself increasingly suffers from chronic congestion: some 7500km, that is some 10 per cent of the road network, is affected daily by traffic jams. And 16,000km of railways,

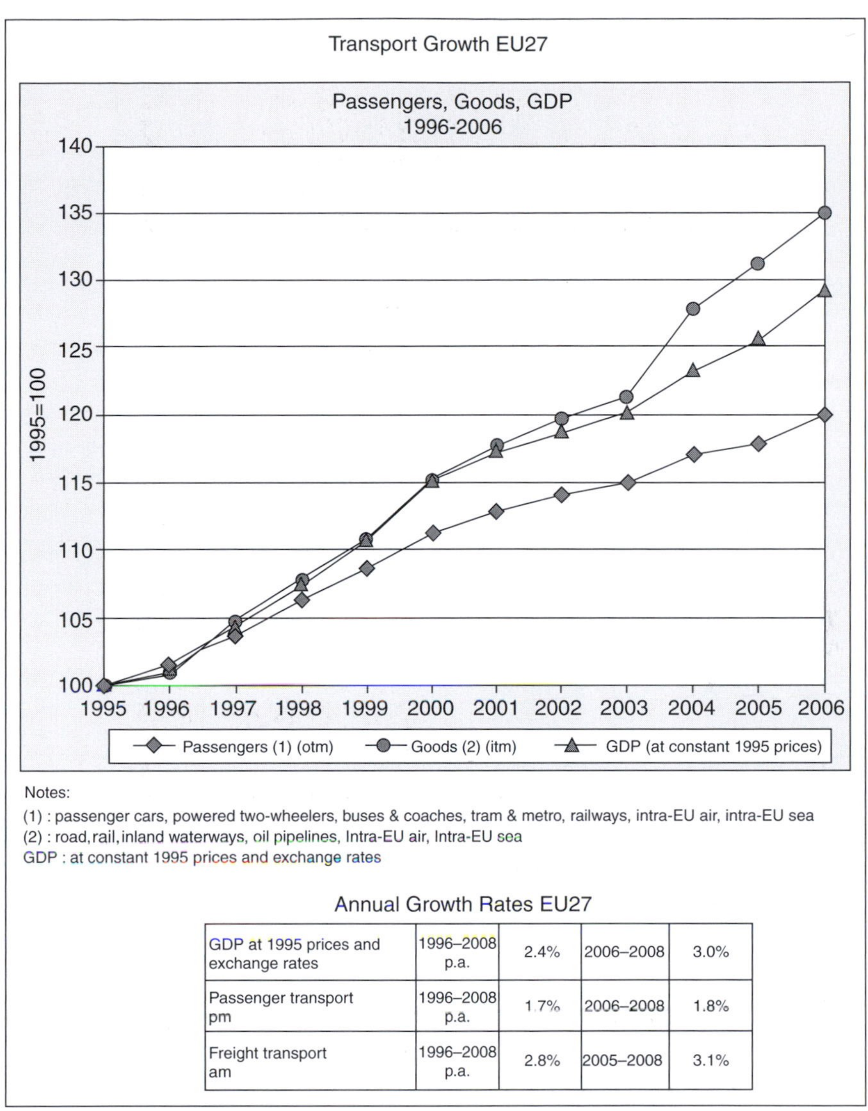


| | | | | |
|---|---|---|---|---|
| GDP at 1995 prices and exchange rates | 1996–2008 p.a. | 2.4% | 2006–2008 | 3.0% |
| Passenger transport pm | 1996–2008 p.a. | 1.7% | 2006–2008 | 1.8% |
| Freight transport am | 1996–2008 p.a. | 2.8% | 2005–2008 | 3.1% |

*Source:* EU Commission, 2007a, sect. 3.1.2

**Figure 4.14** EU transport growth and rates of growth

20 per cent of the network, are classed as bottlenecks. A total of 16 of the EU's main airports recorded delays of more than a quarter of an hour on more than 30 per cent of their flights. Altogether, these delays result in the consumption of an extra 1.9 billion litres of fuel, which is some 6 per cent of annual consumption (EU Commission, 2001). The White Paper proposed almost 60 measures designed to implement a transport system capable of restoring the balance between different modes, revitalizing the railways, promoting sea and waterway transport and controlling the increase in air transport, in response to the sustainable development strategy adopted by the Göteborg European Council in June 2001.

Scenarios produced for the EU Commission into future transport and based on a projection that takes into account transport policy measures,

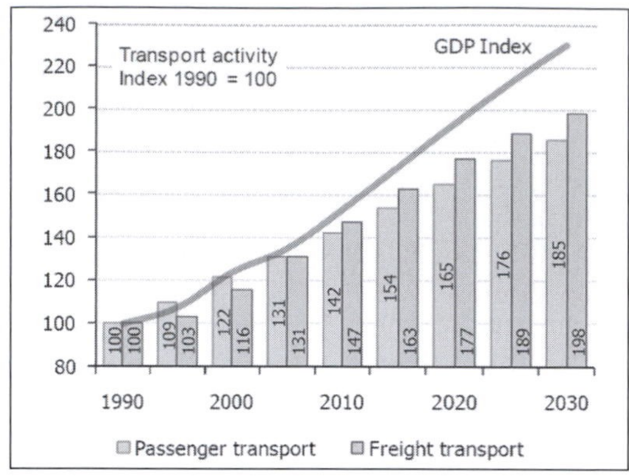


*Source:* Capros et al, 2008, p32

**Figure 4.15** Transport activity growth, 1990–2030

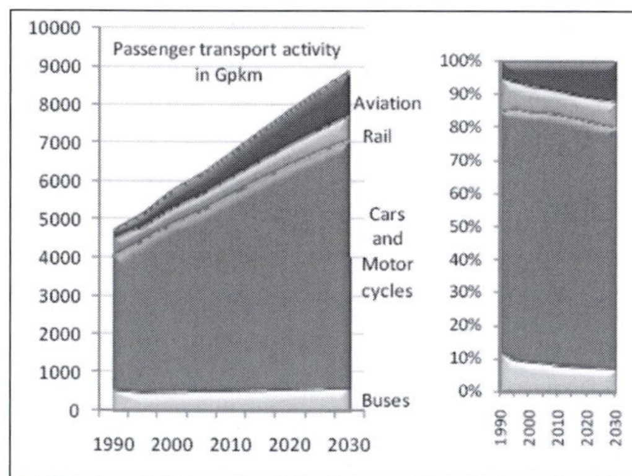


*Source:* Capros et al, 2008, p33

**Figure 4.16** Passenger transport by mode, 1990–2030

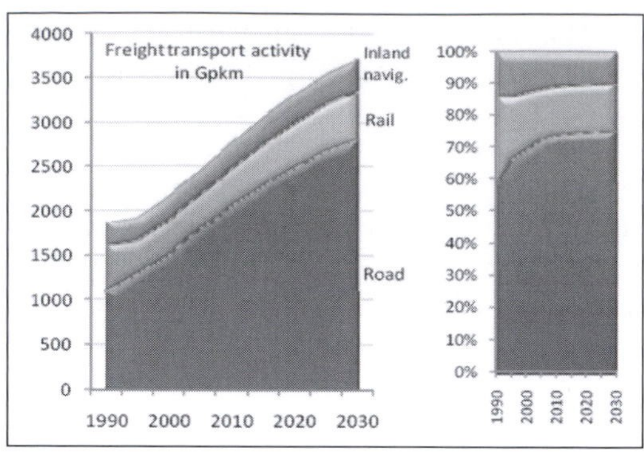


*Source:* Capros et al, 2008, p33

**Figure 4.17** Freight transport activity, 1990–2030

which are in place, or are likely to be implemented before 2010, show continued growth, but with a slowdown in the rate of growth towards 2030 as saturation levels are reached, as shown in Figure 4.15 (Capros et al, 2008).

The volume of passenger transport is projected to increase at a rate of 1.4 per cent per year, between 2005 and 2030, whereas the volume of freight transport is projected to increase by 1.7 per cent per year during the same period of time. In comparison to past trends, the scenario includes a slowdown in the rate of increase of activity, both for passenger and for freight transport. Air, road and rail are all expected to increase whilst public road transport is expected to decline. The results of these scenarios are shown in Figures 4.16 and 4.17.

The thrust of efficiency policy should be seen from two perspectives:

- Those measures aimed at improving the efficiency of the network.
- Those measures aimed at improving the efficiency of transport types (vehicles, vessels, aircraft).

## Network measures

The mid-term review of the White Paper (EU Commission, 2001) identified that road congestion was increasing (costing around 1 per cent of GDP) and there had been a sustained increase in air traffic. Overall, domestic transport accounted for 21 per cent of GHG emissions, which have risen by around 23 per cent since 1990 (COM, 2006). The review does advocate the continued liberalization and expansion of the transport network but does recognize that transport uses a great deal of energy (71 per cent of EU oil consumption: 60 per cent by road transport and approximately 9 per cent by air transport; the remaining 2 per cent is used by rail and inland navigation; rail transport uses 75 per cent electricity and 25 per cent fossil fuels) and that measures are needed to promote improved energy efficiency on a European scale. Essentially the policy framework is aimed at tackling the negative effects of transport, notably through improved logistics and traffic management and the promotion of cleaner, safer vehicles. Technology and the use of fiscal measures are envisaged as mechanisms for optimizing transport systems. Fiscal measures are premised on the external costs of transport systems. Transport demand continues to increase with ever increasing congestion in urban areas and some 60 EU airports predicted as being unable to cope with demand by 2025. The EU has published a handbook on estimates of external costs in the transport sector (Maibach et al, 2008). However, it is uncertain at this time how an EU-wide scheme for infrastructure will be developed. The only proposal at the moment is to allow member states to charge heavy freight vehicles of more than 3.5 tonnes, through the Eurovignette Directive, that will come into force in 2012. The external costs can include congestion costs, environmental pollution, noise, landscape damage, social costs such as health and indirect accident costs which are not covered by insurance. It is difficult to estimate the impact this will have, although congestion charges, for example, as in London, have been shown to reduce traffic levels.

In the aviation sector the EU plans to introduce a Single European Sky (SES) by 2025 that will alleviate the blockages caused by the interaction of 27 different national airspaces under the control of national governments. This, it is argued, will reduce aviation congestion and improve safety by reforming the current Air Traffic Management System. Aviation is currently responsible for 3 per cent of the world's GHG emissions. This is expected to rise as air traffic doubles by 2020. SES is one part of a three part scheme aimed at greening aviation. The other aspects are:

- A Clean Sky Joint Technology Initiative that is aimed at developing breakthrough technologies that will significantly improve the impact of air transport on the environment. The Clean Sky initiative is a €1.6 billion public–private research partnership to help the air-transport industry develop environmentally friendly technology for planes. It aims to reduce noise, fuel consumption and $CO_2$ emissions per passenger kilometre by half, as well as slashing nitrogen oxide ($NO_x$) emissions by 80 per cent by 2020 (Clean Sky, undated).
- The EU will impose a cap on $CO_2$ emissions for all planes arriving at or departing from EU airports from 2012. Airlines will be able to buy and sell 'pollution credits' on the EU 'carbon market' or Emissions Trading Scheme (ETS) (European Parliament, 2008).

## Technology measures

Initially technology measures in the EU have been aimed at reducing the emissions of pollutants from ICEs, for example, the introduction of unleaded fuel and catalytic converters. More recently the EU has developed policy that has been aimed at improving the efficiency of ICEs.

Cars account for around 20 per cent of total European emissions of carbon dioxide. In 1995, the EU heads of state and government set an ambitious goal of reducing emissions of $CO_2$ from new cars to 120 grammes per kilometre (g/km) by 2012 as a measure to combat climate change. This corresponds to fuel consumption of 4.5 litres per 100km (62.7 miles per gallon (mpg))for diesel cars and 5 litres/100km (46.5mpg) for petrol cars.

The EU strategy was based mainly on voluntary commitments from the car industry, which promised to gradually improve the fuel efficiency of new vehicles. Other stratagems, such as raising awareness among consumers and influencing demand through fiscal measures were also expected to contribute to the overall goal.

In 1998 a voluntary agreement between ACEA (the EU's Automobile Manufacturers Association) and the Commission included a commitment by carmakers to achieve a target of 140g/km by 2008. Japanese and Korean car producers, represented by JAMA and KAMA, made a similar commitment for 2009. Although significant progress was made, average emissions only fell from 186g/km in 1995 to 161g/km in 2004.

In 2008 the Commission decided that the voluntary commitments would not achieve their target and that binding legislation was necessary.

The EU has introduced a draft Regulation that sets a binding target for new cars of 120g/km by 2012. The EU has proposed an 'integrated approach' where average emissions are to be brought down to just 130g/km through improvements in vehicle technology. The remaining cuts (10g/km) are to be achieved by complementary measures, such as the further use of biofuels, fuel-efficient tyres and air conditioning, traffic and road-safety management and changes in driver behaviour (eco-driving). Similarly manufacturers of luxury vehicles will be allowed to produce models that exceed this limit provided it is balanced by the production of more efficient models (EU Commission, 2007b).

As the ICE is a heat engine the efficiency that can be obtained is limited by the laws of thermodynamics. Although the efficiency of vehicles can be improved through better design and driver behaviour, the reality is that significant improvements in terms of reducing GHG emissions will be realized through hybrid vehicles and vehicles that do not rely on hydrocarbon fuels, for example, electric vehicles, provided that the electricity is generated from non-hydrocarbon resources such as renewables or nuclear power.

Similar thinking also applies to the aviation sector, where progress has been made in improving efficiency, but again this is limited and a new approach is needed.

# Aviation

Since 1960, global air passenger traffic (expressed as revenue passenger-kms) has increased by nearly 9 per cent per year, which according to the IPCC is 2.4 times the growth rate of the global average GDP (IPCC, 1999). IPCC forecasts that between 1990 and 2015 growth will be 5 per cent per year while fuel use will grow by 3 per cent per year as aircraft have become more efficient. Between 1993 and 2003, world air freight grew at extremely high rates – 6.2 per cent per year according to Boeing (Boeing, 2005). Within Europe, passenger numbers in the EU15 nations increased at 5.3 per cent per year between 1993 and 2002 (Layos, 2005).

Both of the principal aircraft manufacturers, Boeing and Airbus, predict similar growth patterns up to 2023. Airbus predicts that global passenger traffic will grow on average at 5.3 per cent per year between 2004 and 2023 and world passenger kilometres are expected to triple by 2023 and Boeing predicts growth rates of 5.2 per cent per year for passengers and 6.2 per cent per year for cargo. Low cost carriers are likely to continue to drive growth in Europe and for destinations outside of Europe. In 2008 the EU and the US established an Open Skies agreement. This is a two stage programme. The first stage removed all caps on routes, prices, or the number of weekly flights between the two markets. The second stage, to be implemented at the end of 2008, could lead to the removal of limits on services operated by carriers or investors within the other's market. The removal of such limits would normalize transatlantic aviation, bringing it into line with the changes that we have already seen in other sectors of the economy. It is difficult to judge the impact of this agreement (EU Commission, undated-a).

## Impacts of aviation

Aircraft emit gases and particles directly into the upper troposphere and lower stratosphere where they have an impact on atmospheric composition. These gases and particles alter the concentration of atmospheric greenhouse gases, including carbon dioxide ($CO_2$), ozone ($O_3$), and methane ($CH_4$); trigger the formation of condensation trails (contrails); and may increase cirrus cloudiness – all of which contribute to climate change. The IPPC estimates that $CO_2$ emissions from aircraft were 0.14Gt C/year in 1992. This was about 2 per cent of total anthropogenic carbon dioxide emissions in 1992 or about 13 per cent of carbon dioxide emissions from all transportation sources with emissions likely to grow between 1.6 and 10 times this value up to 2050 depending on which scenario is the most likely. For the reference scenario, which is based on mid-range growth and increased technological improvements in efficiency, the IPCC predicts that this would represent 4 per cent of all human produced $CO_2$.

$NO_x$ emitted from aircraft is involved in ozone chemistry and levels are predicted to rise by some 13 per cent, although there is considerable uncertainty. Ozone is a greenhouse gas as well as a shield against harmful ultraviolet radiation but it tends to be localized around flight paths and predominantly in the northern mid-latitudes (unlike $CO_2$ and methane which tend to mix globally). $NO_x$ reduces atmospheric methane (a greenhouse gas) and is forecast to be 5 per cent by 2050 compared to if there were no aircraft.

Contrails are formed by water vapour emitted by aircraft and tend to warm the Earth's surface. In 1992 contrails were estimated to cover about 0.1 per cent of the Earth's surface on an annually averaged basis with larger regional values. Contrail cover is projected to grow to 0.5 per cent by 2050 at a rate which is faster than the rate of growth in aviation fuel consumption due to the increased number of subsonic flights in the upper troposphere (around 13km). Extensive cirrus clouds have been observed to develop after the formation of extensive contrails. However, this phenomenon is poorly understood but may contribute to global warming (IPCC, 1999).

## Mitigating the impacts

Although the industry is aware of the impacts, taking measures to reduce them is not easy. For example, flying at a lower altitude could reduce contrails but would mean increased fuel consumption. The options available to reduce the impact of aviation emissions include changes in aircraft and engine technology, fuel, operational practices, and regulatory and economic measures.

## Fuel

Subsonic aircraft are around 70 per cent more fuel-efficient than 40 years ago, something that has been achieved through improvements to engines and to airframe design. The majority of the efficiency improvement has come from improved engine performance. The IPCC predicts that efficiencies will continue to improve and suggests that by 2015 there will have been a 20 per cent improvement over current designs and a 40 per cent improvement by 2050 (IPCC, 1999). Jet aircraft require a high energy density fuel and it is unlikely that aviation kerosene (also called mineral kerosene) will be replaced in the near future. There is interest in what is termed a kerosene extender. This is where an alternative fuel is added to kerosene, one being biodiesel. However, as a study by the Tyndall Centre notes, there are problems. Performance in cold temperatures, such as those experienced at altitude, can be compromised as biodiesel alters the crystallization properties of aviation fuel at low temperatures. Filtering techniques can be used for mixtures that contain up to 10 per cent biodiesel, so that the fuel continues to meet safety requirements. Research is needed, however, before it is likely that the fuel specifications will be changed to use this kind of approach (Bows et al, 2006). Note that concerns about the increasing use of agricultural land for fuel production are likely to impede progress.

Other approaches include the production of synthetic fuels using the Fischer-Tropsch process. This is typically a three-step procedure:

- *Syngas generation*: the feedstock is converted into synthesis gas composed of carbon monoxide and hydrogen.
- *Hydrocarbon synthesis*: the syngas is catalytically converted into a mixture of liquid hydrocarbons and wax, producing a 'synthetic crude'.
- *Upgrading*: the mixture of Fischer-Tropsch hydrocarbons is upgraded through hydrocracking and isomerization and fractionated into the desired fuels.

This method can manufacture kerosene that is very similar both chemically and physically to aviation kerosene, however, its lack of aromatic molecules and the fact that it is virtually sulphur-free, give it poor lubricity. Additives could be used to improve lubricity but the fuel has a lower energy density than aviation kerosene which could impact long haul flights (Bows et al, 2006).

Hydrogen may be viable as a fuel in the long term, but would require new aircraft designs and new infrastructure for its supply. Hydrogen has a high energy content but its low density will require much larger fuel tanks. There would be a weight advantage due to aircraft carrying lighter fuel, but this would be off-set to some degree by the weight of a larger fuel tank. The volume of hydrogen carried would also be some 2.5 times that of the equivalent kerosene. The airframe would therefore need to be larger, and so would have a correspondingly larger drag (Bows et al, 2006). Hydrogen fuel would eliminate emissions of carbon dioxide from aircraft, but would increase those of water vapour. The overall environmental impacts and the environmental sustainability of the production and use of hydrogen or any other alternative fuels have not been determined (IPCC, 1999).

In summary, there is some potential for biodiesel and synthetic kerosene in the medium term with existing airframe designs. Hydrogen would require significant changes to design and infrastructure. A study by RECP suggests that it is more likely that many of the technically feasible options would be used in surface transport in preference to aviation due to cost factors and ease of implementation (RCEP, 2002). Essentially this means that carbon emissions from the aviation fuels could be offset by the use of low and no-carbon fuels in surface transport systems.

## Airframe design

Although current passenger aircraft still retain the conventional airframe design, there are innovative designs that could be used for either passenger or freight. These are the blended wing-body (BWB) aircraft and the wing-in-ground effect vehicles (WIGs). The Tyndall Centre reports that the BWB idea has a long history. Conceptually the passenger area is blended into the wing configuration, effectively becoming a flying wing. The low drag of the design and the potential for the use of lightweight materials could reduce fuel usage, perhaps by as much as 30 per cent, further reducing aircraft take-off weight. Because of the lower weight and drag, this type of aircraft would have a lower cruise altitude and an extended optimal range.

The WIGs concept relies on a phenomenon known as 'ground effect'. The 'ground effect' occurs as the distance between the ground and the wing decreases to a length less than an aircraft's wing-span and this increases the ratio of lift to drag. This phenomenon has no advantage for small aircraft but for larger aircraft a given payload can be transported much further than with conventional designs. A proposal by Boeing, known as the Pelican aircraft would have a wing-span of 150m, will fly as low as 6m above sea level and carry a load of 750 tonnes of cargo for 18,500km when in 'ground effect' above the sea. At more standard altitude levels, this range for the same fuel burn would be reduced to 12,000km. However there are problems with the noise that would be generated by the large number of separate undercarriages. Current regulations do not permit such low flying vehicles (Bows et al, 2006).

## Airships

The airship is not a new concept but one which has been revisited, especially for the transport of cargo. Despite causing 80–90 per cent less radiative forcing than a conventional jet aircraft, the Tyndall Centre reports that the technology is not promising, mainly because of manoeuvrability difficulties in wind during the loading and unloading stages. There have been promising designs for a cargo lifter such as the Skycat by Airship Technologies Group (UK). To date, no successful large cargo lifter has been built, even though reputable firms such as Lockheed have planned projects (Bows et al, 2006).

## Engine technology

The most fuel-efficient aircraft engines are high bypass, high-pressure ratio gas turbine engines.

These engines have high combustion pressures and temperatures. Although these features are consistent with fuel efficiency, they do increase the formation rates of $NO_x$ especially at high power take-off and at altitude cruise conditions. There is substantial research into reducing $NO_x$ during the Landing and Take-off (LTO) cycle with the goal of reducing $NO_x$ by up to 70 per cent and improving engine fuel consumption by 8–10 per cent by 2010. A reduction of $NO_x$ emissions would also be achieved at cruise altitude, though not necessarily by the same proportion as for LTO (IPCC, 1999).

Currently there are no known alternatives to this type of aircraft engine and it is unlikely that airframe design will change radically. Other methods of improving efficiency are load factors, air-traffic management (ATM), regulatory and market based options.

## Load factors

Load factors are the number of passengers or amount of freight carried on a given aircraft, eliminating non-essential weight, optimizing aircraft speed, limiting the use of auxiliary power (e.g. for heating, ventilation) and reducing taxiing. The IPCC estimates that improvements in these operational measures could reduce fuel burned, and emissions, in the range of 2–6 per cent (IPCC, 1999). For passenger aircraft the charter sector has generally been more successful than scheduled services in optimizing passenger numbers. More effort and research into generating sophisticated ticketing technology, differing pricing bands and demand-focused time-tabling may all lead to load factor improvements (Bows et al, 2006).

## Air-traffic management (ATM)

Air traffic management systems are used for the guidance, separation, coordination and control of aircraft movements. Existing national and international air-traffic management systems have limitations that result, for example, in holding (aircraft flying in a fixed pattern waiting for permission to land), inefficient routings and sub-optimal flight profiles. These limitations result in excess fuel burn and consequently excess emissions (IPCC, 1999).

Inefficient routings have evolved as part of the historic infrastructure when reliable navigation was through ground-based beacons. Global positioning satellites (GPS) and modern on-board flight management systems provide the opportunity to optimize routes. The IPCC estimates that this could result in a 6–12 per cent reduction in fuel use providing the regulatory and institutional frameworks can be established. The EU plans to introduce a Single European Sky (SES) by 2025 that will alleviate the blockages caused by the interaction of 27 different national airspaces under the control of national governments. This will reduce aviation congestion and improve safety by reforming the current ATM system.

## Regulatory and market based options

Essentially this covers issues such as setting targets for efficiency and emission rates for aircraft. The aviation authorities currently use this approach to regulate emissions for carbon monoxide, hydrocarbons, $NO_x$ and smoke. The International Civil Aviation Organization has begun work to assess the need for standards for aircraft emissions at cruise altitude to complement existing LTO standards for $NO_x$ and other emissions (IPCC, 1999).

The aviation industry has itself set research goals for improving fuel efficiency by 50 per cent and to reduce $NO_x$ by 80 per cent. However, it will take a number of years for the whole fleet to be upgraded as the lifetimes of aircraft are long (up to 40 years) and replacement rates are low. The fuel efficiency of the whole fleet is likely to improve slowly given that there is limited fleet renewal. Efficiency improvements over the previous 20 years have been around 1–2 per cent per year, which would in turn lead to around a 1–2 per cent improvement in efficiency per year for the total fleet.

The EU plans to include aviation emissions in the EU ETS. It is unclear what this will mean in terms of greenhouse gas reductions from the sector. At present the costs of aviation fuel are rising and it seems that this is likely to continue in the longer term. Higher costs may modify behaviour, for example, by reducing business travel through the greater use of electronic conference facilities and other forms of communication. Leisure patterns could also change with a greater use of alternatives such as rail or automobiles for shorter journeys. For long haul the use of charter flights could become more popular as their higher load factors will help to keep costs down.

## Summary

Buildings will play a key role in reducing energy. This is particularly important in the EU and OECD areas as much of the energy used in buildings is produced from non-renewable resources. To maximize the efficiency buildings will require changes in building design and this may well mean that efficient buildings in the future look very different to existing types. For ZEBs there will also be a need to embed or install renewable capacity for the production of hot water and electricity within the fabric of the building, which again will give a very different visual aspect. But what this will signal is that there is an intimate connection between the energy we need and the ways in which we gather and manage energy resources and a more efficient approach is sourcing much closer to the point of use. This has some interesting implications for the current structure of energy delivery which is based on a model of a few suppliers and many users. With in-built capacity, buildings have the potential to be both users, producers or autonomous. This will tend to transform the current top-down model to a flatter more integrated and interconnected mode.

There is considerable scope to reduce the energy used by lighting systems and appliances within the domestic sector. Similar arguments apply to the commercial sector as it also requires lighting and many principles for reducing energy use in domestic appliances are also applicable to commercial equipment and machinery. The OECD does suggest that a reduction of one-third in residential use is achievable by 2030 through the use of market measures. The EU sees a more rapid decline in domestic energy with a more interventionist approach. Boardman, using a scenario that aims for an 80 per cent reduction by 2050, identifies that a series of measures can realize that reduction. Improving efficiency of end use has been viewed as the least painful way of reducing energy use. However, this must be coupled to a mindset that sees efficiency as something to be valued. The key policy drivers are climate goals along with energy security. But energy costs also play a role in the wider economy as well as at the household level. Reducing energy costs is important for any business organization or service provider. It is also important in terms of reducing fuel poverty. Improved energy efficiency has a key role in these areas.

Transport is a key issue in both the developed and developing worlds. The road and rail networks are key infrastructures for the movement of people and freight. Although road transport is heavily dependent on fossil fuels, there is the potential to move to more renewable sources such as biofuels and electric vehicles. However, there are concerns that the biofuel route will not be sustainable in the long term, driven by fears that food security could be undermined. Electric vehicles, particularly those powered from a renewable capacity, offer the most sustainable option, but considerable development in terms of battery and fuel cell technologies is needed. The infrastructure challenges are huge. However, the car market is also huge and it is likely that manufacturers will find a way of developing that infrastructure if they believe that the public will make the shift. The high and volatile price of oil may well be a key factor in that transition.

The aviation sector has received considerable attention as the popularity of air travel has continued to grow. A number of environmentalists have expressed concern about the impacts of aviation. The impacts are not just limited to energy and the impact of emissions on the greenhouse effect. Issues such as noise during take off and landing and the amount of land needed for runways, terminals and associated infrastructure are also concerns. However, it is the impact of fuel used that is of growing concern. All projections show that air travel will continue to increase. Unlike in other areas of transport, aviation fuel is not subject to duty, a situation which many feel encourages more air travel. The EU has proposed that the aviation sector becomes part of the ETS. It is not possible to predict the impact that this will have. More recently the sharp rise in the cost of fuel could dampen the rate of growth and hurry the introduction of more efficient engines and airframes. However, an alternative to the jet engine is unlikely. Although the aviation sector has improved fuel efficiency, there are some who believe that the improvement has not been as rapid as portrayed by the industry. A study into efficiency suggests that the benchmark used for the period 1960–2000 does not reflect technological developments. The study determines that improvements of 55 per cent as opposed to 70 per cent were realized and, using this model, projects that future efficiency gains suggested by the industry may be optimistic (Peeters et al, 2005). However, any future efficiency savings could easily be offset by the continued growth in air travel.

If fuel prices continue to rise it may well be the case, as suggested by RCEP, that growth in ground-based forms of transport, such as high speed rail links will grow. This may well impact on the short haul market. At present there appears to be no such alternative for the long haul sector.

## Notes

1 This is essentially the OECD countries, but also includes the industrializing nations such as India and China.

2 For further information on the Motor Challenge Programme see: http://re.jrc.ec.europa.eu/energyefficiency/motorchallenge/index.htm.

## References


ACE (2005) Fact sheet 01 – Key trends in UK domestic sector energy use, October 2005. Available at: www.ukace.org/publications/ACE%20Fact%20Sheet%20(2005-10)%20-%20Key%20Trends%20in%20UK%20Domestic%20Sector%20Energy%20Use

Anderson, B. (2006) Conventions for U-value calculations, BRE Scotland, BRE Press, Watford, UK. Available at: www.bre.co.uk/filelibrary/rpts/uvalue/BR_443_(2006_Edition).pdf

Audenaerta, A., De Cleynb, S. H. and Vankerckhoveb, B. (2007) 'Economic analysis of passive houses and low-energy houses compared with standard houses', *Energy Policy*, vol. 36, pp47–55

BBC (2008a) Work starts on Gulf 'green city' BBC News Online, 10 February 2008. Available at: http://news.bbc.co.uk/1/hi/sci/tech/7237672.stm

BBC (2008b) Technology's low powered future, BBC News Online, 22 August 2008. Available at: http://news.bbc.co.uk/1/hi/programmes/click_online/7576366.stm

BBC (2008c) UN urges biofuel investment halt, BBC News Online 2 May 2008. Available at: http://news.bbc.co.uk/2/hi/7381392.stm

Boardman, B. (2007) Home Truths: A Low-Carbon Strategy to Reduce UK Housing Emissions by 80% by 2050, A research report for The Co-operative Bank and Friends of the Earth, Environmental Change Institute, University of Oxford, UK. Available at: www.eci.ox.ac.uk/research/energy/downloads/boardman07-hometruths.pdf

Boardman, B., Darby, S., Killip, G., Hinnells, M., Jardine, C. N., Palmer, J. and Sinden, G. (2005) 40% house, Environmental Change Institute 2005, Oxford University, UK. Available at: www.eci.ox.ac.uk/research/energy/downloads/40house/40house.pdf

Boeing (2005) World Transport Forecast. Available at: www.boeing.com/commercial/cargo/01_06.html

Bossel, U. (2003) European Fuel Cell Forum Efficiency of Hydrogen Fuel Cell, Diesel-SOFC-Hybrid and Battery Electric Vehicles. Available online at: www.evworld.com/library/fcev_vs_hev.pdf

Bows, A., Anderson, K. and Upham, P. (2006) Contraction & Convergence: UK carbon emissions and the implications for UK air traffic, Technical Report 40, Tyndall Centre for Climate Change Research, UK. Available at: www.tyndall.ac.uk/research/theme2/final_reports/t3_23.pdf

BRE (2005) The Government's Standard Assessment Procedure for Energy Rating of Dwellings 2005 Edition at: http://projects.bre.co.uk/sap2005/pdf/SAP2005.pdf

Business Europe (2007) Energy Efficiency: Reconciling Economic Growth and Climate Protection. Available at: www.bdi-online.de/Dokumente/Energie-Telekommunikation/BUSINESSEUROPE_EnergyEfficiency.pdf

California Hydrogen Highway (2008) California Hydrogen Highway Network, Available at: www.hydrogenhighway.ca.gov/

Capros, P., Mantzos, L., Papandreou, V. and Tasios, N. (2008) European Energy and Transport: Trends to 2030 – Update 2007, Report prepared by the Institute of Communication and Computer Systems of the National Technical University of Athens (ICCS-NTUA), E3M-Lab, Greece, for the Directorate-General for Energy and Transport, Luxembourg. Available at:- http://ec.europa.eu/dgs/energy_transport/figures/trends_2030_update_2007/energy_transport_trends_2030_update_2007_en.pdf

Carrette, L., Friedrich, K. A. and Stimming, U. (2001) 'Fuel cells – fundamentals and applications', *Fuel Cells*, vol. 1, no. 1. Available at: www3.interscience.wiley.com/cgi-bin/fulltext/84502989/PDFSTART

CEC (2006) Communication from the Commission: Action Plan for Energy Efficiency: Realising the Potential Brussels COM(2006)545 final, Commission of the European Communities, Brussels. Available at: http://ec.europa.eu/energy/action_plan_energy_efficiency/doc/com_2006_0545_en.pdf

Chary, R., Correia, P. A., Nagaraj, R. and Song, J. (2004) Low Power Intel Architecture for Small Form Factor Devices. Available at: http://download.intel.com/technology/systems/LowPower_WP.pdf

City of Melbourne (undated) Council House 2. Available at: www.melbourne.vic.gov.au/info.cfm?top=171&pg=1933

Clean Sky (undated). Clean Sky. Available at:- www.cleansky.eu/index.php?arbo_id=83&set_language=en

The Climate Group (2007) In the Black: The Growth of the Low Carbon Economy, London, UK. Available at: www.theclimategroup.org/assets/resources/In_the_Black_full_report_May-06.pdf

COM (2006) Communication from the Commission to the Council and the European Parliament. Keep Europe moving – Sustainable mobility for our continent – Mid-term review of the European Commission's 2001 Transport White Paper (SEC (2006) 768). Available at: http://eur-lex.europa.eu/smartapi/cgi/sga_doc?smartapi!celexplus!prod!DocNumber&lg=en&type_doc=COMfinal&an_doc=2006&nu_doc=314

COM (2007) Communication from the Commission to the European Council and the European Parliament. An Energy Policy for Europe, COM/2007/0001 final (SEC(2007) 12). Available at: http://eur-lex.europa.eu/LexUriServ/site/en/com/2007/com2007_0001en01.pdf

COM (2008) Communication from the Commission. Addressing the challenge of energy efficiency through Information and Communication Technologies, COM(2008) 241 final, Brussels. Available at: http://ec.europa.eu/information_society/activities/sustainable_growth/docs/com_2008_241_1_en.pdf

De Keulenaer, H., Belmans, R., Blaustein, E., Chapman, D., De Almaida, A., De Wachter, B. and Radgen, P. (2004) Energy Efficient Motor Driven Systems. EU-sponsored programme. European Copper Institute. Copyright 2004 European Copper Institute, Fraunhofer-ISI, KU Leuven and University of Coimbra. Available at: http://re.jrc.ec.europa.eu/energyefficiency/pdf/HEM_lo_all%20final.pdf

DEFRA (2007a) Benn launches plan for one stop shop for greener homes. Press release. Available at: www.defra.gov.uk/news/2007/071119a.htm

DEFRA (2007b) UK Energy Efficiency Action Plan 2007. Available at: http://ec.europa.eu/energy/demand/legislation/doc/neeap/uk_en.pdf

Dopp, R. B. (2007) Hydrogen Generation via Water Electrolysis using highly efficient nanometal electrodes, DoppStein Enterprises, QuantumSphere, April. Available at: www.qsinano.com/white_papers/Water%20Electrolysis%20April%2007.pdf

EEW (2007) Screening of National Energy Efficiency Action Plans, EEW – Working paper 01/08, Wuppertal Institute for Climate, Energy and Environment/Ecofys, Wuppertal, Cologne, Berlin.

Available at: www.energy-efficiency-watch.org/fileadmin/eew_documents/Documents/Results/080526EEW_Screening_final.pdf

Elswijk, M. and Kaan, H. (2008) European Embedding of Passive Houses, PEP Promotion of European Passive Houses. The PEP-project is partially supported by the European Commission under the Intelligent Energy Europe Programme: EIE/04/030/S07.39990. See www.europeanpassivehouses.org/. Report Available at: http://erg.ucd.ie/pep/pdf/European_Embedding_of_Passive_Houses.pdf

English, A. (2008) 'Smart car: Think Smart', *Telegraph.co.uk*, 5 July 2008. Available at: www.telegraph.co.uk/motoring/main.jhtml?xml=/motoring/2008/07/05/nosplit/mfsmart105.xml

ESD (2004) Low Carbon Homes: towards zero carbon refurbishment. Feasibility study for the Energy Saving Trust Innovation Programme Reference P00754. Available at: www.generationhomes.org.uk/LowCarbonHomes_ESD.pdf

EST (2006) The rise of the machines. A review of energy using products in the home from the 1970s to today, Energy Savings Trust, London. Available at: www.energysavingtrust.org.uk/uploads/documents/aboutest/Riseofthemachines.pdf

EU (2006) Communication from the Commission, Action Plan for Energy Efficiency: Realising the Potential, COM(2006)545 final. Available at: http://ec.europa.eu/energy/action_plan_energy_efficiency/doc/com_2006_0545_en.pdf

EU Commission (2001) White Paper: European transport policy for 2010: time to decide, EU Commission, Luxembourg. Available at: http://ec.europa.eu/transport/white_paper/documents/doc/lb_texte_complet_en.pdf

EU Commission (2005) Doing More With Less: Green Paper on energy efficiency, Luxembourg: Office for Official Publications of the European Communities. Available at: http://ec.europa.eu/energy/efficiency/doc/2005_06_green_paper_book_en.pdf

EU Commission (2006) Trends to 2030 – Update 2005. Directorate-General for Energy and Transport, European Energy and Transport. EU Commission, Brussels. Available at: http://ec.europa.eu/dgs/energy_transport/figures/trends_2030_update_2005/energy_transport_trends_2030_update_2005_en.pdf

EU Commission (2007a) European Union: Energy and Transport in Figures, 2007, part 3: Transport, Directorate-General for Energy and Transport in co-operation with Eurostat. Available at: http://ec.europa.eu/dgs/energy_transport/figures/pocketbook/doc/2007/pb_3_transport_2007.pdf

EU Commission (2007b) Proposal for a Regulation of the European Parliament and of the Council: setting emission performance standards for new passenger cars as part of the Community's integrated approach to reduce $CO_2$ emissions from light-duty vehicles. COM(2007) 856 final. 0297 (COD), Brussels. Available at:- http://eur-lex.europa.eu/LexUriServ/LexUriServ.do?uri=COM:2007:0856:FIN:EN:PDF

EU Commission (undated-a) Air Transport Portal of the European Commission, EU–US 'Open Skies': The EU and the US start talks on air services agreement to reshape global aviation. Available at: http://ec.europa.eu/transport/air_portal/international/pillars/global_partners/us_en.htm

EU Commission (undated-b) Directorate-General for Energy and Transport, EPBD Building Platform. Available at: www.buildingsplatform.org/cms/index.php?id=8

European Parliament (2008) Aviation to be included in the European Trading System from 2012 as MEPs adopt legislation, Press Release, European Parliament, 8 July 2008. Available at: www.europarl.europa.eu/news/expert/infopress_page/064-33577-189-07-28-911-20080707IPR33572-07-07-2008-2008-false/default_en.htm

Farmery, M. (2006) Future Aviation Fuels: What are the challenges? What are the options? ICAO/Transport Canada Workshop, Montreal 20–21 September. Available at: www.icao.int/env/WorkshopFuelEmissions/Presentations/Farmery.pdf

Federal Energy Management Programme (2001) Low-Energy Building Design Guidelines: Energy-efficient design for new Federal facilities. DOE/EE-0249. Available at: www1.eere.energy.gov/femp/pdfs/25807.pdf

Gallagher, E. (2008) The Gallagher Review of the indirect effects of biofuels production, Renewable Fuels Agency, UK. Available at: www.dft.gov.uk/rfa/_db/_documents/Report_of_the_Gallagher_review.pdf

Gschneidner, K. and Gibson, K. (2001) Magnetic refrigerator successfully tested, Ames Laboratory News Release. Ames Laboratory. Available at: www.external.ameslab.gov/news/release/01magneticrefrig.htm

IEA (2003) Cool Appliances: Policy Strategies for Energy-Efficient Homes, OECD/IEA. Available at: www.iea.org/textbase/nppdf/free/2000/cool_appliance2003.pdf

IEA (2007a) Tracking Industrial Energy Efficiency and $CO_2$ Emissions, IEA/OECD, Paris. Available at: www.iea.org/Textbase/npsum/tracking2007SUM.pdf

IEA (2007b) Outlook for hybrid and electric vehicles V, Hybrid and Electric Vehicle Implementing Agreement. Available at: www.ieahev.org/pdfs/ia-hev_outlook_2008.pdf

IPCC (1999) Aviation and the Global Atmosphere. Summary for Policymakers. A Special Report of IPCC Working Groups I and III in collaboration with the Scientific Assessment panel to the Montreal Protocol on Substances that Deplete the Ozone Layer, (eds) Penner, J. E., Lister, D. H., Griggs, D. J., Dokken, D. J. and McFarland, M. Available at: www.ipcc.ch/pdf/special-reports/spm/av-en.pdf

Irish Energy Centre (undated) What is a U-Value, Information Sheet. Available at: www.sei.ie/uploadedfiles/InfoCentre/whatisauvalue.pdf

ITF (2007) Global transport trends completely at odds with Climate Change aspirations, Press Release 20 November 2007, International Transport Forum, Paris France. Available at: www.internationaltransportforum.org/Press/PDFs/2007-11-20.pdf

Johnston, D., Lowe, R. J. and Bell, M. (2005) 'An exploration of the Technical Feasibility of Achieving $CO_2$ emissions reductions in excess of 60% within the UK housing stock by the year 2050', *Energy Policy*, vol. 33, pp1643–1659

Kahn Ribeiro, S., Kobayashi, S., Beuthe, M., Gasca, J., Greene, D., Lee, D. S., Muromachi, Y., Newton, P. J., Plotkin, S., Sperling, D., Wit, R. and Zhou, P. J. (2007) 'Transport and its infrastructure', in B. Metz, O. R. Davidson, P. R. Bosch, R. Dave, and L. A. Meyer, eds., *Climate Change 2007: Mitigation*. Contribution of Working Group III to the Fourth Assessment Report of the Intergovernmental Panel on Climate Change, Cambridge and New York: Cambridge University Press. Available at: www.ipcc.ch/pdf/assessment-report/ar4/wg3/ar4-wg3-chapter5.pdf

Kavalov, B. and Peteves, S. D. (2005) Status and perspectives of Biomass-to-Liquid fuels in the European Union, European Commission, Directorate General Joint Research Centre (DG JRC), Luxembourg. Available at: www.senternovem.nl/mmfiles/Status_perspectives_biofuels_EU_2005_tcm24-152475.pdf

Langellier, J.-P. and Pedroletti, B. (2006) China to Build First Eco-City, *Guardian Weekly*, China Radio International English. Available at: http://english.cri.cn/811/2006/05/07/301@85444.htm

Lapillonne, B. and Pollier, K. (2007) Energy efficiency trends for households in EU New Member Countries (NMC's) and in the EU 25, Odyssee. Available at: www.odyssee-indicators.org/Indicators/PDF/households_EU_25.pdf

Layos, L.de la F. (2005) Statistics in Focus, Passenger Air Transport, 2002–2003, Eurostat

Lloyd Jones, D., Matson, C. and Pearsall, N. M. (1989) The Solar Office: A Solar Powered Building with a Comprehensive Energy Strategy. Paper Presented at the 2nd World Conference on Photovoltaic Solar Energy Conversion, Vienna, Austria, July 1998. Available at: http://soe.unn.ac.uk/npac/Doxford%20Paper.pdf

Maibach, M., Schreyer, C., Sutter, D., van Essen, H. P., Boon, B. H., Smokers, R., Schroten, A., Doll, C., Pawlowska, B. and Bak, M. (2008) Handbook on estimation of external costs in the transport sector: Internalisation Measures and Policies for All external Cost of Transport (IMPACT) Version 1.1, Delft, CE. Commissioned by: European Commission DG TREN. Available at: http://ec.europa.eu/transport/costs/handbook/doc/2008_01_15_handbook_external_cost_en.pdf

Míguez, J. L., Porteiro, J., López-González ,L. M., Vicuña, J. E., Murillo, S., Morán, J. C. and Granada, E. (2006) 'Review of the energy rating of dwellings in the European Union as a mechanism for sustainable energy', *Renewable and Sustainable Energy Reviews*, vol. 10, no. 1, pp24–45. Available at: www.sciencedirect.com/science?_ob=ArticleURL&_udi=B6VMY-4DHWRKF-1&_user=122879&_rdoc=1&_fmt=&_orig=search&_sort=d&view=c&_acct=C000010138&_version=1&_urlVersion=0&_userid=122879&md5=7010b28175b42a965a689e492e35a5ef

Neese, B. ,Chu, B., Lu, S.-G., Wang, Y., Furman, E. and Zhang, Q. M. (2008) 'Large electrocaloric effect in ferroelectric polymers near room temperature', *Science,* vol. 321, no. 5890, pp821–823. DOI: 10.1126/science.1159655. Abstract available at: www.sciencemag.org/cgi/content/abstract/321/5890/821

OJ (2003) Directive 2002/91/EC of the European Parliament and Council of 16 December 2002 on the energy performance of buildings, *OJ L* 1/65, *Official Journal of the European Communities*. Available at: http://eur-lex.europa.eu/LexUriServ/LexUriServ.do?uri=OJ:L:2003:001:0065:0071:EN:PDF

Olivier, D. (2001) *Building In Ignorance. Demolishing Complacency: Improving the Energy Performance of 21st Century Homes.* Report for the Energy Efficiency Advice Service for Oxfordshire and the Association for the Conservation of Energy. Available at: www.ukace.org/pubs/reportfo/BuildIgn.pdf

Passive House Institute (2006) Definition of Passive Houses. Available at: www.passivhaustagung.de/Passive_House_E/passivehouse_definition.html

Passive House Institute (undated-a) Available at: http://www.passiv.de/

Passive House Institute (undated-b) What is a passive House? Available at: www.passiv.de/

Passive House Solutions (undated) Passive House Solutions Ltd. Available at: www.passivehouse.co.uk/content/view/6/75/

Peeters, P. M., Middel, J. and Hoolhorst, A. (2005) Fuel efficiency of commercial aircraft: An overview of historical and future trends, Amsterdam: National Aerospace Laboratory NLR

PEP Promotion of European Passive Houses (2006) Energy Saving Potential. The PEP-project is partially supported by the European Commission under the Intelligent Energy Europe Programme: EIE/04/030/S07.39990. Available at: http://erg.ucd.ie/pep/pdf/Energy_Saving_Potential_2.pdf

RCEP (2002) The Environmental Effects of Civil Aircraft in Flight, Special Report of the Royal Commission on Environmental Pollution. Available at: www.rcep.org.uk/

SAVE II (2001) Labelling & other measures for heating systems in dwellings. Final Report January 2001. Appendix 5 – Electrical consumption of gas & oil central heating. OMV, Sweden. Available at: http://projects.bre.co.uk/eu_save/pdf/App5Electricalconsumptiono.pdf

SCORE (2007) Score Research Summary. Available at: www.score.uk.com/research/Lists/Announcements/DispForm.aspx?ID=3&Source=http%3A%2F%2Fwww%2Escore%2Euk%2Ecom%2Fresearch%2Fdefault%2Easpx

Scott, D. S. (1995) Interpreting the architecture of the energy system, Proceedings of the World Energy Council 16th Congress, Tokyo, Japan.

Searchinger, R., Heimlich, R. A., Houghton, F., Dong, A., Elobeid, J., Fabiosa, S., Tokgoz, D., Hayes, and Yu, T. (2008) 'Use of U.S. croplands for biofuels increased greenhouse gases through land-use change', *Science Express*, 7 February

Sustainable Development Commission (2006) *Stock Take: Delivering Improvements in Existing Housing*, Sustainable Development Commission, London, UK. Available at: www.sd-commission.org.uk/publications/downloads/SDC%20Stock%20Take%20Report.pdf

Torcellini, P., Pless, S. and Deru, M. (2006) Zero Energy Buildings: A Critical Look at the Definition Preprint. 15 pp.; NREL Report No. CP-550-39833. Available at: www.nrel.gov/docs/fy06osti/39833.pdf

UK Government (2001) English Housing Condition Survey: Building the Picture, Office of the Deputy Prime Minister, London. Available at: www.communities.gov.uk/documents/corporate/pdf/145310.pdf

UK Government (2005) Household Energy Efficiency, Postnote Number 249, Parliamentary Office of Science and Technology, Crown Copyright. Available at: www.parliament.uk/documents/upload/postpn249.pdf

UK Government (2006) Department of Communities and Local Government, Approved Document for Conservation of Fuel and Power (L2A and L2B), NBS, Crown Copyright. L2A available at: www.planningportal.gov.uk/uploads/br/BR_PDF_ADL2A_2006.pdf L2B available at: www.planningportal.gov.uk/uploads/br/BR_PDF_ADL2B_2006.pdf

UK Government (2007) Building a Greener Future: policy statement, Department for Communities and Local Government, Crown Copyright. Available at: www.communities.gov.uk/publications/planningandbuilding/building-a-greener

UK Government (2008) The Code for Sustainable Homes: Setting the standard in sustainability for new homes, Department for Communities and Local Government, Crown Copyright. Available at: www.communities.gov.uk/documents/

planningandbuilding/pdf/codesustainhome-sstandard.pdf

UK Government (undated-a) DEFRA, The Climate Change Bill. Available at: www.defra.gov.uk/Environment/climatechange/uk/legislation/index.htm

UK Government (undated-b) Department of Communities and Local Government, Energy Performance Certificates. Available at: www.communities.gov.uk/planningandbuilding/theenvironment/energyperformance/certificates/energyperformancecertificates/

UK Green Building Council (2008) Zero Carbon Task Group Report, Zero Carbon Task Group. Available at: www.ukgbc.org/site/resources/showResourceDetails?id=180

UN (2005) United Nations Population Division: World Urbanization Prospects: The 2005 Revision. Available at: www.un.org/esa/population/publications/WUP2005/2005wup.htm

WBCSD (2007) *Efficiency in Buildings: Business Realities and Opportunities*, Summary Report. Available at: www.wbcsd.org/DocRoot/kPUZwapTJKNBF9UJaG7D/EEB_Facts_Trends.pdf

WEC (2007) *Transport Technologies and Policy Scenarios to 2050*, World Energy Council, London, UK. Available at: www.worldenergy.org/documents/transportation_study_final_online.pdf

WEC (2008) *Energy Efficiency Policies Around the World: Review and Evaluation*, World Energy Council, London, UK. Available at: www.worldenergy.org/documents/energyefficiency_final_online.pdf

What Car? (2007) 'London to be test-bed for electric Smart', *What Car?* Available at: www.whatcar.com/news-article.aspx?NA=226488

World Coal Institute (undated) 'Coal to Liquid'. Available at: www.worldcoal.org/pages/content/index.asp?PageID=423

Yi, L. and Thomas, H. R. (2007) 'A review of research on the environmental impact of e-business and ICT', Geoenvironmental Research Centre, Cardiff University, UK. Available at: www.aseanenvironment.info/Abstract/41015216.pdf

# 5

# Conventional Fuels

## Introduction

Most of the world's energy comes from fossil fuels. Carbon-based fuels are stores of high quality solar energy accumulated over millions of years, for example, by photosynthesis capturing carbon and storing it in woody biomass, then through a serious of processes eventually storing it as coal. The production of the fossil fuel supply base has been a very slow process. Use of this resource has been profligate with many predicting that we have now reached the point of peak production for oil and shortly that of gas. This point, termed Peak Oil, is where consumption exceeds production and it was first conceptualized by Marion King Hubbert. The combustion of fossil fuels releases carbon dioxide into the atmosphere and this enhances the natural greenhouse effect. It is accepted that the release of carbon is accelerating climate change. The concerns around a diminishing resource base and the impacts of climate change are shaping energy policy and technological development. This chapter will examine the current supply position for oil, gas and coal and predictions of use. It will then look at a number of supply-side technological developments.

## Oil

Crude oil is a mixture of liquid hydrocarbon compounds sometimes found permeating sedimentary rocks. By weight oil is made up of 82.2–87.1 per cent carbon, 11.7–14.7 per cent hydrogen, 0.1–4.55 oxygen, 0.1–1.5 per cent nitrogen and 0.1–5.5 per cent sulphur. Different names, based on the number of carbon atoms in their compounds, are given to products derived from crude oil: gasoline (C4 to C10), kerosene (C11 to C13), diesel fuel (C14 to C18), heavy gas oil (C19 to C25), lubricating oil (C26 to C40) and waxes (over C40). Oil is generated from organic matter in sedimentary rocks at depths of about 800–5000m at temperatures between 66°C and 150°C. Its predominant source material is probably marine organisms, although there is an alternative view that argues that oil is abiotic in origin as discussed in Box 5.1. In the predominant conventional view there are three steps involved in the conversion of organic matter to petroleum.

- *Diagenesis*: During sedimentation organic rich sediment is buried and subjected to slightly increased temperatures and pressures. The organic material is converted to an insoluble solid hydrocarbon called kerogen of which there are three types. The source of type I is mainly marine algal material that yields a light, high quality oil; type II is a mix of various marine organic materials and is the main source of crude oil and some gas, and type III, which produces mainly gas with some oil and waxes is derived from terrestrial material.

## Box 5.1 Abiotic oil

There is an alternative theory about the formation of oil and gas deposits that could change estimates of potential future oil reserves. According to this theory, oil is not a fossil fuel at all, but was formed deep in the Earth's crust from inorganic materials. The theory was first proposed in the 1950s by Russian and Ukrainian scientists. Based on the theory, successful exploratory drilling has been undertaken in the Caspian Sea region, Western Siberia and the Dnieper-Donets Basin. The theory argues that the formation of oil deposits requires the high pressures only found in the deep mantle and that the hydrocarbon contents in sediments do not exhibit sufficient organic material to supply the enormous amounts of petroleum found in very large oil fields.

This notion of abiotic oil was promoted in the West by Thomas Gold. His theory (see T. Gold, *The Deep Hot Biosphere*, Copernicus Books, 1999) is that hydrogen and carbon, under high temperatures and pressures found in the mantle during the formation of the Earth, form hydrocarbon molecules which have gradually leaked up to the surface through cracks in rocks. The biomarkers found in oil are explained by the metabolism of bacteria which have been found in extreme environments similar to those hydrothermal vents and volcanic places where it was formerly believed that life was not possible. Most geologists reject this theory.

Gold did manage to persuade the Swedish government in 1988 to drill a deep hole into non-sedimentary rocks to test his theory. Oil was discovered but only in small quantities, with sceptics arguing that it originated in the drilling mud. There is considerable controversy surrounding abiotic oil. For a good overview see R. Heinberg, 'The "Abiotic Oil" Controversy,' *Energy Bulletin*, 2004; available online at: www.energybulletin.net/node/2423.

- *Catagenesis*: This is the mature stage of the process where further ageing and sedimentation increases temperature and pressure to give a range of petroleum hydrocarbons by thermal cracking.
- *Metagenesis*: Below 5000m or so the increases in temperature and pressure are such as to convert the hydrocarbon material to methane and residual carbon. Oil is seldom found below this depth.

In order for hydrocarbons to accumulate the source or sedimentary rock must have access to a reservoir rock and this must be capped by an impermeable rock so that the hydrocarbons are effectively trapped. These conditions are quite rare. There are some 600 sedimentary basins, although all have not been explored for a variety of reasons such as their location – in deep water or in the polar-regions – or because of political restrictions. Oil deposits are scattered throughout the world and this distribution has led recently to growing concerns about energy security as, in some cases, the deposits are located in areas that are viewed as politically volatile. Figure 5.1 shows the global distribution of oil deposits.

Note that this shows proved reserves. Proved reserves are the amount of oil that it is technically and financially possible to recover from a well. As technology improves it may be possible to extract further oil either by flooding with water or gas. These techniques are used, although not widely, and are termed enhanced oil recovery (EOR). (EOR is actively being tested as a technique for the permanent storage of carbon dioxide produced from the combustion of fossil fuels; see 'Carbon capture and storage' later in this chapter). Proved reserves do not include oil deposits that are thought to exist, but have yet to be discovered. From Figure 5.1 it can be seen that the distribution of oil deposits is very uneven across countries, with some areas, such as Asia Pacific having a low level and others, such as the Middle East, having reserves that are equivalent to 37 per cent of known global reserves.

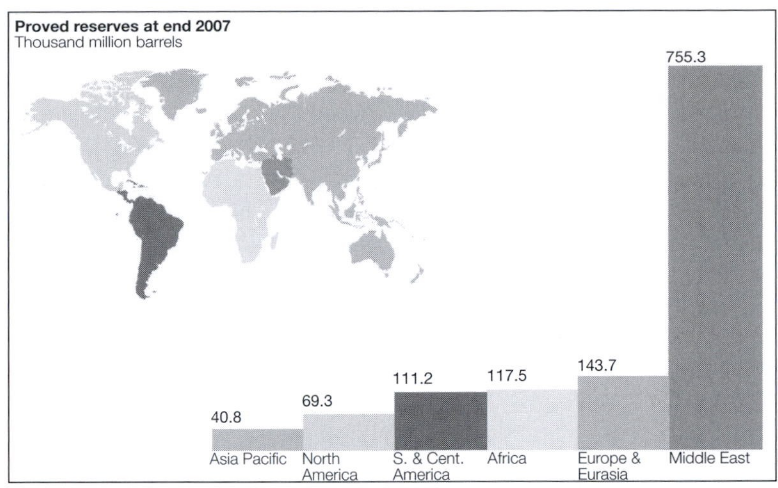


*Source:* BP, 2008

**Figure 5.1** Proved oil reserves as at the end of 2007

There is one further category of reserves, known as unconventional reserves, that could add significantly to total global reserves. These include:

- *Heavy oils*: these can be pumped and refined just like conventional petroleum except that they are thicker and have more sulphur and heavy metal contamination, necessitating more extensive refining. Venezuela's Orinoco heavy oil belt is the best known example of this kind of unconventional reserve. There are estimated reserves of 1.2 trillion barrels, of which about one-third of the oil is potentially recoverable using current technology.
- *Tar-sands*: can be recovered via surface mining or in-situ collection techniques. Again, this is more expensive than lifting conventional petroleum but not prohibitively so. Canada's Athabasca Tar Sands is the best known example of this kind of unconventional reserve. There are estimated reserves of 1.8 trillion barrels of which 280–300bn barrels may be recoverable. Production now accounts for about 20 per cent of Canada's oil supply.
- *Oil shale*: requires extensive processing and consumes large amounts of water and is very environmentally damaging. Oil companies are investing considerable sums in developing appropriate techniques. Reserves are believed to exceed supplies of conventional oil.

In 2000 the US Geological Survey estimated oil reserves at some 3 trillion barrels of oil, which includes unconventional resources. This is very different to the view shown in Figure 5.1 which gives a reserve value of 1.3 trillion barrels of conventional reserves (USGS, 2000). Other authors have suggested that total reserves could be as high as 5 trillion barrels (Odell and Rosing, 1980). These differences are important as it influences the debate over Peak Oil, a debate that reflects concern that at some time oil and gas will run out as they are non-renewable resources. The Peak Oil concept was first advanced by Marion King Hubbert. Hubbert argued, based on an analysis of production and consumption figures in the US, that consumption would exceed production at around 1970 and the world production would peak around 2000, as shown in Figure 5.2 (Hubbert, 1971).

Although Hubbert's estimate for the occurrence of the US peak have proved to be fairly reliable, there is considerable controversy about the date for global Peak Oil as well as for Peak

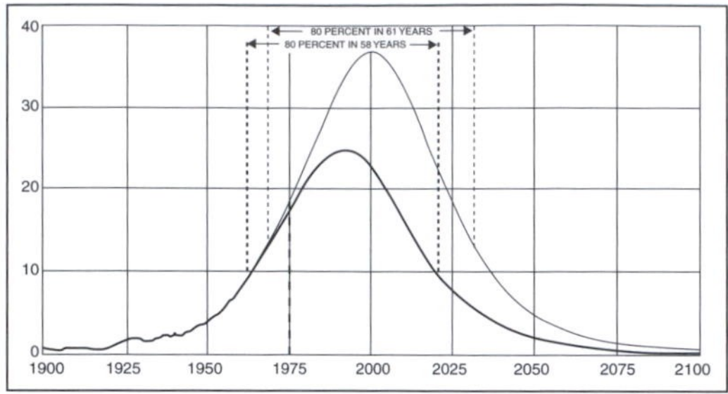


*Source:* Hubbert, 1971, p39

**Figure 5.2** Peak oil

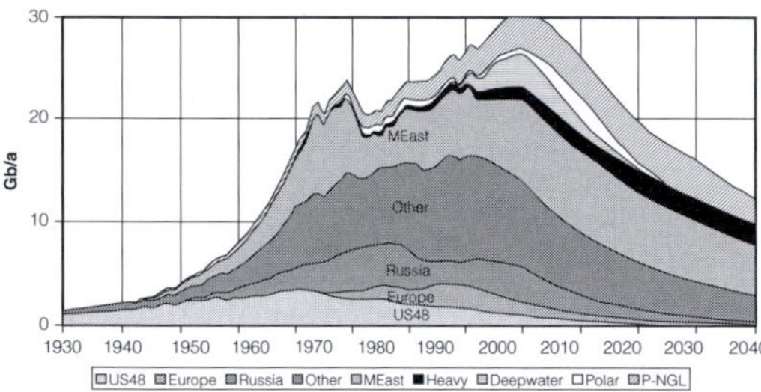


*Source:* Aleklett andCampbell, 2003, p16

**Figure 5.3** Oil and natural gas: 2003 base case scenario

Gas, with USGS claiming that the peak for oil will not occur for some 30 years (USGS, 2000). Others argue that the peak for oil is imminent. One of the most prominent voices in this debate is C. J. Campbell, a geologist and founder of the Association for the Study of Peak Oil, who claims that Peak Oil and Gas is imminent, as shown in Figure 5.3.

The model used to develop this scenario uses published data from public and industry sources but ignores speculative claims about the potential for new sources or increased recovery rates. Even though most unconventional sources are not included, the Association for the Study of Peak Oil does acknowledge these sources and estimates production from them in Canada and Venezuela peaking at only 1.5 million barrels per day. There are clearly very different views on the extent of global resources. Analysis of conventional oil resources using USGS and the Association for the Study of Peak Oil, Campbell gives quite different outcomes; the first suggests a peak using USGS between 2010 and 2030 and the second before 2010 ( Green et al, 2004).

What does appear to be clear is that demand for oil (and other energy resources) is growing and looks likely to continue to do so. The question for energy planners is how much is demand likely to rise over a given period. This is an important issue as it often takes considerable time and investment to develop an energy resource, either a primary resource or the technology and infrastructure to transform a primary resource into an energy service to meet demand. Projecting what might happen in terms of production and consumption over long time periods is complex. The figures shown in the following sections are taken from the Energy Information Administration which is an independent statistical and analytical agency within the US Department of Energy (EIA/DOE, 2008). It should be noted that other bodies also provide energy projections, for example, the International Energy Administration (IEA), which is part of the OECD (see IEA, 2008; Key World Energy Statistics, OECD/IEA, available at:<www.iea.org/textbase/nppdf/free/2008/key_stats_2008.pdf and WETO, 2003, World Energy, Technology and Climate Policy Outlook, European Commission, available at www.ec.europa.eu/research/energy/pdf/weto_final_report.pdf). The important point to note is that the data presented represent predictions based on models of what might happen. These models also indicate uncertainties, as it is impossible to predict with any precision, what may happen in the future. Although models may vary between organizations, in general they all project a business-as-usual case with alternative projections based on possible policy interventions. Similarly they use macro-economic, population and other trends on which to base their projections. The assumptions used by EIA/DOE are briefly discussed in Box 5.2.

World energy consumption is projected to increase by 50 per cent between 2005 and 2030. Demand in the OECD economies is projected to grow slowly at an average annual rate of 0.7

## Box 5.2 Underlying assumptions in the EIA/DOE energy model

The International Energy Outlook 2008 (IEO2008) presents an assessment by the EIA of the outlook for international energy markets through 2030. IEO2008 focuses exclusively on marketed energy. Non-marketed energy sources, which continue to play an important role in some developing countries, are not included in the estimates. The IEO2008 projections are based on US and foreign government laws in effect on 1 January 2008. The potential impacts of pending or proposed legislation, regulations and standards are not reflected in the projections, nor are the impacts of legislation for which the implementing mechanisms have not yet been announced.

The time frame for historical data begins with 1980 and extends to 2005, and the projections extend to 2030. High economic growth and low economic growth cases were developed to depict a set of alternative growth paths for the energy projections. The two cases consider higher and lower growth paths for regional gross domestic product (GDP) than are assumed in the reference case. IEO2008 also includes a high price case and, alternatively, a low price case. In making these projections assumptions are made about macro-economic growth, population trends and changing demand. For example, the projections assume that growth in the OECD area will remain steady but will increase rapidly in India and China. Although growth is projected to increase by 50 per cent by 2030, it is thought that oil will not grow as rapidly (1.2 per cent per year from 2005 to 2030) as renewables and coal (2.0 per cent and 2.1 per cent respectively) mainly because of continued high oil prices and environmental concerns. Coal, however, in areas where it is abundant (India, China and the US) make it an economical choice.

Uncertainties are shown by developing alternative projections based on high and low macroeconomic growth cases and high and low energy price cases. For further discussion see Chapter 1, IEO 2008.

*Source:* EIA/DOE, 2008

per cent, however, in non-OECD countries growth of 2.5 per cent per year is projected, with China and India the fastest growing non-OECD economies. Global projections are shown in Figure 5.4, while Figure 5.5 shows the differential growth projections between OECD and non-OECD areas.

Figure 5.6 illustrates a projection for energy use by fuel type. Note the rapid growth in the use of coal reflecting the increasing use by countries with abundant coal reserves, such as China and India. As coal is a significant producer of carbon dioxide during combustion, there are likely to be significant policy and technology implications if significant reductions in greenhouse gases are to be made.

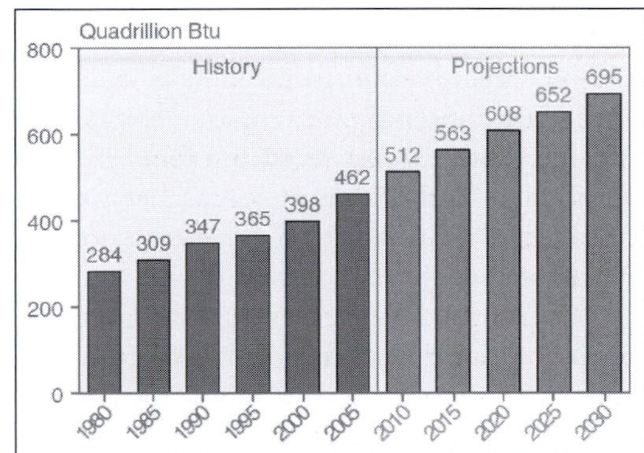


*Source:* Adapted from EIA/DOE, 2008, p7

**Figure 5.4** Growth of world marketed energy consumption 1980–2030

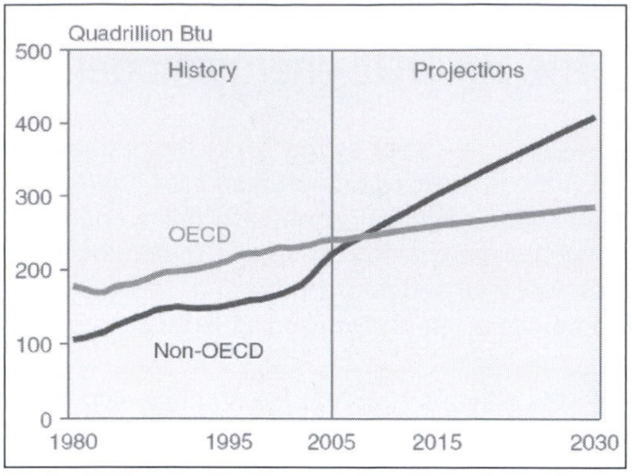


*Source:* EIA/DOE, 2008:8

**Figure 5.5** World marketed energy consumption: OECD and non-OECD, 1980–2030

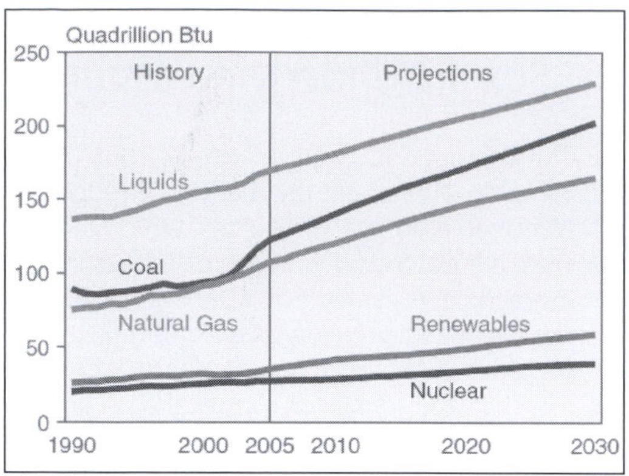


*Source:* EIA/DOE, 2008, p8

**Figure 5.6** World marketed energy use by fuel type, 1990–2030

## Natural gas

The use of natural gas has increased rapidly in recent years for two broad reasons. First, the resource base is much larger than had previously been thought. Second, it is environmentally more benign than other fuels, particularly coal, producing much less carbon in the combustion process. Natural gas consists of hydrocarbons with from one to five carbon atoms, together with small amounts of other gases as impurities. Natural gas is formed under essentially the same kind of conditions as oil; anaerobic decomposition of organic matter under heat and pressure assisted by bacteria. Marine organisms are the primary source material for oil, but natural gas can be formed from both land plants and marine organic material. Gas can be formed in very young deposits, for example, marsh gas in swamps. It can also be formed in association with coal deposits, particularly those of the Permo-Carboniferous. It can be formed with crude oil and as 'thermal' gas below the oil window. This means that the depths and areas of sedimentary basins which may hold gas far exceed those of oil. Gas fields, like oil fields, are not distributed uniformly, and differ in size and geographical concentration, but because of its more diverse origins gas is more widespread. Gas, found on its own in 'dry' wells, is called 'non-associated gas'. It is also found dissolved under pressure in oil in a reservoir or as a 'gas cap' over an oil pool; in these cases it is called 'associated gas'. About 70 per cent of world reserves are non-associated, some 20 per cent are dissolved and about 10 per cent are gas caps (Hill et al, 1995). Figure 5.7 shows the distribution of natural gas resources.

Demand for natural gas is also projected to grow strongly with the strongest growth occurring in the non-OECD area as shown in Figure 5.8 and it remains a key energy source for industrial sector uses (43 per cent in 2030) and electricity generation (35 per cent in 2030) throughout the projection. For electricity production natural gas is an attractive choice for new generating plant because of its relative fuel efficiency and low carbon dioxide intensity.

In order to meet this growth in demand the projection sees an increase in the export of LNG (Liquified Natural Gas) with Africa and the Middle East at the forefront of this trend, In Qatar, for example, export facilities with a total capacity of approximately 3.6 trillion cubic feet of natural gas (77 million metric tons of LNG) are expected to be in operation by 2015, as compared with the country's 2005 LNG exports of 1 trillion

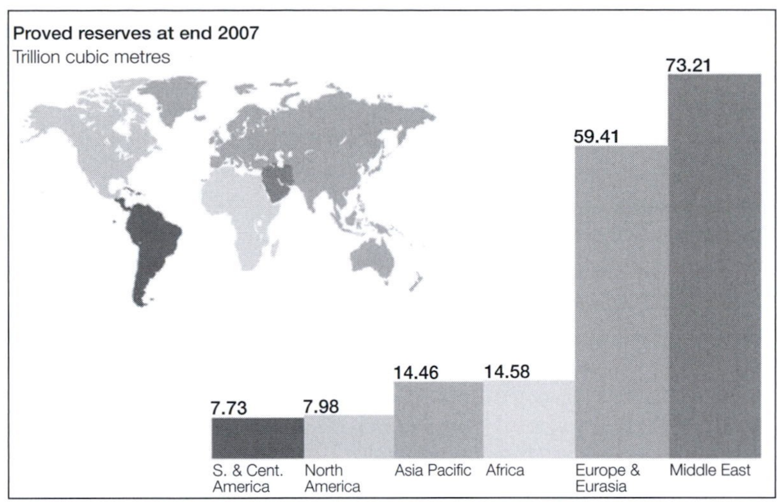


*Source:* BP, 2008

**Figure 5.7** Proved natural gas reserves as at the end of 2006

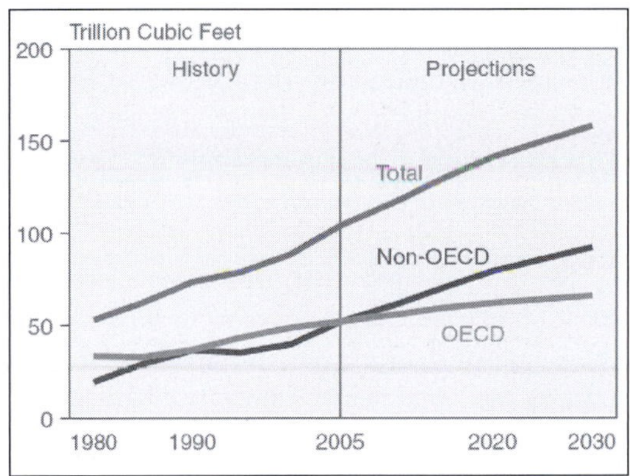


*Source:* EIA/DOE, 2008, p37

**Figure 5.8** World natural gas consumption, 1980–2030

cubic feet. Natural gas liquefies at about minus (–) 163 degrees Celsius, reducing its volume to one sixth-hundredth (1/600), making export by cryogenic sea vessels much more cost efficient than by pipeline.

# Coal

Coal is a complex organic material consisting of fused carbon rings held together by assorted hydrocarbon and other atomic (oxygen, nitrogen and sulphur) linkages. Its average composition is something like C10H8O (this ratio of 10 carbon atoms to 8 of hydrogen can be contrasted with the ratio of 10 carbons to 17.5 hydrogens in crude oil). It is formed from dead plant material which has accumulated in swamps, usually in estuarine deltaic deposits, and which has been consolidated and altered by increasing temperature and pressure. In a similar evolutionary pattern to oil, the first stage in the conversion process is an anaerobic breakdown of the plant material which causes volatile products to be liberated and lost to give a compacted unstructured mass of compounds enriched in carbon. The second stage is the process of coalification which proceeds through the ranks of peat, lignite, sub-bituminous coal, bituminous coal and anthracite to graphite. The proportion of carbon is gradually increased in this progression. Calorific values of the various ranks range from 15–26kJ/g for low rank lignites, through 31–35kJ/g for bituminous coals to

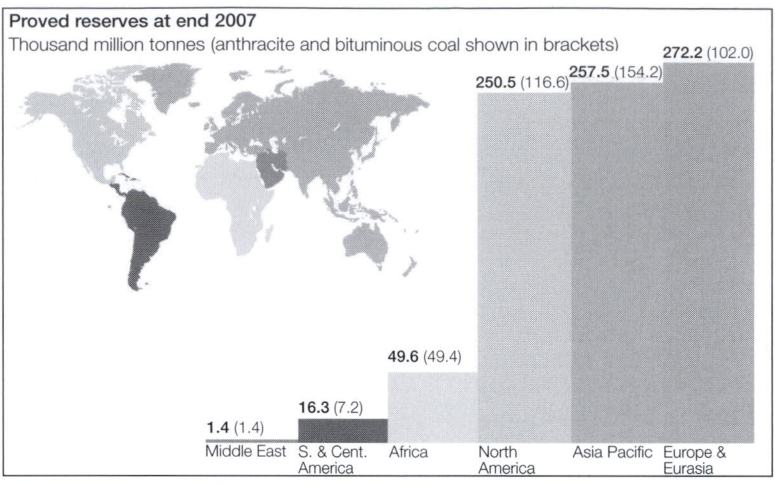


*Source:* BP, 2008

**Figure 5.9** Proved coal reserves as at the end of 2007

30–33kJ/g for anthracite. Coal deposits are not found before the Carboniferous age (about 400 million years ago), and the most important and widespread date from the Carboniferous to the early Triassic age (345–200 million years ago), and from the Jurassic to the early Tertiary age (150–50 million years ago). In general, the older coals have the highest rank but, depending on the geological history of the deposits, this is not necessarily so. Although coal is widespread the major deposits are unevenly distributed, as shown in Figure 5.9, the bulk of the deposits are to be found in North America, Europe and Eurasia and Asia Pacific.

Altough coal had a major role in the Industrial Revolution, the rapid changes in OECD countries in the latter half of the twentieth century brought about changes in the market for coal. For example, in the UK, following the Clean Air Act, coal was no longer used in domestic households. Coal has been replaced in transport systems and the decline in energy intensive industries such as iron and steel has seen the market for coal shrink. Coal remains an important fuel for electricity production, but climate concerns are beginning to influence how coal will be used in the future. In terms of carbon emissions, coal emits slightly more than oil, and about double that of natural gas. In the rapidly industrializing nations such as India and China, coal is increasingly used to generate power and this is reflected in non-OECD growth, as shown in Figure 5.10.

No one disagrees that nations, developed or developing, have a right to strive to improve the well-being of their citizens and that they have a right to use whatever resources available to do that task. However, coal produces a large amount of carbon during combustion and this has global consequences. Finding a way of either developing a new and cleaner resource or finding ways of minimizing or eliminating the adverse impacts of coal is an urgent policy and technological challenge.

## Policy context

There is no global energy policy. Energy policy is determined at the level of the state. Energy is so fundamental to development that governments throughout the world have acted to ensure that

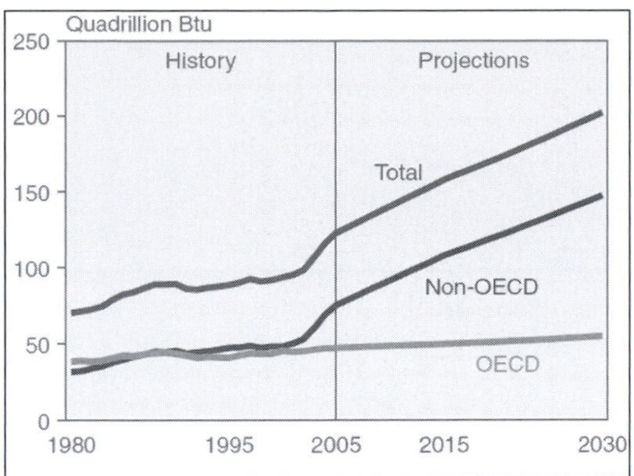


*Source:* EIA/DOE, 2008: 47

**Figure 5.10** World coal consumption by country grouping, 1980–2030

sufficient supplies of fuel are available to meet the needs of their citizens. The history of energy is complex, but generally energy demands in many nations have evolved from a reliance on coal to a reliance on oil, gas and nuclear resources. The pattern of the energy mix is very varied with nations using a combination of indigenous resources and imported supplies to meet needs. There are a variety of models of ownership and control of energy infrastructure throughout the world. In general though the trend has been away from state ownership and control to a market based system with regulatory oversight. The exception is oil where typically the infrastructure for exploring, extracting, refining and distributing petroleum products has been the remit of the oil companies.

The contribution to accelerated climate change made by the use of conventional fuels is now firmly established (IPCC, 2007). The policy context for energy is now increasingly influenced by climate concerns. The United Nations Framework Convention on Climate Change (UNFCCC), which came into force in 1994, established a framework for reducing greenhouse gas emissions. The initial target of reducing emissions by 2000 to 1990 levels was not realized, mainly because the targets were poorly devised and the Convention was not legally binding. The Kyoto Protocol, ratified in 2005, committed the signatories to reducing greenhouse emissions and provided differential targets that recognized differential emission rates and capacities to reduce emissions. The first commitment period for the Kyoto Protocol is 2008–2012. The Annex 1 countries of the Kyoto Protocol, primarily the OECD nations, have agreed targets for emission reductions. They can do so by implementing domestic policies and by engaging in Joint Implementation (JI) projects, the Clean Development Mechanism (CDM) and emissions trading. The Convention and the Protocol have, in essence, provided a framework for energy policy. Policy now has the twin aims of providing secure and affordable energy supplies whilst reducing greenhouse gas emissions. Meeting these aims is having significant impact on the future of energy. On the supply side of the energy systems this means introducing technologies that do not emit greenhouse gases, or at the very least emit a minimal amount of greenhouse gases. For conventional fuels this is a significant challenge. It is being addressed in a number of ways: first, by improving the efficiency of existing systems; second, by fuel switching to fuels with a lower carbon content, for example, from coal to gas; and third, by developing technologies that can capture carbon emissions.

In summary, the use of all fossil fuels is likely to increase and despite the interest in renewable technologies, it is unlikely that they will meet expected demand levels. In short it appears that fossil fuels will be around for some time and this means that although efforts must continue to develop renewable capacity and to improve end use efficiency, efforts to improve the supply-side of conventional fuels must also continue.

## Supply-side strategies

The supply side of the energy system is that part of the energy system which transforms primary energy resources into secondary energy resources, for example, from coal to electricity, and distributes these to the point of use as shown in Figure 5.11.

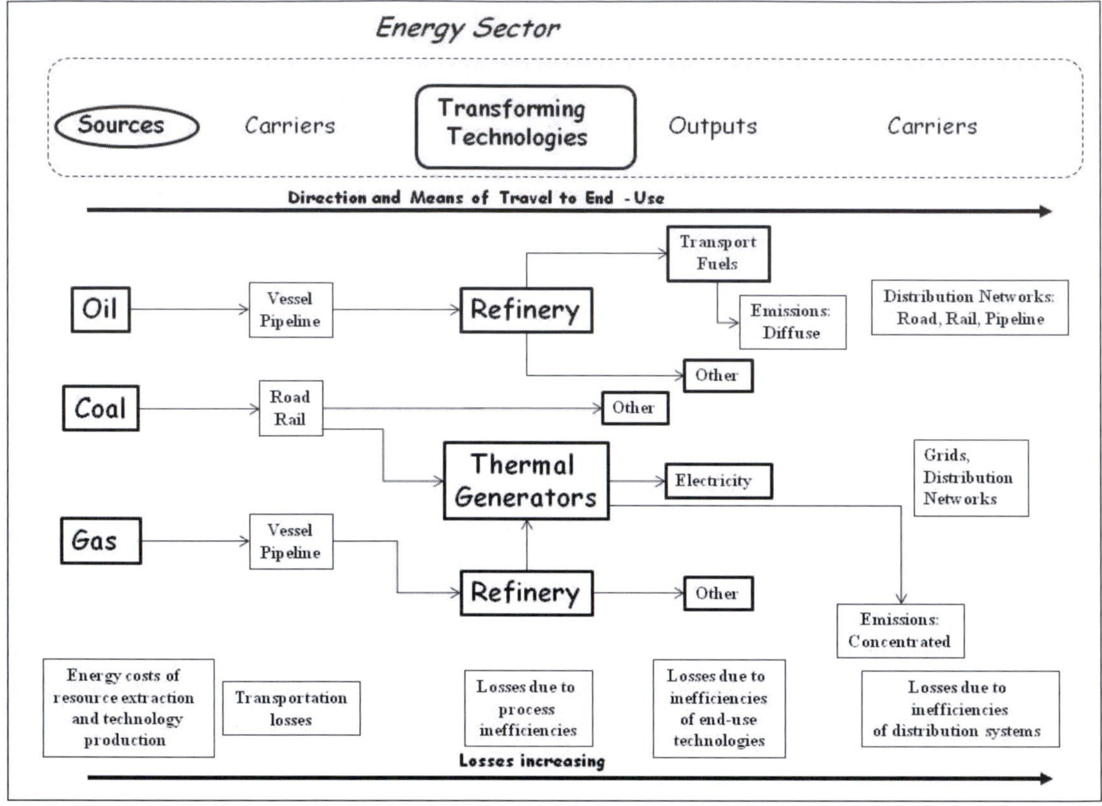


**Figure 5.11** Supply-side: Conventional energy resources

From Figure 5.11 the losses in the system increase as resources are transformed into usable energy services. There are losses or energy costs associated with each step, for example, energy is required both to build a pipeline and to transport oil and gas through it. Losses could occur if the pipeline is damaged. This is also the case for products leaving a refinery and electricity leaving a power station. However, the largest areas of loss are in the transformation of coal and gas into electricity and the use of transport fuels. In electricity production the losses are realized as heat and these can be captured and used for other purposes. Carbon emissions are realized at the point of production and because they are concentrated the emissions can be captured and processed for storage. The losses in the transport systems are much more diffuse, as are the emissions.

Improving efficiency is an important component of reducing greenhouse production. That requires efforts to improve supply-side and end use efficiencies. End use efficiency has been covered in Chapter 4. On the supply side of conventional energy, improving the efficiency of electricity production offers the largest scope. However, power station efficiency is governed by the laws of thermodynamics and only marginal improvements are possible (see Box 5.3).

By combining both the Rankin and Brayton Cycles, by using the waste heat from one as the heat source for the other, then significant efficiency improvements can be made. This is known as the Combined Cycle Gas Turbine (CCGT) and is shown in Figure 5.12. The overall efficiency of the CCGT is about 60 per cent. By using the waste heat for other purposes such as space heat-

## Box 5.3 Heat engines

A heat engine operates by transferring energy from a warm region to a cool region of space and, in the process, converting some of that energy to mechanical work. For example, a hot gas when introduced into a piston will cause the piston to move. The gas expands and its temperature (a measure of the energy in the gas) will decrease. In short the energy in the gas has been converted to work, or motion, in this case. The process can be reversed. By applying external work (force) thermal energy can be transferred from a cool place to a warmer one. This is the basis of refrigeration. This cycle is known as the Carnot Cycle and is the most efficient cycle for transferring work and energy. In the Carnot heat engine the efficiency of a system is 1 – TC/TH where TC is the temperature of the sink and TH is the temperature of the source. Temperature is expressed in degrees Kelvin. At absolute zero (–273°C) TC would be zero and theoretically the efficiency would be 100 per cent. In reality efficiencies for heat engines are less than 50 per cent and typically for coal fired stations that generate steam to drive a turbine, less than 40 per cent. Heat engines of all kinds operate with modified versions of the Carnot Cycle.

- *The Steam Turbine*: this operates on the Rankin Cycle, a thermodynamic cycle that converts heat into work. This is a closed loop system that uses water as the working fluid. It is the most commonly used cycle for electricity production. Superheated steam (high pressure and temperature) is introduced to an expansion device, the turbine, where, as it expands and cools, it exerts force on the turbine blades. The blades are mounted perpendicular to the turbine axis and the force causes it to rotate. The temperature difference between the input and the output is a measure of the work done by the turbine. At the turbine exhaust the steam enters a condenser where it is cooled and then re-circulated by means of a pump. Ideally the efficiency of the Rankine Cycle is 63 per cent. In reality the overall thermal efficiency is between 35–40 per cent. The exhaust heat from the condenser is very energetic and could be used for other purposes.
- *The Gas Turbine*: this operates on the Brayton Cycle, a constant pressure cycle used in gas turbines and jet engines. It has three components; a compressor, a combustion chamber and a turbine. Compressed air from the compressor is heated either by directly burning fuel in it or by burning fuel externally in a heat exchanger. The heated air with or without products of combustion is expanded in a turbine resulting in work. More than 60 per cent of the work produced is used to drive the compressor and the balance, up to 40 per cent, is available as useful work output. The exhaust stream from a gas turbine is also very energetic.

ing or industrial processes efficiency can exceed 90 per cent. This is known as Cogeneration or Combined Heat and Power (CHP). These cycles are shown in Figure 5.12. The fuel normally used for CCGT systems is natural gas. In response to climate and energy security concerns, replacing natural gas with Oxyfuel generated from coal and then capturing and storing the carbon offers the opportunity to extract the maximum energy from coal and avoid greenhouse gas emissions. This is known as the Integrated Gasification Combined Cycle – see the Vattenfall pilot in Carbon Capture and Storage in Box 5.5.

## Combined heat and power (CHP)

Combined heat and power (CHP), also known as Cogeneration, is the simultaneous generation of usable heat, either for industrial use or space heating, and power, usually electricity, in a single process. CHP systems cover a wide range of sizes, applications, fuels and technologies. In terms of scale it can range from large systems through to micro CHP systems for use at the household level. Fuels that can be used include coal, gas, biomass and waste. Some schemes use municipal

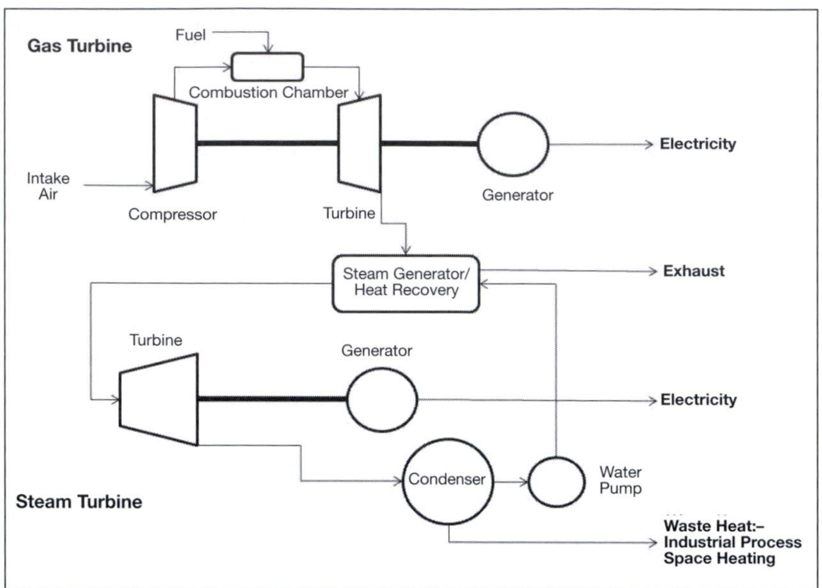


**Figure 5.12** The combined cycle gas turbine (CCGT)

waste as a fuel, although this is somewhat controversial. The heat can be used for industrial purposes or space heating. Space heating can range from the municipal level, known as district heating, through to the household level.

The concept of district heating is not new and is used in Europe and North America. In London, for example, the Pimlico District Heating Undertaking (PDHU), constructed in 1950, used waste heat from Battersea Power Station. Although Battersea Power Station is now closed, the PDHU uses other systems to provide heat. In 1903 the municipality of Frederiksberg in Denmark constructed a municipal incinerator that provided heat and power. In 2006 in Europe CHP systems produced about 11 per cent (143GW) of overall electricity production and some 3100PJ of heat. About 68 per cent of electricity and 32 per cent of heat production is supplied from dedicated facilities and some 32 per cent of electricity and 68 per cent of heat is produced by organizations for their own use (Eurostat, 2008).

In 1997 the EU Commission's cogeneration strategy set an overall target of 18 per cent of electricity production from cogeneration by 2010. In 2004 the Commission introduced a Directive on the promotion of cogeneration (CHP) based on useful heat demand in the internal energy market. Its purpose is to increase energy efficiency and improve security of supply by creating a framework for promotion and development of high efficiency cogeneration. It defines high efficiency cogeneration as cogeneration providing at least 10 per cent energy savings compared to separate production. The Directive does not specify a target but focuses on providing a framework that will promote cogeneration (EU Commission, 2004).

Improving the efficiency of the conventional supply-side can bring efficiency improvements and reduce greenhouse gas production. However, the supply side will continue to produce significant amounts of greenhouse gases and to make significant inroads into reducing these emissions will require a different approach.

## Carbon capture and storage and carbon sequestration

Given the dependence on fossil fuels and the increasing demand, another approach is to capture and store carbon emissions before they enter the atmosphere. Greenhouse gas emissions can be reduced through, for example, the use of renewables and improved efficiencies, but the scale of use and growth in conventional fuels has led to an increasing interest in capturing carbon emissions from existing production technologies and storing these emissions in ways that do not interfere with the climate system. Broadly these fall into two categories. The first is the enhancement of natural carbon sinks such as forests. This is termed sequestration (IPCC, 2000). The second is carbon capture and storage from existing production sites, for example, coal fired electricity plants. This is termed carbon capture and storage (CCS) (IPCC, 2005a).

## Carbon sequestration

Carbon sequestration refers to the enhancement of natural carbon sinks such as forests, soils and oceans. Vast amounts of carbon are naturally stored in forests by trees, plants and the soil. Through photosynthesis, plants absorb carbon dioxide from the atmosphere and store the carbon as sugar, starch and cellulose, whilst releasing oxygen into the atmosphere. Young forests with rapidly growing trees absorb carbon dioxide and act as a sink. Mature forests can prove to be carbon neutral as they house dead and decaying matter that releases carbon to the atmosphere. The gradual build up of soil slows the decay process and carbon gradually accumulates. Most forests are a mix of both growing and mature tress and carbon is stored and released continuously. Figure 5.13 gives an overview of the global carbon cycle.

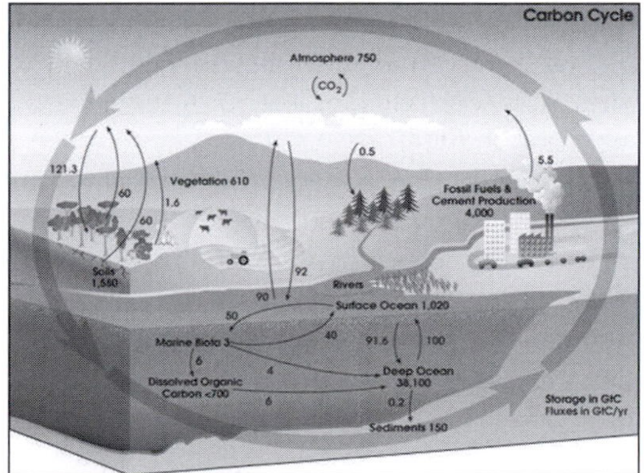


*Source:* NASA, undated

**Figure 5.13** The global carbon cycle

## Soils

Soils contain about three times more carbon than that stored in vegetation and about twice as much carbon as the atmosphere. Carbon storage in soils is the balance between the input of dead plant material (leaf and root litter) and losses from decomposition and mineralization processes. Increasing the amount of carbon naturally stored in soils could provide a short-term bridge to reduce the impacts of increasing carbon emissions until low-carbon and sustainable technologies are implemented. For example, FAO (the Food and Agriculture Organization of the United Nations) estimates that the carbon content of soils ranges between 7 and 24 tonnes in normal, non-degraded soils, depending on the climate zone and vegetation. Soil degradation is a global problem, particularly the desertification of drylands. The dynamics of carbon storage in soils is complex because of the variability of composition and environmental factors and the real potential for terrestrial carbon storage is not known because of the lack of a reliable database and fundamental understanding of the dynamics of soil organic carbon at the molecular, landscape, regional and global scales. Speculative estimates suggest that improved terrestrial management over the next 50–100 years could sequester up to 150Pg of carbon, the amount released to the atmosphere since the mid-19th century as a result of past agricultural conversion of grasslands, wetlands and forests.

**Table 5.1** Agricultural practices for enhancing productivity and increasing the amount of carbon in soils

| Traditional practices | Recommended |
|---|---|
| Plough till | Conservation till or no-till |
| Residue removal or burning | Residue return as mulch |
| Summer fallow | Growing cover crops |
| Low off-farm input | Judicious use of fertilizers and integrated nutrient management |
| Regular fertilizer use | Soil-site specific management |
| No water control | Water management/conservation, irrigation, water table management |
| Fence-to-fence cultivation | Conversion of marginal lands to nature conservation |
| Monoculture | Improved farming systems with several crop rotations |
| Land use along poverty lines and political boundaries | Integrated watershed management |
| Draining wetland | Restoring wetlands |

*Source:* FAO, 2004, p4

Mechanisms to enhance carbon sequestration in soils include conservation tilling, cover cropping and crop rotation. A more detailed listing is shown in Table 5.1. This suggests that carbon sequestration in soils through improved management should be a component in mitigation. At the very least it could help to lengthen the time needed to implement other mitigation technologies.

## Forests and peatlands

The dead trees, plants and moss in peatlands undergo slow anaerobic decomposition below the surface. This process is slow enough that in many cases the peatlands grow rapidly and fixes more carbon from the atmosphere than is released. Peatlands cover approximately 3 per cent of the Earth's land area and are estimated to contain 350–535Gt of carbon, or between 20–25 per cent of the world's soil organic carbon stock (Gorham, 1991).

Forests and peatlands can also be a source of carbon dioxide, for example, forest fires can quickly release absorbed carbon into the atmosphere. In addition, forest flooding, for example, by the construction of a hydroelectric dam would allow the rotting vegetation to become a source of carbon dioxide and methane. These can be comparable in magnitude to the amount of carbon released from a fossil fuel powered plant of equivalent power.

Peat material has traditionally been used as a fuel and more recently as a soil enhancement material. Although the peat cycle has been disturbed through such activities, other interventions such as fire can result in considerable releases of carbon. The peatlands of Borneo and the neighbouring territories of Sumatra and Irian Jaya are up to 20m deep and cover more than 200,000km$^2$. They contain 50 billion tonnes or more of carbon – far more than the forests above. As farmers clear the forests by burning, the peat can catch fire and release carbon for months afterwards. During 1997 and 1998 smouldering peat beneath the Borneo forests was estimated to have released between 0.8 and 2.6 billion tonnes of carbon into the atmosphere. That is equivalent to 13–40 per cent of all emissions from burning fossil fuels in 1998 (Page et al, 2002).

Forests are estimated to contain about 20 per cent of the carbon stored in soils. Forests have an important role in capturing and storing carbon. However, these systems need to be carefully managed. Disturbance of the forest system, either though natural causes or human intervention, can release carbon into the atmosphere. For example, at present, temperate forests are considered to act as a carbon sink because of reduced harvest levels, increased regeneration efforts and administrative set-asides. However, tropical forests are still reported to be a net carbon emitter as a result of mainly human-induced land use change. Globally, the clearing of forests has reduced the forest area by almost 20 per cent in the last 140 years. By far the greatest sources of forestry-related emissions are clear-cutting and logging in forests. These activities are responsible for about 20 per cent of global, human-induced emissions (Streck and Schloz, 2006).

## Oceans

Oceans are natural carbon dioxide sinks and the level of carbon dioxide rises within the oceans with increases in atmospheric concentrations of carbon dioxide. Consequently, increasing atmospheric emissions could generate potentially disastrous acidic oceans. Phytoplankton and other marine animals absorb $CO_2$ from the water to build their skeletons and shells, which removes the $CO_2$ from the water enabling more to be absorbed. These skeletons and shells eventually die and decay. To be sequestered for 1000 years they must sink to the bottom of deep waters, 2000–4000m.

A promising method of increasing carbon sequestration efficiency is to add micrometer-sized iron particles called hematite or iron sulphate to the water. This stimulates growth in plankton. Natural sources of ocean iron have been declining in recent decades, contributing to an overall decline in ocean productivity. The application of iron nutrients in select parts of the oceans, at appropriate scales, could have the combined effect of restoring ocean productivity while simultaneously mitigating the effects of anthropogenic emissions of carbon dioxide to the atmosphere. Critics argue the effect of periodic small scale phytoplankton blooms on ocean ecosystems is unclear and further research is required (Coale, undated).

## Carbon capture and storage

Carbon capture and storage (CCS) is an approach to mitigating climate change through separating and capturing carbon dioxide from the production, processing and burning of oil, gas, coal and biomass from power plants and industrial processes and subsequently transporting and storing it instead of releasing it into the atmosphere. There are two distinct new dimensions to this process. The first is the capture of carbon dioxide and the second is its long-term storage. Issues such as the transportation of carbon, for example, by pipeline, are well understood. The capture and storage of carbon dioxide is not a wholly new concept, for example, the technology for capturing carbon dioxide is commercially available and is used by the oil industry for EOR. Carbon dioxide is injected commercially into oil reservoirs for enhanced oil recovery in many parts of the world. The carbon is essentially 'stored' (see Box 5.4).

### Box 5.4 Enhanced oil recovery (EOR)

EOR is a particular type of CCS where $CO_2$ pumped into a near depleted field dissolves in the oil, making it more mobile and easier to extract. This can lengthen the life of the field and increase the overall yield of oil. EOR is an established onshore technology but it has not so far been used commercially offshore. Although some of the injected $CO_2$ returns to the surface with the oil, this is recaptured and added back to the $CO_2$ being injected. The climate change benefit of EOR arises if the $CO_2$ has been captured from fossil fuel combustion and if most of it is left in the reservoir at the end of its productive life.

*Source:* House of Commons Science and Technology Committee, 2006, p8

**Table 5.2** Profile by process or industrial activity of worldwide large stationary $CO_2$ sources with emissions of more than 0.1 million tonnes of $CO_2$ per year ($MtCO_2\ yr^{-1}$)

| Process | Number of sources | Emissions ($MtCO_2\ yr^{-1}$) |
|---|---|---|
| Fossil fuels | | |
| Power | 4942 | 10,539 |
| Cement production | 1175 | 932 |
| Refineries | 638 | 798 |
| Iron and steel industry | 269 | 646 |
| Petrochemical industries | 470 | 379 |
| Oil and gas processing | n/a | 50 |
| Other sources | 90 | 33 |
| Biomass | | |
| Bioethanol and bioenergy | 303 | 91 |
| Total | 7887 | 13,466 |

*Source:* Adapted from IPCC, 2005, Table SPM.1

Around 33 million tons of $CO_2$ have been captured and stored in over 70 projects (IEA, 2004). Many are experimental, but there are large-scale commercial projects in operation in Salah (Algeria), Weyburn (Canada), the North Sea (Sleipner) and forthcoming in the Barents Sea, Gorgon (Australia), Gassi Touil (Algeria) and other fields. CCS, however, is a relatively untried concept in terms of industrial and power production processes. Up to 2007 no power plant had been developed that operated with a full carbon capture and storage system (WEC, 2007a). The scope, however, is considerable and in 2008 a pilot scheme using coal as the fuel was developed in Germany. This is discussed later in this section. The potential for CCS technology is considerable as shown in Table 5.2. Note this table refers to stationary sources. CCS for mobile sources such as vehicles is not realizable. Further power production accounts for some 75 per cent of carbon emissions making this sector very attractive in terms of CCS.

There are a number of technological approaches to carbon capture:

- *Pre-combustion capture*: Currently used in the industrial manufacture of hydrogen and ammonia. Natural gas, coal, oil residuals or biomass is reacted with oxygen, air and/or steam to generate a synthesis gas or 'syngas' consisting mainly of carbon monoxide and hydrogen. The carbon monoxide is reacted with steam to give $CO_2$ and more hydrogen. The resulting gas mixture contains predominantly hydrogen gas and $CO_2$ (15–40 per cent) at a high pressure. The carbon is separated from the hydrogen, usually by physical solvent absorption although membranes may be a promising option for the future.
- *Post-combustion capture*: This process is used to separate $CO_2$ from power station exhaust streams for use in the food industry. Similar technology has been in use since 1996 separating 1Mt/y of $CO_2$ from a natural gas stream for injection into an aquifer beneath the Norwegian North Sea (the Sleipner project). $CO_2$ is captured from flue gas by separating it from nitrogen and oxygen gases; the $CO_2$ content is low (3–13 per cent) and the separation is done at low pressure. The leading technology in post-combustion capture is chemical solvent absorption using amine based solvents (commonly referred to as 'amine scrubbing') although other solvents are being developed.
- *Oxyfuel capture*: This technique is still at the pilot stage. Fossil fuels, especially coal, are burnt in oxygen rather than air, producing a flue gas comprised mainly of $CO_2$ and water. This greatly facilitates the separation of $CO_2$. Pure oxygen is produced by cryogenic separation of air into mainly oxygen and nitrogen; this part of the process uses significant amounts of energy and is therefore costly. Burning fuel in pure oxygen results in an extremely high flame temperature so part of the flue gas is recycled to the combustion chamber to control the flame temperature. Finally, water is condensed from the flue gas that is not recycled. Some additional clean up of the $CO_2$ may also be required (Adapted from House of Commons Science and Technology Committee, 2006, p15).

These approaches are illustrated in Figure 5.14.

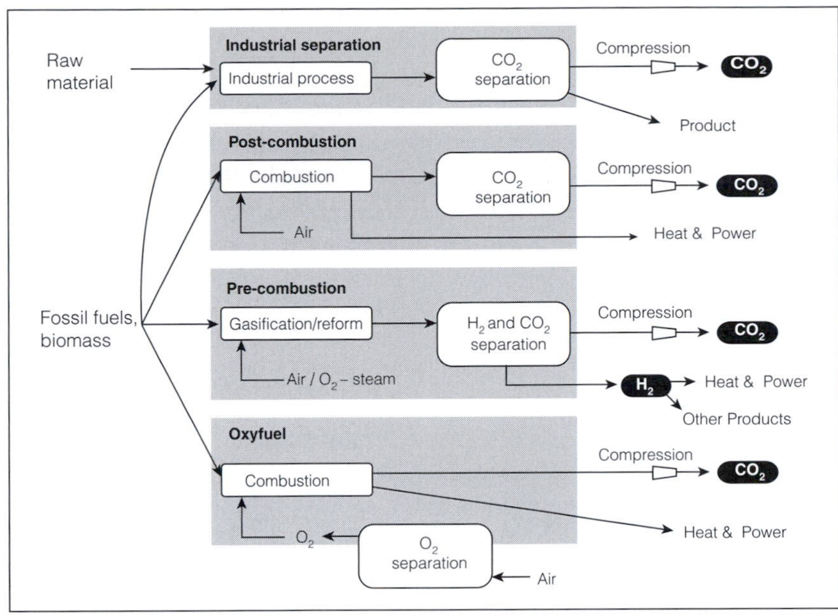


Source: IPCC, 2005b, Figure SPM.3:4

**Figure 5.14** Overview of carbon capture systems

Given that power production is the largest source of carbon emissions (see Table 5.2) the focus of development will be on the pre-, post- and oxyfuel capture technologies for fossil fuelled power stations. There is no single approach which is better that the other. Although there is experience of pre- and post-combustion technologies, oxyfuel, considered as having a promising future, has only recently been developed into a pilot stage project (see Box 5.5).

There are also considerations on how best to implement the technologies, for example, should existing power stations be retrofitted or should proposed new stations have the technology included or, at a minimum, be designed so that the technology can be retrofitted? These are complex issues to which there is no single answer. Two of the main considerations will be the economic and efficiency implications. CCS projections indicate a reduction in atmospheric carbon dioxide emissions within a modern conventional power plant by 80–90 per cent compared to a plant without CCS. However, capturing and

## Box 5.5 Vattenfall Oxyfuel Pilot Project

The Vattenfall pilot project will use coal burned in pure oxygen. This is a 30MW project that will produce heat, water vapour and about 9 tonnes of carbon dioxide per hour. The heat is used to raise steam that would normally drive a turbine. In this pilot the steam is being supplied to a nearby industrial estate. To avoid pollution the flue gases are cleaned to remove particles and sulphur dioxide. The remaining gas stream is almost pure carbon dioxide which is then cooled and compressed to one 500th of its volume which liquefies the gas. The liquid gas is then transported to a geological storage site.

*Source:* Vattenfall, undated; Harrabin, 2008

compressing $CO_2$ requires much energy. This would increase the energy needs of a plant with CCS by approximately 10–40 per cent. This and other system expenses would increase the costs of energy from a power plant with CCS by 30–60 per cent depending on the specific circumstances (IPCC, 2005a)

The IEA estimated that the costs of CCS could be US$50–100 per tonne of $CO_2$ captured and stored depending on the power plant fuel and the technology used. The bulk of the cost is on the capture side. By 2030, costs could fall to US$25–50 per tonne. Using CCS with new power plants would increase electricity production costs by 2–3 US cents/kWh falling to 1–2 US cents by 2030 including capture, transportation and storage. Scenarios developed by the IEA suggest that CCS potentials are between 3Gt and 7.6Gt $CO_2$ in 2030 and between 5.5Gt and 19.2Gt $CO_2$ in 2050. This compares to 38Gt $CO_2$ emissions by 2030 under the WEO Reference Scenario (The IEA World Energy Outlook (WEO) Reference Scenario projects that, based on policies in place, by 2030 $CO_2$ emissions will have increased by 63 per cent from today's level, which is almost 90 per cent higher than 1990 levels). For 2030 this gives a range of 8–20 per cent of carbon emissions. The scale of the range reflects the uncertainty in technology development and the rate of implementation. This has an impact on the longer term potential of CCS. The Stern Review, for example, estimates that CCS could contribute up to 28 per cent of global carbon dioxide mitigation by 2050, while IPCC estimates that it could be 50 per cent by 2050 (Stern, 2006; IPCC, 2005a). But the fact that all scenarios show a potential on a Gt scale suggests that CCS technologies constitute a robust option for emissions reduction.

Although CCS has considerable potential, countries such as the UK, which has a fleet of coal fired power stations that were designed some 30 years ago and are inefficient, the value of retrofitting a CCS technology that will further decrease efficiency is questionable (House of Commons Science and Technology Committee, 2006). Evidence presented to the Committee suggested that retrofitting to improve efficiency should be undertaken prior to fitting CCS technology. With ageing equipment, retrofitting poses some difficult challenges. However, it would seem that building a new plant that either has the capability of retrofitting or has capture technology in-built would be a more sensible method. The committee concluded that CCS has considerable potential and that a long-term incentive framework and a policy signal from government would be needed to give industry the confidence to proceed. But Haszeldine and Yaron (2008) find that to date the picture from the UK government is muddled and argue that CCS, like other low carbon technologies, should be incentivized and they suggest a number of options:

- create a Decarbonized Renewable Obligation Certificate band similar to that for wind energy;
- introduce long-term purchase contracts for decarbonized fossil fuel electricity;
- allocate free EU Emission Trading Scheme allowances after 2012 to reward $CO_2$ stored.

This illustrates some of the policy difficulties for any government. Which of the competing technologies is the most appropriate and how best to support that technology? There are no clear answers.

## Storage

Storage of the carbon dioxide is envisaged either in deep geological formations, deep oceans or in the form of mineral carbonates. Geological formations are currently considered the most promising. WEC estimates that there are considerable underground depositories for $CO_2$. For example, global capacities in saline formations are estimated at 1000–10,000Gt $CO_2$ and in depleted oil and gas fields at 1100Gt $CO_2$. This corresponds to 90–480 years of current world emissions at 23–24Gt $CO_2$/year. In addition $CO_2$ can be stored in abandoned or unminable coal seams or glacial clathrates (WEC, 2007b). IPCC estimates the economic potential of CCS could reach between

10–50 per cent of the total carbon mitigation effort until the year 2100 (IPCC, 2005a). Carbon storage is a relatively new concept in terms of climate change mitigation and the main options are discussed briefly below.

## Geological storage

Also known as geo-sequestration, this method involves injecting carbon dioxide directly into underground geological formations where various physical and geochemical trapping mechanisms prevent the carbon dioxide from escaping to the atmosphere. Possible formations include:

- Depleted oil and gas reservoirs at depths below 800m: This is a proven technology where carbon dioxide is injected to increase recovery. This option ensures the storage cost is offset by the sale of the additional oil and gas that is recovered. Various physical and geochemical trapping mechanisms would prevent it from migrating to the surface. In general, an essential physical trapping mechanism is the presence of a caprock.
- Saline formations, both on- and offshore at depths below 800m: These formations contain highly mineralized brines and to date have been considered of no benefit to humans. Saline aquifers have been used for the storage of chemical waste. Saline aquifers possess a large potential storage volume and are commonly found. This will reduce the distance over which $CO_2$ must be transported. Unfortunately, little is known of saline aquifers and to maintain acceptable storage costs (as there are no side products to offset this cost) the geophysical exploration may be limited, resulting in larger uncertainty about the aquifer structure. Leakage is a further issue, but research indicates several trapping mechanisms immobilize the carbon dioxide underground, reducing this risk.
- Unminable coal seams: These can be used to store $CO_2$ as it is absorbed readily in the coal's surface and storage can take place at depths of less than 800m. The technical feasibility depends on the permeability of the coal bed. This process releases methane previously absorbed in the coal's surface, which can be recovered. This is known as Enhanced Coal Bed Methane (ECBM) recovery. The sale of the methane can offset the cost of the storage. The feasibility of this approach has not been tested (IPCC, 2005a). Figure 5.15 illustrates these techniques.

## Ocean storage

Two concepts exist for ocean storage of carbon dioxide. The 'dissolution' type injects carbon dioxide by ship or pipeline into the water column at depths of 1000m or more, and the $CO_2$ subsequently dissolves. The 'lake' type deposits carbon dioxide directly onto the sea floor at depths greater than 3000m, where carbon dioxide is denser than water and is believed to form a 'lake' that will delay dissolution of carbon dioxide into the environment. These methods are illustrated in Figure 5.16.

Both dissolution and lake methods possess a range of issues to be overcome if they are to be widely accepted and effective in mitigating climate change. Large concentrations of $CO_2$ kill ocean organisms and react with the water

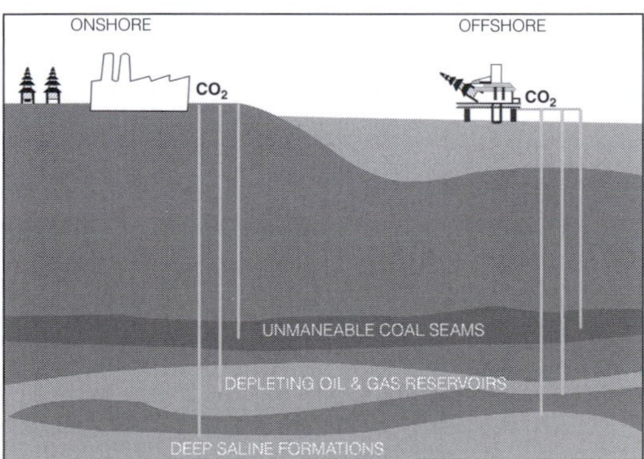


*Source:* World Coal Institute, undated

**Figure 5.15** Geological storage options for carbon dioxide

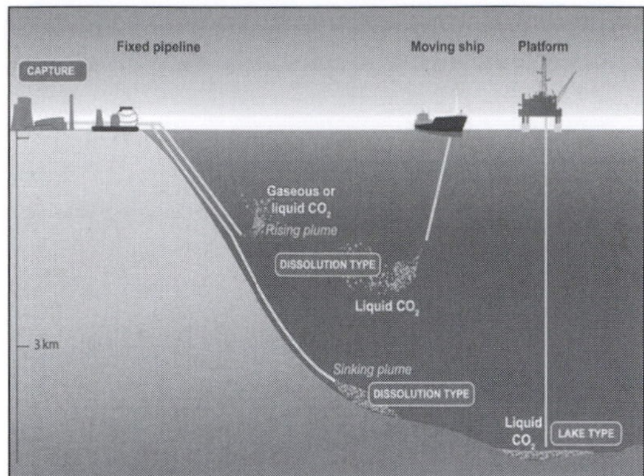


*Source:* IPCC, 2005b, Fig. SPM:6

**Figure 5.16** Overview of ocean storage concepts

to form carbonic acid, increasing the acidity of the ocean water, and consequently pose a serious threat to aquatic ecosystems (Greenpeace, 1999). The environmental effects on life-forms living at great depths (bathypelagic, abyssopelagic and hadopelagic zones) are poorly understood. Hence, further research is necessary to determine the full effects of carbon injections. Furthermore, the dissolved $CO_2$ would eventually equiliberate with the atmosphere and, therefore, storage would not be permanent. IPCC estimates 30–85 per cent of the injected $CO_2$ would be retained after 500 years for depths of 1000–3000m (IPCC, 2005a).

The IPCC estimates that $CO_2$ could be trapped for millions of years, retaining over 99 per cent of the injected $CO_2$ over 1000 years for well-selected, designed and managed geological storage sites (IPCC, 2005a). Leakage, however, remains a major concern with CCS and its resulting ability to mitigate climate change is greatly debated. In both geological and ocean storage, further research is necessary.

## Policy context

Establishing the appropriate policy framework is important if CCS is to play a role in mitigation strategies. In terms of establishing an appropriate policy framework for carbon capture technologies, that is the role of national governments, as is policy for storage if the storage area is within national boundaries. However, ocean storage and deep geological storage that requires the use of international waters or access to geological formations that cross borders, must complie with international law. Treaties, both international and regional, that have been established and are applicable to ocean and deep geological storage were originally intended to prevent dumping at sea with the purpose of avoiding transboundary environmental damage and protecting the marine environment. These obligations had been specified in a number of legally binding global and regional international instruments, established before CCS became an environmental and climate mitigation option and may need to be updated (some have) to take into account CCS. For example:

- The UN Convention on the Law of the Sea, 1982. This does not specifically regulate or prohibit CCS activities, but calls on states to protect the marine environment from human activities such as dumping.
- The London Convention on the Prevention of Marine Pollution by Dumping of Wastes and Other Matter, 1972. This prohibits the dumping of 'waste' into the sea.
- The London Protocol to the above Convention, 1996, allows, as of 10 February 2007, the injection of $CO_2$ streams from $CO_2$ capture processes and incidental associated substances in sub-seabed geological formations.
- The Basel Convention on the Control of Transboundary Movements of Hazardous Waste, 1989, which might be applicable if $CO_2$ contained toxic substances.
- The UN Framework Convention on Climate Change, 1994, under which CCS could be considered as an option to mitigate climate change.
- The Kyoto Protocol, 2005, excludes CCS from the Clean Development Mechanism.

At present, the above major conventions are being reconsidered to distinguish $CO_2$ injections from dumping. Parties to the London Protocol have defined the conditions under which $CO_2$ can be stored in sub-seabed geological formations as noted above (IPCC, 2005a; WEC, 2007a).

The G8 Summit of Gleneagles in 2005 mandated the IEA and CSLF (Carbon Sequestration Leadership Forum) to submit recommendations to the G8 Summit in Japan in 2008. IEA pointed out that establishing legal and regulatory frameworks must be one of the priorities for the G8 community in order to promote CCS (IEA, 2008). The Summit Leaders Declaration of G8 agreed to support IEA in developing a roadmap for innovative technologies, including CCS, to mitigate climate change. G8 also gave its support to the development of 20 large-scale CCS demonstration projects globally by 2010, with a view to beginning broad deployment of CCS by 2020 (G8, 2008, p31).

## EU context

A proposal for a directive on CCS was introduced in 2008 as part of the Commission's Climate Change and Energy Package. This originated from an agreement reached in the European Council in March 2007 that the EU should aim for all fossil fuel power stations built beyond 2020 to be equipped with CCS technology, subject to the necessary technical, economic and regulatory frameworks. The draft CCS Directive contains proposals designed to take this forward. It requires that all new combustion plants over 300MW have the capacity to have CCS retrofitted and that all carbon that is stored will not be considered as having been emitted and therefore not subject to the EU ETS. By 2015 the EU intends to have 12 large scale demonstration projects for coal and gas fired power plants (EU Commission, 2008). These should help CCS technology to become commercially viable and publicly acceptable. Although little research has been conducted into public perception, a study undertaken by the Tyndall Centre in 2003 suggests that acceptability is likely, providing that the purpose is fully explained (as a climate change mitigation technology), the key risks are acknowledged and that CCS is part of a range of mitigation measures (Shackley et al, 2004).

To support the development of CCS capacity in 2007, the EU adopted the European Technology Platform for Zero-Emission Fossil Fuels Power Plant (Zero Emissions Technology Platform, ZETP) and started working on the design of a mechanism to stimulate the construction and operation by 2015 of up to 12 large scale demonstration CCS plants (EU Commission, undated).

## Summary

The supply side of the energy system will continue to use fossil fuels for some time. Although efficiency improvements can be made, making significant cuts in greenhouse gas emissions is likely to require innovative approaches. CCS and sequestration have a clear role in climate mitigation. This is even more the case if alternative sources to fossil fuels are not developed sufficiently to meet current and future needs. Coal reserves are plentiful and the use of CCS could mean that fossil fuels will continue to have a significant role in the energy mix, particularly in those parts of the world where coal reserves are plentiful. It could also see the reactivation of mining in countries such as the UK where this activity has only recently ceased.

## References

Aleklett, K. and Campbell, C. J. (2003) 'The Peak and Decline of World Oil and Gas Production', *Minerals and Energy*, vol. 18, pp5–20

BP (2008) Statistical Review of World Energy, BP. Available at: www.bp.com/liveassets/bp_internet/globalbp/globalbp_uk_english/reports_and_publications/statistical_energy_review_2008/STAGING/local_assets/downloads/pdf/statistical_review_of_world_energy_full_review_2008.pdf

Coale, K. (undated) *Open Ocean Iron Fertilization for Scientific Study and Carbon Sequestration*, Moss Landing Marine Laboratories, Moss Landing,

California 95039, US. Available at: www.netl.doe.gov/publications/proceedings/01/carbon_seq/6b1.pdf

EIA/DOE (2008) *International Energy Outlook*, EIA/DOE, Washington DC, US. Available at: www.eia.doe.gov/oiaf/ieo/pdf/0484(2008).pdf

EU Commission (undated) *European Technology Platform for Zero Emission Fossil Fuel Power Plants (ETP-ZEP)*. Available at: www.zero-emissionplatform.eu/website/

EU Commission (2004) Directive 2004/8/EC, 2004, The promotion of cogeneration based on a useful heat demand in the internal energy market, OJ- l 52/50. Available at: http://eur-lex.europa.eu/LexUriServ/LexUriServ.do?uri=OJ:L:2004:052:0050:0060:EN:PDF

EU Commission (2008) Proposal for a Directive of the European Parliament and of the Council on the Geological Storage of Carbon Dioxide 2008/0015 (COD). Available at: http://ec.europa.eu/environment/climate/ccs/pdf com 2008 18.pdf)

Eurostat (2008) Data in Focus, Combined Heat and Power (CHP) in the EU, Turkey,

Norway and Iceland – 2006 data, Eurostat. Available at: http://epp.eurostat.ec.europa.eu/cache/ITY_OFFPUB/KS-QA-08-022/EN/KS-QA-08-022-EN.PDF

FAO (2004) 'Carbon sequestration in dryland soils', *World Soil Resources Reports:* 102, Rome. ISBN 92-5-105230-1. Available at: ftp://ftp.fao.org/agl/agll/docs/wsrr102.pdf

G8 (2008) G8 Hokkaido Toyako Summit Leaders Declaration. Available at: www.g8summit.go.jp/eng/doc/doc080714__en.html

Gorham, E. (1991) 'Northern peatlands: Role in the carbon cycle and probable responses to climatic warming', *Ecological Applications*, vol. 1, no. 2, pp182–195

Green, D. L., Hopson, J. L. and Li, J. (2004) 'Running out of and into oil: Analyzing global oil depletion and transition through 2050', *Transportation Research Record: Journal of the Transportation Research Board*. No. 880. TAB. National Research Council, Washington. DC Available at: http://cta.ornl.gov/cta/Publications/Reports/RunningOutofandIntoOil_TRB1880.pdf

Greenpeace (1999) 'Ocean Disposal/Sequestration of Carbon Dioxide from Fossil Fuel Production and Use: An Overview of Rationale, Techniques and Implications'. http://archive.greenpeace.org/politics/co2/co2dump.pdf

Harrabin, R. (2008) 'Germany's clean coal pilot', BBC News Online, 8th September 2008. Available at: http://news.bbc.co.uk/1/hi/sci/tech/7603694.stm

Haszeldine, S. and Yaron, G. (eds) with Singh, T. and Sweetman, T. (2008) *Six Thousand Feet Under: Burying the Carbon Problem*, London: Policy Exchange. Available at: www.policyexchange.org.uk/images/libimages/390.pdf

Hill, R., O'Keefe, P. and Snape, C. (1995) *The Future of Energy Use*, London: Earthscan

House of Commons Science and Technology Committee (2006) *Meeting UK Energy and Climate Needs: The Role of Carbon Capture and Storage First Report of Session 2005–06*, Volume I, HC 578-I, The Stationery Office Limited, UK. Available at: www.geos.ed.ac.uk/research/subsurface/diagenesis/CCS_energy_climate_578i_report_S_T_ctte_06.pdf

Hubbert, M. K. (1971) *The Energy Resources of the Earth*, A Scientific American Book. Available at: www.hubbertpeak.com/hubbert/energypower/

IEA (2004) Prospects for CO2 Capture and Storage, OECD/IEA, Paris. Available at: www.iea.org/textbase/nppdf/free/2004/prospects.pdf

IEA (2008) The International Energy Agency, supporting the Gleneagles Plan of Action. Available at: www.g8summit.go.jp/doc/pdf/0708_06_en.pdf

IPCC (2000) *Land Use, Land-Use Change and Forestry. IPCC Special Report*, R. T. Watson, I. R. Noble, B. Bolin, N. H. Ravindranath, D. J. Verardo and D. J. Dokken (eds), Cambridge: Cambridge University Press

IPCC (2005a) *IPCC Special Report on Carbon Dioxide Capture and Storage. Prepared by Working Group III of the Intergovernmental Panel on Climate Change*, B. Metz, O. Davidson, H. C. de Coninck, M. Loos, and L. A. Meyer (eds), Cambridge and New York: Cambridge University Press. Available at: www.mnp.nl/ipcc/pages_media/SRCCS-final/SRCCS_WholeReport.pdf

IPCC (2005b) *Carbon Capture and Storage, Summary for Policymakers and Technical Summary*, B. Metz, O. Davidson, H. de Coninck, M. Loos and L. Meyer (eds), IPCC. Available at: www.mnp.nl/ipcc/pages_media/SRCCS-final/ccsspm.pdf

IPCC (2007) *Climate Change 2007: Synthesis Report*, Contribution of Working Groups I, II and III to the Fourth Assessment Report of the Intergovernmental Panel on Climate Change Core Writing Team, Pachauri, R. K. and Reisinger, A. (Eds.) IPCC, Geneva

NASA (undated) The Earth Observatory, NASA (National Aeronautics and Space Administration), The Carbon Cycle: The Human Role. Available at: http://earthobservatory.nasa.gov/Library/CarbonCycle/carbon_cycle4.html

Odell, P. R. and Rosing, K. E. (1980) *The Future of Oil*, London: Kogan Page and New York; Nichols Publishing Company

Page, S. E., Siegert, F., Rieley, J. O., Boehm, H. D. V., Jaya, A. S. and Limin, S. (2002) 'The amount of carbon released from peat and forest fires in Indonesia during 1997', *Nature*, No. 420, pp61–65

Shackley, S., McLachlan, C. and Gough, C. (2004) *The Public Perceptions of Carbon Capture and Storage*, Tyndall Centre Working Paper No. 44, Tyndall Centre for Climate Change Research, Manchester UK.

Stern, N. (2006) *The Economics of Climate Change*, Cambridge: Cambridge University Press

Streck, C. and Scholz, S. M. (2006) 'The role of forests in global climate change: Whence we come and where we go', *International Affairs*, vol. 82, no. 5, pp861–879. Available at: www.gppi.net/fileadmin/gppi/Streck_Scholz__2006__Forests_Global_Climate_Change__3_.pdf

USGS (2000) US Geological Survey World Petroleum Assessment. Available at: http://pubs.usgs.gov/dds/dds-060/

Vattenfall (undated) Vattenfall's project on CCS. Available at: www.vattenfall.com/www/co2_en/co2_en/index.jsp

WEC (2007a) Carbon Capture and Storage: a WEC 'interim balance', World Energy Council. Available at: www.worldenergy.org/documents/ccsbrochurefinal.pdf

WEC (2007b) Survey of Energy Resources, World Energy Council, London. Available at: www.worldenergy.org/documents/ser2007_final_online_version_1.pdf

World Coal Institute (undated) Carbon Capture and Storage. Available at: www.worldcoal.org/pages/content/index.asp?PageID=414

# 6

# Nuclear Energy

## Introduction

Electricity has been generated using nuclear energy since 1954. Although initially viewed as a cutting edge technology that promised to deliver virtually limitless power, it has had a very controversial history. The industry has had many problems, but Chernobyl in 1986, represents what many opponents of nuclear would view as its worst moment. Interest in nuclear energy was declining at that time through a combination of problems and public concerns compounded by competition from fossil fuels and it seemed likely that nuclear power would be consigned to the scrapheap of history. More recently nuclear power has experienced a renewed surge of interest with many countries either building new plants or planning to do so. Driving this resurgence of interest is a number of factors. These are discussed below. It should be noted that there is no single issue that is driving this resurgence, but combinations of these that are influencing the debate.

### Climate change

This is the most significant threat facing humankind. It is generally accepted that anthropogenic activities are enhancing the greenhouse effect and the principal culprit is the use of fossil fuels. Advocates of nuclear energy point out that, at point of use, nuclear power stations do not emit greenhouse gases. But nuclear power stations only produce electrical power, unlike fossil fuels, which produce a variety of power. Even though at point of production, nuclear stations do not emit greenhouse gases, it should be noted that over their whole lifecycle, that is, from fuel production, construction, decommissioning and waste storage, nuclear power does add to the greenhouse effect. The level of contribution is subject to debate, but it is less than the contribution from fossil fuels on a like-for-like basis.

### Energy security

Humankind is approaching what has been termed the end of oil. Fossil fuels have a finite life and 'peak oil', where use exceeds production, is thought to be either very near or actually happening. The developed world is highly dependent on fossil fuels and many of these resources are located in areas that could be considered politically unstable. Ensuring security of supply is a key challenge. Advocates of nuclear power argue that its use can enhance energy security, as nuclear fuel is plentiful and reprocessing strategies can be used to maximize the life of spent fuel. Others argue that the fast breeder technology can produce fuel. Taken together these factors argue that more nuclear capacity would lessen reliance on imported resources as well as diversifyng the energy mix.

## Costs

The history of nuclear power shows that costs have often been grossly underestimated and there is still some uncertainty over the real costs of the long-term storage of nuclear waste, particularly high level waste. In comparison with fossil fuels, nuclear power generation has been slightly more expensive. This has changed recently with the surge in fossil fuel prices. Advocates argue that with growing experience in nuclear build programmes and more standardization, costs will be better managed and as fuel plays a less significant role in operating costs when compared to fossil fuel generation, nuclear power does become a cost effective option.

## Safety

Advocates of nuclear power argue that safety has improved radically since Chernobyl and the growing international experience of nuclear power will further enhance the safety culture of the nuclear industry. The new generation of reactors is argued to be inherently safer than earlier designs, again enhancing safety. However, there are still issues around the disposal of nuclear waste.

In summary, although nuclear energy is a proven technology, it is still very controversial. This is likely to remain the case.

# Technology overview

A nuclear power station can be described as a conventional method for generating electricity. It works in much the same way as a coal or gas fired power station. The reactor generates heat which is used to raise steam which then drives a conventional steam turbine; quite simply it boils water. Like a coal or gas fired station the overall efficiency is governed by the laws of thermodynamics. The significant difference in the technologies is the way in which heat is generated in a nuclear reactor. Coal and gas fired stations generate heat from combustion. Combustion takes place when fuel reacts with the oxygen to produce heat. Heat or energy in a nuclear reaction is generated very differently. Nuclear energy is produced in two ways:

1. *Fission*: where the splitting of a heavy nucleus into two or more radioactive nuclei is accompanied by the emission of gamma rays, neutrons and a significant amount of energy. Fission is usually initiated by the heavy nucleus absorbing a neutron, but it also can occur spontaneously.
2. *Fusion*: this is a process in which two nuclei literally fuse together, and in doing so release considerable energy. Fusion processes power the sun.

All nuclear reactors use the fission process to generate the energy to raise steam to drive a turbine to produce electricity. Although there is considerable interest and ongoing research into a fusion reactor, it seems likely that an operational system will take many decades to finalize.

# Nuclear reactions

Atoms are made up of a nucleus and electrons. The nucleus can contain protons and neutrons. These particles are known as nucleons. Protons have a positive electric charge, neutrons do not have a charge and electrons have a negative charge. The number of protons and neutrons determine the mass of an atom. Different elements are defined by their atomic number, which is the number of protons in the nucleus. Naturally occurring elements range in mass from the lightest, hydrogen, which consists of one proton and one electron, to uranium, which has an atomic number of 92. Uranium, like other elements can have slightly different forms known as isotopes. Different isotopes of the same element have different numbers of neutrons in the nucleus. Natural uranium is largely a mixture of two isotopes: uranium-238 (U-238), which has 92 protons, 92 electrons and 143 neutrons, and uranium-235 (U-235), which has 146 neutrons. U-238 accounts for 99.3 per cent of uranium and U-235 for 0.7 per cent.

## Nuclear forces

There are four basic forces in nature, the strong nuclear force, the electromagnetic force, the weak nuclear force and gravity. The first three forces have roles in forming the nucleus of the atom. The strong nuclear force binds together the sub-atomic particles that make up the nucleons. In science a nucleon is represented by the Standard Model which comprises 16 sub-atomic particles: 12 matter particles and 3 force-carrier particles. It is the effect of the force-carrier particles acting just beyond the boundary of the nucleon that create the residual strong force that can bind together a neutron and a proton. Two protons, because of the electromagnetic force, would tend to repel each other. The neutron and the strong residual force play a key role in holding the nucleus together. As neutrons have no charge they do not add to the repulsion already present and help separate the protons. This lessens the strong repulsive force from any other nearby protons. Neutrons are a source of more of the strong residual force for the nucleus. This balance of forces helps to keep the nucleus stable. As the nucleus increases in size, the numbers of nucleons increase and the repulsive force increases. To counterbalance this, the strong forces must be increased. The number of neutrons associated with the protons therefore increases and the ratio of neutrons to protons gradually rises from 1 for a small nucleus to more than 1.5 for the heaviest nucleus. Eventually, a point is reached beyond which the nucleus becomes unstable. The bismuth nucleus, with 83 protons and 126 neutrons, is the largest stable nucleus. Nuclei with more than 83 protons are all unstable, and will eventually decay, that is break up into smaller pieces; this is known as radioactivity. Uranium, remember, has 92 protons.

## Decay

An unstable nucleus will eventually decay by emitting a particle, transforming the nucleus into another nucleus, or into a lower energy state. A chain of decay takes place until a stable nucleus is reached. There are three common types of radioactive decay: alpha, beta and gamma. The difference between them is the particle emitted by the nucleus during the decay process.

Alpha decay occurs when the nucleus has too many protons. This causes excessive repulsion and a helium nucleus is emitted to reduce the repulsion. Alpha particles do not travel far in air before being absorbed.

Beta decay occurs when the neutron to proton ratio in the nucleus is too great. In basic beta decay, a neutron is turned into a proton and an electron. The electron is then emitted. There is also positron emission when the neutron to proton ratio is too small. A proton turns into a neutron and a positron, and the positron is emitted. A positron is basically a positively charged electron. The final type of beta decay is known as electron capture and it also occurs when the neutron to proton ratio in the nucleus is too small. The nucleus captures an electron which basically turns a proton into a neutron. Beta particles have a higher penetration than alpha particles.

Gamma decay occurs because the nucleus is at too high an energy. The nucleus falls down to a lower energy state and, in the process, emits a high energy photon known as a gamma particle. Gamma emissions are very penetrating, but they can be most efficiently absorbed by a relatively thick layer of high-density material such as lead.

## Nuclear energy

When fossil fuels burn, their hydrocarbons react with the oxygen in air to produce carbon dioxide and water. This is a chemical reaction involving the outermost electrons of the atoms and the process releases energy through the rearrangement of chemical bonds. Nuclear energy results from rearrangements of the components of the nucleus. As the forces between these are very much greater than those between the outer electrons, the energy released in nuclear reactions is very much greater. Typically 100 million times more energy is released in a nuclear reaction than in a chemical reaction. It is this reaction that powers the nuclear reactor.

The nucleus is composed of protons and neutrons, but the mass of a nucleus is always less than the sum of the individual masses of the protons and neutrons which constitute it. The difference is a measure of the nuclear binding energy which holds the nucleus together. The difference in mass is known as the mass defect. When a high mass material such as uranium fissions, it creates two lower-mass and more stable nuclei while losing mass in the form of kinetic and/or radiant energy. The amount of energy released can be determined from Einstein's equation that relates energy and mass:

$$E = mc^2,$$

where $E$ is the energy, $m$ is the mass and $c$ is the velocity of light.

The mass defect of a nucleus is the difference in mass between its separated components and its own mass. If $\Delta m$ denotes the mass defect,

$$\Delta m = (A - Z)m_n + Zm_p - M$$

where $M$ is the mass of the nucleus, $m_n$ the mass of a neutron and $m_p$ the mass of a proton, expressed in mass units.

When $E = mc^2$ is used to convert the mass defect into an equivalent amount of energy, the energy is the binding energy (BE) of the nucleus and will be given by:

$$\text{BE (Joules)} = \Delta mc^2$$

$$= [(A - Z)m_n + Zm_p - M]uc^2$$

($u$ is the unified mass unit = $1.66 \times 10^{-27}$kg). Or, it can be expressed in MeV using

$$\text{BE (MeV)} = 931\ [(A - Z)m_n + Zm_p - M]$$

The binding energies for a range of nuclei have been measured. The results are shown in Figure 6.1. The curve has a flattish maximum over the range of mass numbers from about 50 to about 110. This means that these nuclei are the most stable. It also indicates the two possible ways of releasing energy from the nucleus. Previously, we discussed producing energy from nuclear fission where the process of creating stable nuclei results in mass being converted into energy. This occurs when materials that have a very high atomic mass are used. The heavier the mass, the greater the yield. The left hand side of the graph shows that the binding energy curve is much steeper for nuclei with low mass numbers. Fusing two light nuclei together to form a nucleus will release far greater levels of energy than that in the fissioning of a heavy nucleus. Although the fusion reaction has been achieved in the thermonuclear or hydrogen bomb, fusion has not yet been harnessed in a controlled reaction suitable for power generation. Progress to date will be discussed later in this chapter.

As stated earlier, the amount of energy that can be released by a nuclear reaction as opposed to the chemical reactions in combustion is greater by orders of magnitude. Using Figure 6.1 we can calculate the energy output from a fission reaction.

From the curve of the graph in Figure 6.1, a nucleus with a mass number of 220 can be seen to have nucleons with an average binding energy of 7.8MeV, while one with a mass number of 110 has an average binding energy per nucleon of 8.6MeV.

If the heavy nucleus is split so that two of the lighter nuclei are produced, the average binding energy of the nucleons increases and there is an energy release of:

$$(8.6 - 7.8) \times 220 = 176\text{MeV}$$

Obviously the exact amount of energy released, which is predominantly the kinetic energy of the two lighter nuclei, depends on the mass of the nucleus that is fissioned and the masses of the nuclei which result. A figure of 200MeV per fission is often used as a rough order of magnitude when estimating the energy released in fissioning a large mass of material. This amount of energy is equivalent to a mass loss of about 0.2 mass units per fission. Box 6.1 gives an example of estimating

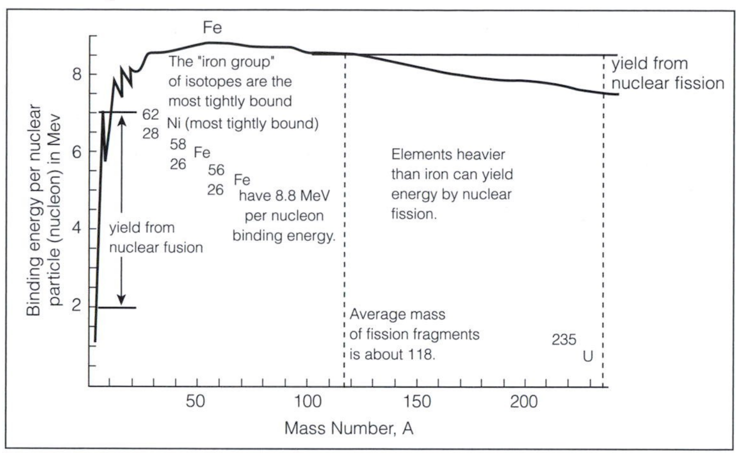


*Source:* Adapted from Hill et al., 1995

**Figure 6.1** Graph showing binding energies

## Box 6.1 Energy output of 1kg of U-235

We can estimate the energy released when 1kg of U-235 is fissioned as follows. Avogadro's number, $6 \times 10^{23}$, is the number of nuclei in 1 gram atomic weight of an element and this can be used to determine the number of nuclei in 1kg of U-235:

$(6 \times 10^{23}) \times 1000/235$

In their fission the energy released can be found by the number of nuclei divided by the amount of energy released per fission:

$(6 \times 10^{23}) \times 1000/235 \times 200$MeV,

This can be converted into Joules:

$(6 \times 10^{23}) \times 1000/235 \times 200 \times 106 \times 1.6 \times 10^{-19}$ joules,

that is:

$(6 \times 2 \times 1.6)/235 \times 10^{15} = 8.2 \times 10^{13}$ joules of energy released.

This amount of energy could be used to operate a 1000MW power station, which can convert thermal energy to electrical energy with an efficiency of 33 per cent, for:-=

$(6 \times 2 \times 1.6 \times 10^{15})/(235 \times 3 \times 1000 \times 106 \times 60 \times 60 \times 24)$days = 0.35 days

the energy released and how that relates to the operation of a 1000MW power station.

The amount of U-235 used when the power station is run continuously for a year is therefore 1158kg (i.e. approximately 1.2 tonnes). For comparison, a coal fired power station of similar output might need to burn 2.5 million tonnes of coal (Hill et al, 1995). A typical 1000 megawatt (MWe) reactor can provide enough electricity for a modern city of up to one million people (WNA, 2006a).

## Nuclear reactions for a nuclear power station

Although U-258 is an unstable atom, it requires the right conditions for it to split and release energy. When the nucleus of a U-235 atom captures a moving neutron it splits in two (fissions) and releases some energy in the form of heat and two or three additional neutrons are thrown off. If enough of these expelled neutrons cause the nuclei of other U-235 atoms to split, releasing further neutrons, a fission 'chain reaction' can be achieved. When this happens over and over again, many millions of times, a very large amount of heat is produced from a relatively small amount of uranium. The role of the neutron is important. As discussed earlier, the electromagnetic force would mean that considerable kinetic energy would be needed to propel a proton into a nucleus and cause it to split, because of the repulsive effect.

For nuclear reactions involving the neutron, the problem of repulsion does not exist as the neutron has no electric charge, it does not experience the repulsive force and can contact a nucleus more easily. Fission reactions, and indeed all of the major processes which need to be considered to understand nuclear reactors, involve the interaction of neutrons with nuclei. In essence the conditions inside a reactor need to ensure a supply of neutrons of the optimum kinetic energy for the fission of U-235 to take place at a reasonable rate. It must also be possible to maintain control of the reactor by reducing or increasing the availability of neutrons as necessary.

When a neutron impacts U-235 a number of fission products are generated as well as a number of neutrons, typically 2.4 neutrons per reaction. Provided that there are sufficient atoms of U-235 in close proximity, the neutrons emitted will interact with these atoms generating further fission products and neutrons. This is essentially the chain reaction that lies at the heart of nuclear power that produces the energy that is used to generate steam for electricity production. Chain reactions would not be achievable if, on average, one or fewer fission neutrons were produced per fission because inevitably some of them would not interact with further U-235 nuclei. The fission neutrons can take part in all types of neutron reactions. In broad terms these neutrons are involved in three kinds of process. They either leak out of the fissionable material and escape, are absorbed by it or surrounding material to react in ways other than fission, or are absorbed by U-235 and provoke further fissions. If, on average, one of the fission neutrons produces a further fission, the reaction is self-sustaining; if more than one produces a further fission the number of fissions increases with time; and if less than one produces a further fission the number of fissions gradually decreases with time. These three possibilities are described by a factor, known as k, the reaction multiplication constant. If $k > 1$ the reaction is supercritical and the number of fissions increases with time; if $k = 1$ the reaction is critical and the reaction rate is constant; and if $k < 1$ the reaction is sub-critical and the number of fissions decreases with time. To produce a working, controllable nuclear reactor it must be possible to initiate a chain reaction, allow the number of nuclei involved per second to increase gradually until the fission rate is sufficient to generate the required power level and then to maintain this rate steady. In the event of wanting to close down the reactor, for example, for refuelling or to repair a fault, it must be possible to decrease the reaction rate in a controlled fashion. Control of the reactor is achieved by controlling the number of neutrons which are present in the core.

## The reactor core

There are a number of conditions that must be satisfied if a stable nuclear reactor is to be built. These are as follows.

1 The core must contain sufficient fissionable material for a chain reaction of the required magnitude to be maintained. This depends on the amount of U-235, the configuration of the core, the moderator and the coolant (see below). Some reactors, for example, the Magnox and Candu types, use natural uranium as a fuel, others, for example, advanced gas cooled reactors (AGRs) and pressurized water reactors (PWRs), use enriched uranium as a fuel. In enriched uranium the proportion of U-235 is increased to around 3 per cent (reactor types are discussed later in this chapter). As the life of a fuel load progresses, some of the neutrons are absorbed by U-238 in reactions which produce plutonium. U-238 may also fission occasionally.

2 At least one fission neutron from each fission event must produce a further fission. This depends on the factors mentioned in 1 and on the profile, or reaction cross-section, of U-235. Because the fission neutrons can be energetic, they must be slowed down to

enhance their chance of inducing further fissions. This is achieved by a moderator. The moderator is a material where fast neutrons collide with its nuclei and lose energy in each collision and are subsequently slowed down to thermal speeds where they can react with U-235 atoms. Moderator materials must have a low mass number and low neutron absorption. Magnox and AGR reactors use a graphite moderator, PWR reactors use water and Candu reactors heavy water.

3 An uncontrolled supercritical chain reaction must not be possible. Control rods are made of materials which strongly absorb neutrons. These rods constrain the number of neutrons in the core to the required operating levels. Additional control rods are incorporated for use in emergency. The materials which are suitable for this purpose include steel with cadmium, boron or indium dispersed through it.

4 The heat which is generated must be removed to prevent the core overheating. A coolant which is compatible with the moderator must be used. In Magnox and AGR reactors carbon dioxide gas is the coolant, whereas in PWRs water is both the coolant and the moderator. The coolant is circulated through a heat exchanger where steam is produced to drive the turbines.

5 Radiation must be contained. Containment is incorporated at various levels. The fuel is sealed within cylindrical tubes of an appropriate metal such as magnox, zircalloy or stainless steel, so that the fission products are not in contact with the coolant. The core and coolant are contained in a sealed system and the whole of this system is encased in reinforced concrete which serves as a biological shield. There are numerous radiation detectors at critical parts of the system looking for leaks. Any containment must also be effective under fault conditions. Overheating and partial melting of the core due to a failure of the cooling systems is the condition most likely to lead to a serious release of radioactivity. The containment arrangements must prevent radioactivity from reaching the environment.

## Reactor technology

The important point to note is that nuclear power stations differ from conventional power stations only by using the heat from a nuclear reactor to raise steam for their steam turbines. Their steam turbines and electrical generating sets are identical with those of conventional fossil fuel fired stations.

Underpinning the approach to any reactor design is the fuel used, the moderator, the control systems and the medium used to transfer heat from the reactor. There are two basic approaches. The first uses different materials for the moderator and coolants, for example, the Magnox and Advanced Gas Cooled reactors (AGR). The second uses the same material as both moderator and coolant, for example, the Pressurized Water Reactor (PWR).

## Magnox and AGR reactors

One of the first generation of commercial nuclear stations was the Magnox reactor type, named after the magnesium alloy used to make the fuel can containing the uranium fuel. Magnox reactors used natural uranium metal as the fuel, had a graphite moderator and used pressurized $CO_2$ as the coolant. They are now obsolete. They have been succeeded in the UK by AGR (see Figure 6.2). The AGR uses enriched uranium clad stainless steel cans and a graphite moderator and pressurized $CO_2$ as the coolant. AGRs operate at a higher temperature than the Magnox reactor. The AGR is encased in a steel-lined pre-stressed concrete pressure vessel several metres thick which acts as the biological shield, with the boilers inside. The coolant conveys heat from the reactor to the boilers which, in turn, heat water in an isolated steam circuit. This steam is then used to turn the turbines, just as in coal, oil or gas fired stations.

One other type of design that uses different material for moderator and coolant is the Soviet designed RBMK (Reactor Bolshoy Moshchnosty Kanalny – a high-power channel reactor).

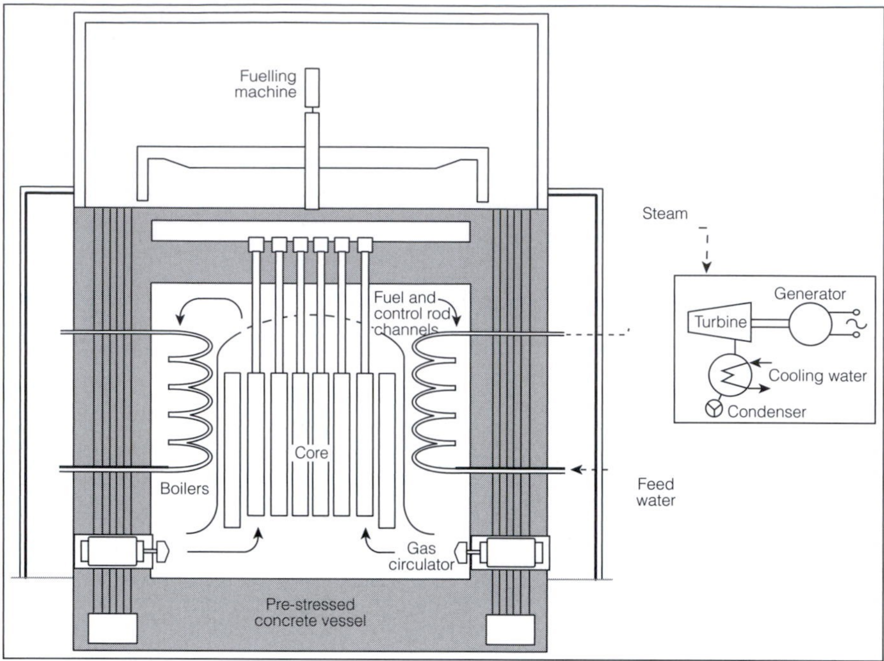


*Source:* Hill et al, 1995, p116

**Figure 6.2** AGR reactor

This is a pressurized water reactor with individual fuel channels using ordinary water as its coolant and graphite as its moderator. It is very different from most other power reactor designs as it was intended and used for both plutonium and power production. The combination of graphite moderator and water coolant is found in no other power reactors. This is the design of the Chernobyl reactor.

## Water reactors

Water reactors are the most common type of nuclear design throughout the world and it is based on the strategy that uses the same material as both moderator and coolant method. There are a number of types, falling into two main categories:

1 *Heavy Water Reactors*: the moderator and coolant is heavy water. A molecule of water contains two atoms of hydrogen and one of oxygen. Most water is comprised of hydrogen/oxygen but a small percentage has another hydrogen isotope, deuterium and oxygen. Deuterium differs from hydrogen by having one neutron in the nucleus of each atom. This is known as heavy water. The deuterium in heavy water is slightly more effective in slowing down the neutrons from the fission reactions, meaning that it can use natural uranium as the fuel. The Canadian style reactors of this type are commonly called CANDU reactors.

2 *Light Water Reactors*: There are two types of the light water reactor. The first is the boiling water reactor (BWR). In this design the water which passes over the reactor core to act as moderator and coolant is also the steam source for the turbine. The disadvantage of this is that any fuel leak might make the water radioactive and that radioactivity would reach the turbine and the rest of the loop. The sec-

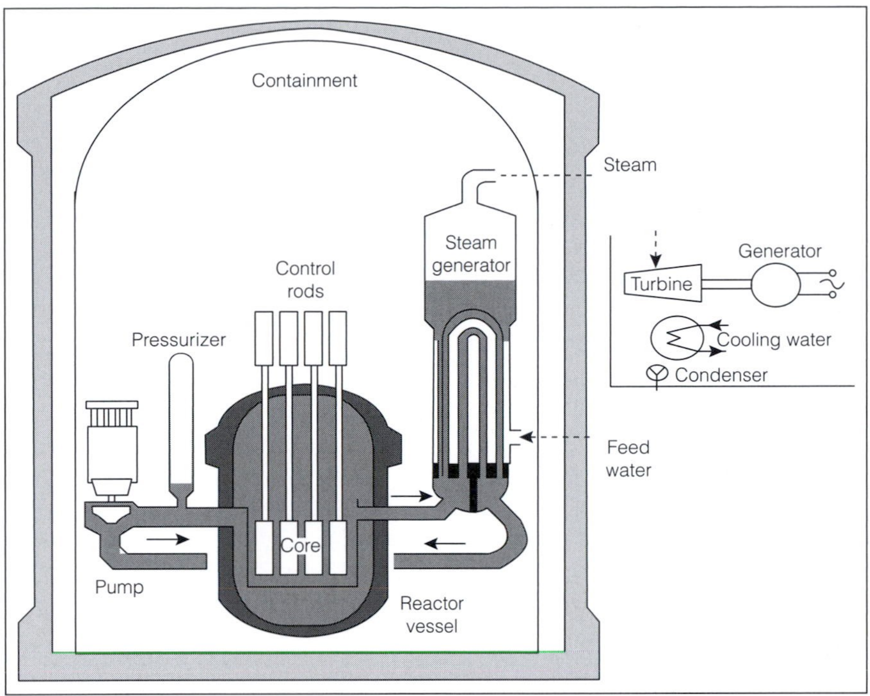


*Source:* Hill et al, 1995, p117

**Figure 6.3** The pressurized water reactor

ond is the Pressurized Water Reactor (PWR). The PWR design is based on US technology and is the most common reactor type used in the world (Figure 6.3). The reactor is contained in a steel pressure vessel. Pressurized water, which acts as both moderator and coolant, is pumped around the reactor and through the boilers. The pressure vessel, boilers and connecting pipework form a sealed primary pressurized circuit, which is contained within a steel-lined pre-stressed concrete containment building, which also acts as a biological shield. The remainder of the generation process is similar to that for other power stations.

## Fast Breeder Reactors (FBR)

Under appropriate operating conditions, the neutrons given off by fission reactions can 'breed' more fuel from otherwise non-fissionable isotopes. The most common breeding reaction is that of plutonium-239 from non-fissionable uranium-238. The term 'fast breeder' refers to the types of configurations which can actually produce more fissionable fuel than they use, such as the Liquid Metal Fast Breeder (LMFBR). This scenario is possible because the non-fissionable uranium-238 is 140 times more abundant than the fissionable U-235 and can be efficiently converted into Pu-239 by the neutrons from a fission chain reaction.

The FBR was originally conceived to extend the world's uranium resources, and could do this by a factor of about 60. When those resources were perceived to be scarce, several countries embarked upon extensive FBR development programmes. However, significant technical and materials problems were encountered and also geological exploration showed by the 1970s that scarcity was not going to be a concern for some

**Table 6.1** Nuclear power plants in commercial operation

| Reactor type | Main countries | Number | GWe | Fuel | Coolant | Moderator |
|---|---|---|---|---|---|---|
| PWR | US, France, Japan, Russia | 264 | 250.5 | enriched $UO_2$ | water | water |
| BWR | US, Japan, Sweden | 94 | 86.4 | enriched $UO_2$ | water | water |
| PHWR 'CANDU' | Canada | 43 | 23.6 | natural $UO_2$ | heavy water | heavy water |
| AGR & Magnox | UK | 18 | 10.8 | natural U (metal), enriched $UO_2$ | $CO_2$ | graphite |
| RBMK | Russia | 12 | 12.3 | enriched $UO_2$ | water | graphite |
| FBR | Japan, France, Russia | 4 | 1.0 | $PuO_2$ and $UO_2$ | liquid sodium | none |
| Other | Russia | 4 | 0.05 | enriched $UO_2$ | water | graphite |
| Total | | 439 | 384.6 | | | |

GWe = capacity in thousands of megawatts (gross)

*Source: Nuclear Engineering International Handbook*, 2007

time. As a result of these two factors, it was clear by the 1980s that FBRs would not be commercially competitive with existing light water reactors. Although there has been progress on the technical front, the economics of FBRs still depend on the value of the plutonium fuel which is bred, relative to the cost of fresh uranium. Also there is international concern over the disposal of ex-military plutonium, and there are proposals to use fast reactors for this purpose (WNA, 2008d). In both respects the technology is important to long-term considerations of world energy sustainability and there have been calls for a more concerted approach to FNR technology and the associated reprocessing issues (American Nuclear Society, 2005).

Although there is considerable research interest in this technology, it is likely that the price of uranium fuel will be a key determinant. This is rising at present, but there is no global shortage of fuel. Table 6.1 shows the distribution of reactor types in commercial use throughout the world.

## New fission technologies

Since 1996 research and development has continued into what are termed third-generation reactors. Third-generation reactors have:

- a standardized design for each type to expedite licensing, reduce capital cost and reduce construction time;
- a simpler and more rugged design, making them easier to operate and less vulnerable to operational upsets;
- higher availability and longer operating life – typically 60 years;
- reduced possibility of core melt accidents;
- resistance to serious damage that would allow radiological release from an aircraft impact;
- higher burn-up to reduce fuel use and the amount of waste;
- burnable absorbers ('poisons') to extend fuel life.

The greatest departure from second-generation designs is that many incorporate passive or inherent safety features which require no active controls or operational intervention to avoid accidents in the event of malfunction, and may rely on gravity, natural convection or resistance to high temperatures.

Many of this new generation are evolutionary designs, for example, the advanced boiling water reactor (**ABWR**) derived from a General Electric design which builds on experience from the LWRs. In Europe several designs are being developed to meet the European Utility

Requirements (EUR) of French and German utilities, which have stringent safety criteria. The European Pressurized Water Reactor (EPR) has been designed as the new standard for France and received design approval in 2004. The first EPR unit is being built at Olkiluoto in Finland, the second at Flamanville in France (ENS, 2005). However, recent reports have uncovered a series of faults in the construction of these reactors (Lean and Owen, 2008).

Table 6.2 lists the main third-generation concepts. All, except the PMBR (pebble bed modular reactor) use either heavy or light water as the moderator and coolant. The PMBR uses a graphite moderated gas cooled nuclear reactor.

In 2000 a group of international nations with a significant interest in nuclear power formed the Generation IV International Forum (GIF). The role of GIF, formally chartered in 2001, is to explore the joint development of the next generation of nuclear technology. Led by the US, its members are Argentina, Brazil, Canada, France, Japan, South Korea, South Africa, Switzerland and the UK, along with the EU. Russia and China were admitted in 2006 (GIF, 2007).

In 2002 GIF announced the selection of six reactor technologies which they believe represent the future shape of nuclear energy. These are selected on the basis of being clean, safe and cost-effective means of meeting increased energy demands on a sustainable basis, while being resistant to diversion of materials for weapons proliferation and secure from terrorist attacks. They will be the subject of further development internationally. These are shown in Table 6.3.

The goals for the fourth generation of nuclear reactors were established by GIF in its roadmap in terms of sustainability, economics, safety and reliability and proliferation resistance and physical protection. These are set out in Box 6.2.

Whether or not any of these designs will come into commercial operation remains to be seen. Issues such as public perception, costs and safety are likely to influence the debate.

## Fusion

Fusion is the combination of light atoms into a heavier atom. This reaction produces a considerable amount of energy. It also requires a considerable amount of energy to make the light atoms fuse. This has been one of the main constraints to fusion power. Fusion fuel, which comprises different isotopes of hydrogen, must be heated to extreme temperatures of over 10 million degrees Celsius, and must be kept dense enough, and confined for long enough (at least one second) to trigger the energy release. The aim of the controlled fusion research programme is to achieve 'ignition' which occurs when enough fusion reactions take place for the process to become self-sustaining, with fresh fuel then being added to continue it.

The problem is in developing a method of heating the fuel to a high enough temperature and confining it long enough so that more energy is released through fusion reactions than is used to get the reaction going. At present, two different experimental approaches are being studied: fusion energy by magnetic confinement (MFE) and fusion by inertial confinement (ICF). The first method uses strong magnetic fields to trap the hot plasma. The second involves compressing a hydrogen pellet by smashing it with strong lasers or particle beams.

Magnetic confinement operates by using magnetic fields. Fuel is heated into a plasma. A gas becomes a plasma when the addition of heat or other energy causes a significant number of atoms to release some or all of their electrons. The remaining parts of those atoms are left with a positive charge, and the detached negative electrons are free to move about. Those atoms and the resulting electrically charged gas are ionized. When enough atoms are ionized to significantly affect the electrical characteristics of the gas, it is a plasma. And it is this electrical property of the plasma that is used in magnetic confinement. The purpose is to prevent contact with any physical surfaces that could reduce the temperature of the plasma. Additional energy is required to raise the

**Table 6.2** Advanced thermal reactors being marketed

| Country and developer | Reactor | Size MWe | Design progress | Main features (improved safety in all) |
|---|---|---|---|---|
| US-Japan (GE-Hitachi, Toshiba) | ABWR | 1300 | Commercial operation in Japan since 1996–1997. In US: NRC certified 1997, FOAKE. | Evolutionary design. More efficient, less waste. Simplified construction (48 months) and operation. |
| USA (Westinghouse) | AP-600 AP-1000 (PWR) | 600 1100 | AP-600: NRC certified 1999, FOAKE. AP-1000 NRC certification 2005. | Simplified construction and operation. 3 years to build. 60-year plant life. |
| France-Germany (Areva NP) | EPR US-EPR (PWR) | 1600 | Future French standard. French design approval. Being built in Finland. US version developed. | Evolutionary design. High fuel efficiency. Low cost electricity. |
| US (GE) | ESBWR | 1550 | Developed from ABWR, under certification in US | Evolutionary design. Short construction time. |
| Japan (utilities, Mitsubishi) | APWR US-APWR EU-APWR | 1530 1700 1700 | Basic design in progress, planned for Tsuruga US design certification application 2008. | Hybrid safety features. Simplified construction and operation. |
| South Korea (KHNP, derived from Westinghouse) | APR-1400 (PWR) | 1450 | Design certification 2003, First units expected to be operating c. 2012. | Evolutionary design. Increased reliability. Simplified construction and operation. |
| Germany (Areva NP) | SWR-1000 (BWR) | 1200 | Under development, pre-certification in US | Innovative design. High fuel efficiency. |
| Russia (Gidropress) | VVER-1200 (PWR) | 1200 | Replacement for Leningrad and Novovoronezh plants | High fuel efficiency. |
| Russia (Gidropress) | V-392 (PWR) | 950–1000 | Two being built in India, bid for China in 2005. | Evolutionary design. 60-year plant life. |
| Canada (AECL) | CANDU-6 CANDU-9 | 750 925+ | Enhanced model Licensing approval 1997 | Evolutionary design. Flexible fuel requirements. C-9: Single stand-alone unit. |
| Canada (AECL) | ACR | 700 1080 | Undergoing certification in Canada | Evolutionary design. Light water cooling. Low-enriched fuel. |
| South Africa (Eskom, Westinghouse) | PBMR | 170 (module) | Prototype due to start building (Chinese 200 MWe counterpart under const.) | Modular plant, low cost. High fuel efficiency. Direct cycle gas turbine. |
| US-Russia et al (General Atomics – OKBM) | GT-MHR | 285 (module) | Under development in Russia by multinational joint venture | Modular plant, low cost. High fuel efficiency. Direct cycle gas turbine. |

*Source:* WNA, 2008e

**Table 6.3** Overview of Generation IV Systems

| Type | Neutron spectrum (fast/ thermal) | Coolant | Temp. (°C) | Pressure | Fuel | Fuel cycle | Size(s) (MWe) |
|---|---|---|---|---|---|---|---|
| Gas-cooled fast reactors | fast | helium | 850 | high | U-238 + | closed | 288 |
| Lead-cooled fast reactors | fast | Pb-Bi | 550–800 | low | U-238 + | closed | 50–150<br>300–400<br>1200 |
| Molten salt reactors | fast/ thermal | fluoride salts | 700–800 | low | UF in salt | closed | 1000 |
| Sodium-cooled fast reactors | fast | sodium | 550 | low | U-238 & MOX | closed | 150–500<br>500–1500 |
| Supercritical water-cooled reactors | thermal/ fast | water | 510–550 | very high | $UO_2$ | open/ closed | 1500 |
| Very high temperature gas reactors | thermal | helium | 1000 | high | $UO_2$ prism or pebbles | open | 250 |

*Source:* Adapted from GIF, 2007

temperature of the plasma to about 10 million degrees Celsius.

The most promising design for magnetic confinement has been the tokamak. The Joint European Torus (JET) is the largest tokamak operating in the world today. It is located at Culham in Oxfordshire, UK. Experiments have been conducted there since 1983. To date up to 16MW of fusion power for one second has been achieved in D-T plasmas using the device. JET conducts many experiments to study different heating schemes and other techniques. JET has been very successful in operating remote handling techniques in a radioactive environment to modify the interior of the device and has shown that the remote handling maintenance of fusion devices is realistic.

In 2006 an international partnership of the EU, the US, Russia, Japan, South Korea and China agreed to develop a research project to investigate nuclear fusion. The project called ITER (International Thermonuclear Experimental Reactor) is being constructed at Cadarache in France. The programme is anticipated to last for 30 years, 10 for construction, and 20 for operation, and cost approximately €10 billion. ITER is designed to produce approximately 500MW of fusion power sustained for up to 1000 seconds (compared to JET's peak of 16MW for less than one second) by the fusion of about 0.5g of a deuterium/tritium (D-T) mixture in its approximately 840m$^3$ reactor chamber. Although ITER is expected to produce (in the form of heat) 5–10 times more energy than the amount consumed to heat up the plasma to fusion temperatures, the generated heat will not be used to generate any electricity.

Inertial confinement (ICF) is a newer line of research. In this method laser or ion beams are focused very precisely onto the surface of a target, which is a sphere of D-T ice, a few millimetres in diameter. This evaporates or ionizes the outer layer of the material to form a plasma crown which expands, generating an inward-moving compression front or implosion which heats up the inner layers of material. The core or central hot spot of the fuel may be compressed to one thousand times its liquid density, and ignition occurs when the core temperature reaches about

## Box 6.2 Goals for Generation IV nuclear energy systems

**Sustainability–1** Generation IV nuclear energy systems will provide sustainable energy generation that meets clean air objectives and promotes long-term availability of systems and effective fuel utilization for worldwide energy production.
**Sustainability–2** Generation IV nuclear energy systems will minimize and manage their nuclear waste and notably reduce the long-term stewardship burden, thereby improving protection for the public health and the environment.
**Economics–1** Generation IV nuclear energy systems will have a clear lifecycle cost advantage over other energy sources.
**Economics–2** Generation IV nuclear energy systems will have a level of financial risk comparable to other energy projects.
**Safety and reliability–1** Generation IV nuclear energy systems operations will excel in safety and reliability.
**Safety and reliability–2** Generation IV nuclear energy systems will have a very low likelihood and degree of reactor core damage.
**Safety and reliability–3** Generation IV nuclear energy systems will eliminate the need for offsite emergency response.
**Proliferation resistance and physical protection–1** Generation IV nuclear energy systems will increase the assurance that they are a very unattractive and the least desirable route for diversion or theft of weapons-usable materials, and provide increased physical protection against acts of terrorism.

*Source:* GIF, 2002

100 million degrees Celsius. Thermonuclear combustion then spreads rapidly through the compressed fuel, producing several times more energy than was originally used to bombard the capsule. The time required for these reactions to occur is limited by the inertia of the fuel (hence the name), but is less than a microsecond. The aim is to produce repeated micro-explosions.

Recent work at Osaka in Japan suggests that fast ignition may be achieved at lower temperatures with a second very intense laser pulse through a millimetre-high gold cone inside the compressed fuel, and timed to coincide with the peak compression. This technique means that fuel compression is separated from hot spot generation with ignition, making the process more practical.

So far most inertial confinement work has involved lasers, although their low energy makes it unlikely that they would be used in an actual fusion reactor. The world's most powerful laser fusion facility is the NOVA at Lawrence Livermore Laboratory in the US, and declassified results show compressions to densities of up to 600 times that of the D-T liquid. Various light and heavy ion accelerator systems are also being studied, with a view to obtaining high particle densities.

In summary there are many lines of research into fusion power. However, it is impossible to predict when, if ever, a commercially viable system will be available.

## Radiation risks

Chernobyl in 1986 is the most serious nuclear accident to have occurred to date. The accident itself has been thoroughly documented. However, the long-term effects on human and environmental health are unclear. We do know that to date some 50 people have died directly as a result of radiation from the accident and at least 9 children and young people out of about 4000 cases contracted thyroid cancer as a result of the accident's contamination. An international team of more than 100 scientists concluded that up to 4000 people could eventually die of exposure to the radiation from the accident. A 2005 study of the event came to the conclusion that the health effects have been far smaller than expected

(Chernobyl Forum, 2005). It was first thought that there would be a greater loss of life.

However, others dispute the findings of the Chernobyl Forum, claiming that the figures could be higher. In response to the Chernobyl Forum, *The Other Report on Chernobyl* (TORCH) was produced on behalf of the European Greens. It predicted 30,000–60,000 excess cancer deaths, and urged more research, stating that large uncertainties made it difficult to properly assess the full scale of the disaster (Fairlie and Sumner, 2006).

In short, it is unlikely that we will ever know that real impact of Chernobyl on human health. However, the accident and the conflicting reports in its aftermath do serve to highlight the great levels of uncertainty and the fear that many have of the nuclear industry. The reality is that radiation is a natural phenomenon which occurs widely, at low levels, in nature. Soil is radioactive, houses are radioactive, people are radioactive and food is radioactive. Cereals have a comparatively high radioactive content, as do Brazil nuts; whereas milk, fruit and vegetables have a low content. Evolution has occurred in the presence of this background of radiation. In addition to the natural background, human activity has introduced more radiation, principally due to the diagnostic use of X-rays and the use of radionuclides in the treatment of cancer. Figure 6.4 shows sources of radiation.

Concern about the use of nuclear power arises from the knowledge that nuclear radiation can harm living tissue. The effects are usually considered in three categories:

1 *Genetic*: risks of radiation exposure to the reproductive organs that can be passed on to progeny.
2 *Somatic*: effects caused if radiation strikes, and damages, molecules of living matter, perhaps causing a cell to grow in an abnormal manner and eventually being manifest as cancer. Not all cancers are associated with exposure to radiation. The risk of dying from radiation induced cancer is about one half the risk of getting the cancer.
3 *In-utero*: spontaneous risks of foetal abnormalities. The risk of childhood cancer from exposure in-utero is about the same as the risk to adults exposed to radiation. By far, medical practice is the largest source of in-utero radiation exposure (USNRC, 2003).

The damage potential of radiation is a function of the level and length of the exposure. Box 6.3 discusses how radiation is measured.

Knowledge of radiation effects has been derived from groups of people who have received high doses. There is an assumption in setting radiation protection standards that any dose of radiation, no matter how small, involves a possible risk to human health. However, available scientific evidence does not indicate any cancer risk or immediate effects at doses below 100mSv a year. At low levels of exposure, the body's natural repair mechanisms seem to be adequate to repair radiation damage to cells soon after it occurs. Table 6.4 sets out some comparative radiation doses and their effects.

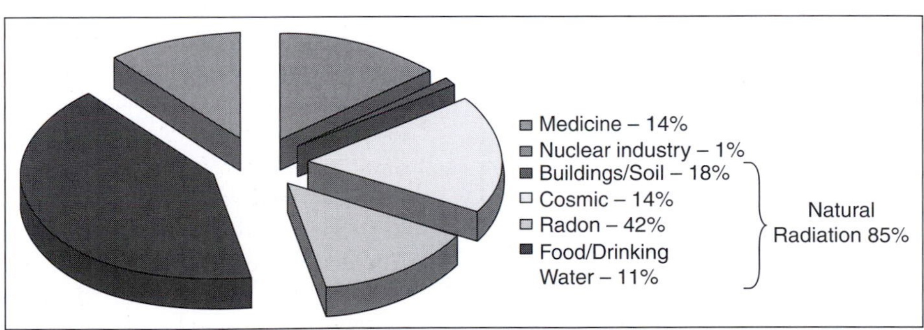


*Source:* WNA, 2007b

**Figure 6.4** Sources of radiation

## Box 6.3 Measurement units for radiation

The basic unit of radiation dose absorbed in tissue is the gray (Gy), where one gray represents the deposition of one joule of energy per kilogram of tissue.

As neutrons and alpha particles cause more damage per gray than gamma or beta radiation, another unit, the sievert (Sv) is used in setting radiological protection standards. This unit of measurement takes into account the biological effects of different types of radiation.

One gray of beta or gamma radiation has one sievert of biological effect, one gray of alpha particles has 20Sv effect and one gray of neutrons is equivalent to around 10Sv (depending on their energy). Since the sievert is a relatively large value, a dose to humans is normally measured in millisieverts – mSv – one thousandth of a sievert.

The average dose received by all of us from background radiation is around 2.4mSv/y. This can vary depending on the geology and altitude where people live, ranging between 1 and 10mSv/y. The maximum annual dose allowed for radiation workers is 20mSv/yr, although, in practice, doses are usually kept well below this level. In comparison, the average dose received by the public from nuclear power is 0.0002mSv/y and corresponds to less than 1 per cent of the total yearly dose received by the public from background radiation.

The becquerel (Bq) is a unit or measure of actual radioactivity in material (as distinct from the radiation it emits, or the human dose from that), with reference to the number of nuclear disintegrations per second (1Bq = 1 disintegration/sec.). Quantities of radioactive material are commonly estimated by measuring the amount of intrinsic radioactivity in becquerels – 1Bq of radioactive material is that amount which has an average of one disintegration per second, that is an activity of 1Bq.

Older units of radiation measurement continue in use in some literature:

1 gray = 100 rads
1 sievert = 100 rem
1 becquerel = 27 picocuries or 2.7 × 10–11 curies

One curie was originally the activity of one gram of radium-226, and represents 3.7 × 1010 disintegrations per second (Bq).

*Source:* WNA, 2007b

Even so, the radiation protection community conservatively assumes that any amount of radiation may pose some risk for causing cancer and hereditary effect, and that the risk is higher for higher radiation exposures. A linear, no-threshold (LNT) dose–response relationship is used to describe the relationship between radiation dose and the occurrence of cancer. This dose–response model suggests that any increase in dose, no matter how small, results in an incremental increase in risk. The LNT risk model is shown in Figure 6.5. The LNT hypothesis is accepted by the United States Nuclear Regulatory Commission (USNRC) as a conservative model for determining radiation dose standards, recognizing that the model may over estimate radiation risk (USNRC, 2004).

Epidemiological studies of the survivors of the atomic bombings of Hiroshima and Nagasaki show that high levels of exposure, ranging up to more than 5000mSv do increase the death rate from cancer above levels that would be normally found in any population. The assumption in the LNT hypothesis is that the linear relationship between exposure and risk is true for lower levels of exposure.

There has been extensive research into the effects of low-level radiation and to date a

**Table 6.4** Some comparative radiation doses and their effects

| Dose | Effect |
|---|---|
| 2mSv/year | Typical background radiation experienced by everyone (av 1.5 mSv in Australia, 3 mSv in North America). |
| 1.5–2.0mSv/year | Average dose to Australian uranium miners, above background and medical. |
| 2.4mSv/year | Average dose to US nuclear industry employees. |
| up to 5mSv/year | Typical incremental dose for aircrew in middle latitudes. |
| 9mSv/year | Exposure by airline crew flying the New York–Tokyo polar route. |
| 10mSv/year | Maximum actual dose to Australian uranium miners. |
| 20mSv/year | Current limit (averaged) for nuclear industry employees and uranium miners. |
| 50mSv/yea | Former routine limit for nuclear industry employees. It is also the dose rate which arises from natural background levels in several places in Iran, India and Europe. |
| 100mSv/year | Lowest level at which any increase in cancer is clearly evident. Above this, the probability of cancer occurrence (rather than the severity) increases with dose. |
| 350mSv/lifetime | Criterion for relocating people after Chernobyl accident. |
| 1000mSv/cumulative | Would probably cause a fatal cancer many years later in 5 of every 100 persons exposed to it (i.e. if the normal incidence of fatal cancer were 25%, this dose would increase it to 30%). |
| 1000mSv/single dose | Causes (temporary) radiation sickness such as nausea and decreased white blood cell count, but not death. Above this, severity of illness increases with dose. |
| 5000mSv/single dose | Would kill about half those receiving it within a month. |
| 10,000mSv/single dose | Fatal within a few weeks. |

*Source:* WNA, 2007b

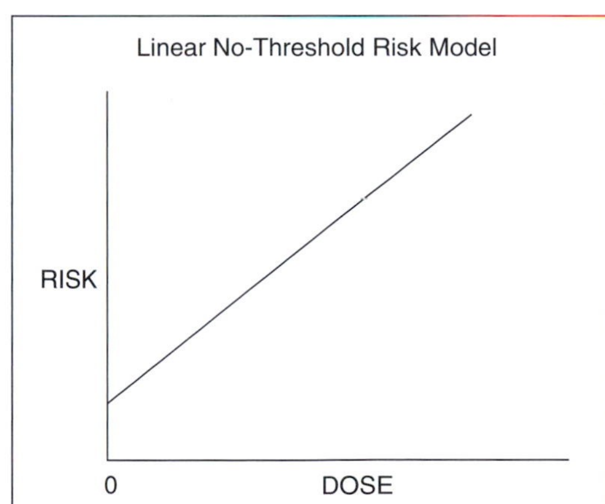


*Source:* USNRC, 2003

**Figure 6.5** Linear no-threshold risk model

number of findings have failed to support the LNT hypothesis. The causal relationship between the body and its response to low-level exposure is poorly understood. This level of uncertainty means that the conservative approach to standard setting will remain (WNA, 2007b)

## Proliferation

Nuclear power is often associated with nuclear proliferation. Nuclear proliferation is a term now used to describe the spread of nuclear weapons, fissile material, and weapons-applicable nuclear technology and information, to nations which are not recognized as 'nuclear weapon States' by the Treaty on the Nonproliferation of Nuclear Weapons, also known as the Nuclear Nonproliferation Treaty or NPT.

Globally the nuclear sector is regulated through the IAEA (International Atomic Energy Agency). The IAEA was set up by unanimous resolution of the United Nations in 1957 to help nations develop nuclear energy for peaceful purposes. Allied to this role is the administration of safeguards arrangements. This provides assurance to the international community that individual countries are honouring their treaty commitments to use nuclear materials and facilities exclusively for peaceful purposes. The IAEA undertakes regular inspections of civil nuclear facilities to verify the accuracy of documentation supplied to it. The agency checks inventories and undertakes sampling and analysis of materials. Safeguards are designed to deter diversion of nuclear material by increasing the risk of early detection. They are complemented by controls on the export of sensitive technology from countries such as the UK and the US through voluntary bodies such as the Nuclear Suppliers' Group. They are backed up by the threat of international sanctions.

Uranium processed for electricity generation cannot be used for weapons. The uranium used in power reactor fuel for electricity generation is typically enriched to about 3–4 per cent of the isotope U-235, compared with weapons-grade which is over 90 per cent U-235. For security purposes uranium is deemed to be highly-enriched when it reaches 20 per cent U-235. Few countries possess the technological knowledge or the facilities to produce weapons-grade uranium.

Plutonium is produced in the reactor core from a proportion of the uranium fuel. Plutonium contained in spent fuel elements is typically about 60–70 per cent Pu-239, compared with weapons-grade plutonium which is more than 93 per cent Pu-239. Weapons-grade plutonium is not produced in commercial power reactors but in a production reactor operated with frequent fuel changes to produce low burn-up material with a high proportion of Pu-239.

The only use for 'reactor grade' plutonium is as a nuclear fuel, after it is separated from the high-level wastes by reprocessing. It is not and has never been used for weapons, due to the relatively high rate of spontaneous fission and radiation from the heavier isotopes such as Pu-240 making any such attempted use fraught with great uncertainties.

## Status of the nuclear power industry

Nuclear power stations are sites designed specifically for producing electricity from nuclear fuels. The IAEA has a broad definition for nuclear sites or facilities that reflects a range of uses for radioactive materials. IAEA defines a nuclear facility as:

> *a facility and its associated land, buildings and equipment in which radioactive materials are produced, processed, used, handled, stored or disposed of on such a scale that consideration of safety is required.* (IAEA, 2003, 4.12)

This is a broad definition of the nuclear industry that incorporates all of the pre-generation process such as mining and enrichment, the power production phase and post-generation phases of waste management and safe disposal and storage. The nuclear power industry has had a troubled history. Initially claimed as being cheap to meter, the industry experienced successive increases in the building costs as well as time delays in construction. Chernobyl in 1986 led to questions from many about the future viability of nuclear power. A survey conducted by the IAEA in 2005 of 18,000 people in 18 countries shows the diversity of views on nuclear power (Figure 6.6).

Despite these problems, from 1975 through 2006 global nuclear electricity production increased from 326 to 2661TWh. Installed nuclear capacity rose from 72 to 369.7GW(e) due to both new construction and upgrades at existing facilities. In 2006 nuclear power supplied about 15.2 per cent of the world's electricity. In October 2007, there were 439 operating nuclear power plants (NPPs) around the world totalling 371.7GWe of installed capacity. Table 6.5 gives a complete listing of national nuclear capacity.

In addition there were also five operational units in long-term shutdown with a total net capacity of 2.8GWe. There also were 31 reactor units with a total capacity of 23.4GWe under construction.

*Source:* IAEA, 2005, p18

**Figure 6.6** Aggregate results of a global public opinion poll

The ten countries with the highest reliance on nuclear power in 2006 were: France, 78.1 per cent; Lithuania, 72.3 per cent; Slovakia, 57.2 per cent; Belgium, 54.4 per cent; Sweden, 48.0 per cent; Ukraine, 47.5 per cent; Bulgaria, 43.6 per cent; Armenia, 42.0 per cent; Slovenia 40.3 per cent; and Republic of Korea 38.6 per cent.

In North America, where 121 reactors supply 19 per cent of electricity in the US and 16 per cent in Canada, the number of operating reactors has increased in the last three years due to reconnection of two long-term shutdown reactor units in Canada (Bruce-3 in 2004 and Pickering-1 in 2005) and one in the US (Browns Ferry-1 in 2007).

In Western Europe, with 130 reactors, overall capacity has declined by 1966GWe because of the shutdown of 11 ageing reactor units. In Eastern Europe the same number of shutdowns and new grid connections (4) resulted in an unchanged number of operating units (68). In Asia, with a total of 111 reactors at present, the number of operating reactors has increased by 10 since the beginning of 2004 (WEC, 2007a).

Despite continued concern about nuclear power, for example, with the recent problems around proliferation involving North Korea and Iran and worries that nuclear materials could be used by terrorists to produce a dirty bomb (a dirty bomb is one where conventional explosives are used to scatter nuclear materials making the area effected unusable – if deployed in a city, for example, this would probably lead to it having to be abandoned) interest in expanding the nuclear sector has recently increased, primarily for two reasons.

The first is that, at the point of production, nuclear generation does not produce greenhouse gases. However, it is not carbon free over the whole lifecycle. For example, the energy used to extract, process and transport the fuel and the energy embedded in the buildings and structures needed for power stations and storage facilities, will have both used and embedded energy derived from fossil fuels. The second reason is that the availability of fuel increases energy security by diversifying the supply chain.

## The EU and nuclear power

Europe has a mature nuclear industry. Countries such as France produce some 75 per cent of electricity from nuclear reactors. However, historically there has been considerable public opposition to nuclear power, particularly in Germany and Sweden. But more recently public opinion has shifted and in those countries that have implemented nuclear phase out programmes support for nuclear power is still quite strong (Sweden 62 per cent, Germany 46 per cent, Belgium 50 per cent). Since 2005 support for nuclear power has grown with 44 per cent in favour compared to 45 per cent against. In 2005, 37 per cent were in favour and 55 per cent against. Although waste remains a concern, four out of ten would change their mind if an effective solution were found, giving a majority of 61 per cent of EU citizens in

**Table 6.5** Nuclear energy annual capacity (MWe), 2005

| Country | Capacity MWe | Units number | Per cent electricity | Country | Capacity MWe | Unit number | Per cent electricity |
|---|---|---|---|---|---|---|---|
| Argentina | 935 | 2 | 6.9 | Mexico | 1310 | 2 | 5.0 |
| Armenia | 376 | 1 | 42.7 | Netherlands | 449 | 1 | 3.9 |
| Belgium | 5801 | 7 | 55.6 | Pakistan | 425 | 2 | 2.8 |
| Brazil | 1901 | 2 | 2.5 | Romania | 655 | 1 | 9.3 |
| Bulgaria | 2722 | 4 | 44.1 | Russian Fed. | 21,743 | 31 | 15.8 |
| Canada | 12,500 | 18 | 15.0 | Slovak Rep. | 2460 | 6 | 56.1 |
| China | 6572 | 9 | 2.0 | Slovenia | 656 | 1 | 42.4 |
| Czech Rep. | 3368 | 6 | 30.5 | South Africa | 1800 | 2 | 4.9 |
| Finland | 2696 | 4 | 32.9 | Spain | 7588 | 9 | 19.6 |
| France | 63,363 | 59 | 78.3 | Sweden | 8961 | 10 | 46.6 |
| Germany | 20,303 | 17 | 31.0 | Switzerland | 3220 | 5 | 38.0 |
| Hungary | 1755 | 4 | 37.2 | Taiwan | 4904 | 6 | 17.7 |
| India | 3040 | 15 | 2.8 | Ukraine | 13,107 | 15 | 48.5 |
| Japan | 47,839 | 56 | 29.3 | UK | 12,144 | 23 | 19.0 |
| Korea, Rep. | 16,810 | 20 | 44.7 | US | 99,988 | 104 | 19.3 |
| Lithuania | 1185 | 1 | 69.6 | | | | |
| **Region** | | | | **Region** | | | |
| Africa | 1800 | 2 | | South America | 2836 | 4 | |
| North America | 113,798 | 124 | | Asia | 79,996 | 109 | |
| Europe | 172,176 | 204 | | World | 370,576 | 443 | |

*Source:* Adapted from WEC, 2007b, p250

favour of nuclear power compared to 57 per cent in 2005. The reasons for this shift lie in the more open debate that has taken place, leading people to feel better informed about nuclear issues; for example:

- 64 per cent of EU citizens believe that nuclear energy enables European countries to diversify their energy sources;
- 63 per cent believe that using more nuclear energy would help reduce Europe's dependency upon oil;
- 62 per cent agree that one of the main advantages of nuclear energy is that it produces less greenhouse gas emissions than coal and oil.

In general the survey shows that levels of awareness are rising, but, on average, EU citizens do not feel well informed about the nuclear issue and radioactive waste in particular (Eurobarometer, 2008).

The EU, in its Energy Policy for Europe, supports nuclear power. Many states have opted to extend the life of existing reactors and others, such as Finland, are developing new capacity. The UK has also decided to develop new capacity on existing sites and Italy has announced it will recommence its nuclear programme. This recent interest is driven by climate and energy security concerns. But there are issues around security associated with nuclear power plants that typically are not associated with conventional power plants. Security costs could become prohibitive. Little research has been done into the long-term terrorist threat and nuclear power stations.

Other potential problems relate to climate change. Nuclear plants (as do other types of conventional power stations) require considerable amounts of cooling water. For example, in 2003 many reactors in France were threatened by a lack of cooling water as river flows dropped. Some plants had to shut down and others were given an exceptional exemption from legal requirements to return water to the water course at a temperature that does not exceed environmental safety limits. Six nuclear reactors and a number of conventional power stations were granted these exemptions. The nuclear power plants of Saint-Alban (Isère), Golfech (Tarn-et-Garonne), Cruas (Ardèche), Nogent-sur-Seine (Aube), Tricastin (Drôme) and Bugey (Ain) continued functioning, although the upper legal limits were exceeded (UNEP, 2003).

Other long-term climate problems such as sea-level rise could be an issue for plants that are located near to the sea, for example, Sizewell B in the UK. A study by Middlesex University Flood Hazard Research Centre commissioned by Greenpeace into the nuclear plants at Bradwell, Dungeness, Hinkley Point and Sizewell found them to be very vulnerable to the threat of sea-level rise and storm surges. All four sites have been identified as candidates for replacement nuclear plants (Greenpeace, 2007).

In its latest publication 'Energy Policy for Europe' published in January 2007, the European Commission stressed that nuclear power production must be considered as an option to reduce $CO_2$ emissions and to meet the targets of the Kyoto protocol. In 2004, there were 148 nuclear power reactors in operation in the EU member states, with a total net capacity of 131 Gigawatts (GWe). France has the highest number with 58 units (63.4GWe), followed by the UK with 23 units (11.9GWe) and Germany with 18 units (20.3GWe). Nuclear power is used for electricity production in 13 of the 25 EU member states.

The majority of nuclear reactors, which comprise 107 units, are the pressurized light water type (PWR), with an absolute capacity of 103GWe, which accounts for 79 per cent of the total nuclear power in the EU. This type of reactor is used in all the EU member states apart from Lithuania where the LWGR type reactor is exclusively operated. The boiling light water reactor (BWR) has the second largest quota with 18 units and a capacity of 16.3GWe. The BWR generates approximately 12 per cent of the total nuclear power in the EU and is operated throughout Sweden, Germany, Spain and Finland. With 14 units (8.4GWe) and 8 units (2.3GWe) the advanced gas cooled (AGR) and the gas cooled (GCR) type reactors come in third and fourth position respectively. The gas cooled reactors are operated solely in the UK (WEC, 2007c).

## The economics of nuclear power

The issue of the cost of nuclear power is fraught with difficulty. This is an important issue as plants, once built, will operate for many years and will need to recoup the investment made in them. There are a number of areas to consider; these are shown in Figure 6.7 and discussed below:

1. *Capital costs*: this is the actual cost of the plant, land, infrastructure, fees, etc. and is termed the overnight cost – that is the cost at today's prices if the plant were built in one night. In reality the build time can be ten years or more. Capital costs can vary considerably, for example, if the plant is a replacement plant on an existing site, then the costs of land and infrastructure will not be included, meaning this is likely to be lower. If it is a greenfield site, the costs could be very high.
2. *Financing costs*: this is the cost associated with financing the capital costs. In effect, money would have to be borrowed to finance the project before it produces any power and starts to generate an income. This can be as high as 50 per cent or more of the project cost and will depend on the debt–equity ratio and the rate of interest on the debt.
3. *Operating costs*: these are the costs associated with operating and maintaining the plant, fuel costs, a return for investors and an element to generate funds for decommissioning.
4. *Operational waste and spent fuel management*: these are the costs associated with the safe disposal of waste materials from the operational phase of the station.
5. *Decommissioning and long-term storage*: these are the costs associated with closing the plant and storing the radioactive waste.

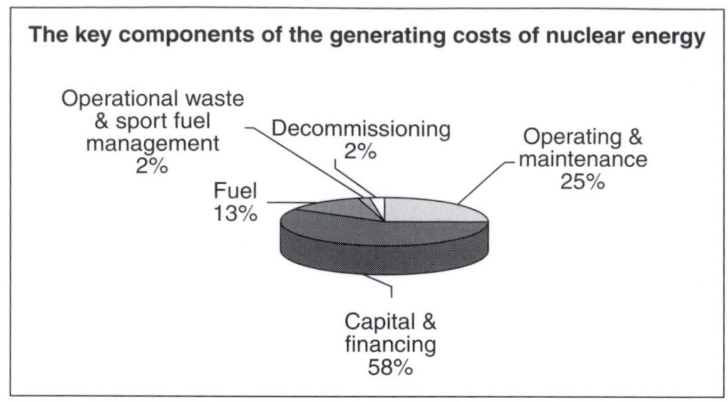


*Source:* WEC, 2008

**Figure 6.7** An overview of the key components of the generating costs of nuclear energy

Establishing a clear cost profile is important. In general, for conventional power stations the cost of fuel is important. For example, coal and gas, the predominant fuel for electricity generation, have been rising in price. Further, the concern about the greenhouse gas content of these fuels, ignored in the calculation of the cost of power produced, has led in Europe to these plants being included in the EU ETS, meaning that prices could rise further. Conventional stations do not have some of the costs that are associated with nuclear power, such as decommissioning. Comparisons are therefore not straightforward.

A 2005 OECD comparative study showed that nuclear power had increased its competitiveness over the previous seven years. The principal changes since 1998 were increased nuclear plant capacity factors and rising gas prices. The study did not factor in any costs for carbon emissions from fossil fuel generators, and focused on over 100 plants able to come on line in 2010–2015, including 13 nuclear plants. Nuclear overnight construction costs ranged from US$1000/kW in the Czech Republic to $2500/kW in Japan, and averaged $1500/kW. Coal plants were costed at $1000–1500/kW, gas plants $500–1000/kW and wind capacity $1000–1500/kW. These figures are shown in Table 6.6.

At a 5 per cent discount rate nuclear, coal and gas costs are as shown in Table 6.6 and wind is around 8 cents. Note that the discount rate is the interest rate or the amount charged to borrow the capital. Nuclear costs were highest by far in Japan. Nuclear is comfortably cheaper than coal in seven of ten countries, and cheaper than gas in all but one. At a 10 per cent discount rate nuclear was 3–5 cents/kWh (except Japan: near 7 cents, and The Netherlands), and capital becomes 70 per cent of power cost, instead of the 50 per cent with a 5 per cent discount rate. Here, nuclear is

**Table 6.6** OECD electricity generating cost projections for year 2010 at a 5 per cent discount rate

| Country | Nuclear | Coal | Gas |
|---|---|---|---|
| Finland | 2.76 | 3.64 | — |
| France | 2.54 | 3.33 | 3.92 |
| Germany | 2.86 | 3.52 | 4.90 |
| Switzerland | 2.88 | — | 4.36 |
| Netherlands | 3.58 | — | 6.04 |
| Czech Rep. | 2.30 | 2.94 | 4.97 |
| Slovakia | 3.13 | 4.78 | 5.59 |
| Romania | 3.06 | 4.55 | — |
| Japan | 4.80 | 4.95 | 5.21 |
| Korea | 2.34 | 2.16 | 4.65 |
| US | 3.01 | 2.71 | 4.67 |
| Canada | 2.60 | 3.11 | 4.00 |

US 2003 cents/kWh, discount rate 5 per cent, 40 year lifetime, 85 per cent load factor.

*Source:* OECD/IEA NEA, 2005

again cheaper than coal in 8 of 12 countries and cheaper than gas in all but 2. Among the technologies analysed for the report, the new EPR, if built, in Germany would deliver power at about 2.38c/kWh – the lowest cost of any plant in the study.

Many dispute that nuclear stations can compete with other sources. For example, a number of studies suggest that the overnight cost per KW for nuclear stations is very much higher than for other fuels. A study in the US by the Keystone Center, which was funded by several nuclear plant operators as well as other interested parties such as General Electric, estimates overnight costs of $2950/kWe (in 2007 dollars). With interest, this figure translates to between $3600/kWe and $4000/kWe (The Keystone Center, 2007). Other commentators suggest that these figures may be too low with final construction costs in real 2007 dollars in the range of $4300–4550/kWe (Harding, 2007).

In a report commissioned for the Greens-EFA Group of the European parliament cites Moody's, a US-based capital market service company, gave a low estimate for new nuclear capacity in the US at $5,000/kW and its high estimate was $6,000/kW. Some of the reasons for this are higher material and labour costs. It is likely that there will continue to be a range of uncertainty for the capital or overnight price, particularly if demand rises rapidly compared to supply, in which case it is possible the overnight prices may rise further (Schneider and Froggart, 2007).

Couple this to the uncertainties in the global financial markets, and borrowing capital to finance the projects may also be high cost. Given the high proportion of the costs, up to 70 per cent, then arguably the costs of servicing the capital of a nuclear power plant over its lifetime is the most sensitive parameter to overall costs (Postnote, 2003).

The load factor is another determinant of the cost effectiveness of different energy sources. The most effective operating mode for nuclear power stations is base load. This is where the station operates at maximum rated output. Typically load factors for nuclear stations are around 85 per cent. This recognizes that at times the station will need maintenance and re-fuelling. The load factors for wind turbines are determined by the prevailing wind conditions. Typically they can range from 20 to 30 per cent. For nuclear, long-term contracts with predictable prices offer the best scenario. But in the UK for example, the electricity market, requires much shorter contract times. This could be problematic and, for nuclear to achieve the most optimal load factor, special market conditions will be needed.

There are two further areas that need to be considered in costing nuclear power: waste storage and decommissioning.

## Nuclear waste

Radioactive wastes are waste types that contain radioactive chemical elements that have no practical purpose. Radioactive waste typically comprises a number of radioisotopes. These are unstable configurations of elements that decay, emitting ionizing radiation that can be harmful to human health and to the environment. Those isotopes emit different types and levels of radiation, which last for different periods of time.

The management and disposal of spent nuclear fuel remains a challenge for the nuclear power industry. In the nuclear cycle there are two phases where waste is produced while producing usable fuel for reactors. The first is related to the production of fuel for the reactors, this is called front end waste; the second is waste that is generated from the fission process, this is spent fuel and is called back end waste. Globally some 12,000 tonnes of spent nuclear fuel are produced every year (WEC, 2007b).

There are three categories of nuclear waste associated with nuclear power stations; low, intermediate and high. Nuclear power stations are not the only source of nuclear waste. It is generated in hospitals and industry as well as in the processing of uranium ore (uranium mill tailings). However, the costs associated with nuclear power have to be related to the overall cost of a nuclear station (IAEA, 1997).

The safe treatment of waste is a crucial factor in the nuclear industry. In 1997 the IAEA developed the Joint Convention on the Safety of Spent Fuel Management and on the Safety of Radioactive Waste Management (IAEA, 1997). This came into force for signatories in 2001. In December 2007 there were 42 signatories to the Convention. A full list is available at: www.iaea.org/Publications/Documents/Conventions/jointconv_status.pdf.

Effectively the Convention helps to establish guidelines for the safe management of materials defined to be radioactive waste by appropriate national authorities. Note that the responsibility for waste disposal is the responsibility of those nations that produce the waste. The IAEA has established criteria for defining different levels of wastes with suggested disposal options as shown in Table 6.7. Note, however, that typically classifications are determined by national governments.

### Low level waste (LLW)

This comprises paper, rags, tools, clothing, filters, etc., which contain small amounts of mostly short lived radioactivity. Commonly, LLW is designated as such as a precautionary measure if it originated from any region of an 'Active Area', which frequently includes offices where there is only a remote possibility of being contaminated with radioactive materials. Such LLW typically exhibits no higher radioactivity than one would expect from the same material disposed of in a non-active area, such as a normal office block. Some high activity LLW requires shielding during handling and transport but most LLW is suitable for shallow land burial. To reduce its volume, it is often compacted or incinerated before disposal.

### Intermediate level waste (ILW)

Intermediate-level waste (ILW) contains higher amounts of radioactivity and does require shielding in the form of lead, concrete or water. It is further categorized into short lived and long lived. The former is dealt with in a similar way to LLW and the latter to HLW.

### High level waste (HLW)

This is perhaps the most controversial area in waste disposal from nuclear plants, with no obvious solution. High level waste (HLW) is highly radioactive, contains long lived radioactivity and generates a considerable amount of heat. Storage solutions involve disposal in underground sites that have been specially prepared for that purpose.

LLW and ILW account for 90 per cent by volume of radioactive waste generated and contain about 1 per cent of the total radioactivity. HLW accounts for 10 per cent by volume of radioactive waste generated and contains about 99 per cent of the total radioactivity. This includes fission products and spent fuel. Fission products, residual waste that occurs from reprocessing, are first extracted in liquid form (after acid has dissolved them). They are then stored in stainless steel tanks that have cooling systems. The products transform into solids and are incorporated into solid blocks of borosilicate glass (also known as vitrification).

Spent fuel can be packaged in containers made of steel or concrete for shielding purposes. It must be stored underwater or in a space with a cooling system; the heat the fuel generates needs to be removed. There is usually a cooling period of 20–50 years before removal of the spent fuel from the reactor site and its long-term disposal.

## Long-term storage

Currently waste is incinerated, compacted, encapsulated or goes through the process of vitrification. Spent fuel is also reprocessed to achieve two objectives:

- the recovery of reusable materials, uranium and plutonium, which reduces the need for natural uranium extraction;
- the reduction of waste toxicity and volumes.

Reprocessing reduces waste volume by 80 per cent. A long-term option that is being looked

**Table 6.7** The IAEA's proposed waste classification scheme

| Waste classes | Typical characteristics | Disposal options |
|---|---|---|
| 1. Exempt waste (EW) | Activity levels at or below clearance levels … based on an annual dose to members of the public of less than 0.01mSv | No radiological restrictions |
| 2. Low and intermediate level waste (LILW) | Activity levels above clearance levels … and thermal power below about 2kW/m$^3$ | |
| 2.1. Short lived waste (LILW-SL) | Restricted long lived radionuclide concentrations (limitation of long lived alpha emitting radionuclides to 4000Bq/g in individual waste packages and to an overall average of 400Bq/g per waste package) | Near surface or geological disposal facility |
| 2.2. Long lived waste (LILW-LL) | Long lived radionuclide concentrations exceeding limitations for short lived waste | Geological disposal facility |
| 3. High level waste (HLW) | Thermal power above about 2kW/m$^3$ and long lived radionuclide concentrations exceeding limitations for short lived waste | Geological disposal facility |

*Source:* IAEA, 2002

into by many countries is geological disposal. This involves disposing waste in rock, clay or salt 500–1000m deep. The waste is first immobilized through the process of encapsulation or vitrification, then sealed in a canister made from stainless steel or copper (which is corrosion resistant) and finally buried in one of the three geological structures. The final disposal of HLW is regulated by governments with the support of the nuclear industry. Examples of geological repositories are Olkiluoto in Finland and Yucca Mountain in Nevada, US.

The most developed programmes for deep geological storage are those of Finland, Sweden and the US. None is likely to have a repository in operation much earlier than 2020. In France new legislation on spent-fuel management and waste disposal, which established spent-fuel reprocessing and recycling of usable materials as French policy, also established deep-geologic disposal as the reference solution for high-level long-lived radioactive waste. The legislation sets goals of applying for a licence for a reversible deep geological repository by 2015 and of opening the facility by 2025. In 2006, the UK's Committee on Radioactive Waste Management concluded that the best disposal option for the UK is deep geological disposal, with robust interim storage until a repository site is selected (WEC, 2007b).

There are many uncertainties about deep geological storage as little is understood about the impacts of storing radioactive materials for very long periods. The IAEA reports a number of research projects that have been trying to model the capacity of different storage strategies, for example, embodiment in glass and ceramics, to immobilize waste in deep geologic storage sites. The report finds that this type of investigation is far from being finished (IAEA, 2007).

Nuclear waste is the responsibility of national governments. One problem for deep geologic storage is finding geological formations that will remain stable for a long time – many thousands of years. The IAEA supports the concept of waste storage in other countries, principally because many parts of the world do not have suitable geological conditions for deep storage. A major research programme in the 1990s by Pangea Resources has identified Australia, southern Africa, Argentina and western China as having the appropriate geological credentials for a deep geological repository, with Australia being favoured on economic and political grounds. It would be located where the geology has been stable for several hundred million years, so that there need not be total reliance on a robust engineered barrier system to keep the waste securely isolated for thousands of years.

For a number of reasons Western Australia was judged the most suitable location. After consideration, the Western Australian parliament passed a Bill to make it illegal to dispose of foreign high-level waste in the state without specific parliamentary approval. Russia has passed legislation to allow the import of high-level wastes, but appears unlikely to proceed with this. The European Commission has funded studies to assess the feasibility of European regional waste repositories (WNA, 2006b).

It seems unlikely that international repositories for nuclear waste will be developed for the foreseeable future. In the interim national schemes will have to be developed and the costs associated with these will become part of the economics of nuclear energy.

## Decommissioning

The IAEA defines decommissioning as:

> *Administrative and technical actions taken to allow the removal of some or all of the regulatory controls from a facility. This does not apply to a repository or to certain nuclear facilities used for mining and milling of radioactive materials, for which closure is used.* (IAEA, 2003, 4.12)

The IAEA has defined three options for decommissioning, the definitions of which have been internationally adopted:

- Immediate Dismantling (or Early Site Release/ Decon in the US): This option allows for the facility to be removed from regulatory control relatively soon after shutdown or termination of regulated activities. Usually, the final dismantling or decontamination activities begin within a few months or years, depending on the facility. Following removal from regulatory control, the site is then available for re-use.
- Safe Enclosure (or Safestor): This option postpones the final removal of controls for a longer period, usually in the order of 40–60 years. The facility is placed into a safe storage configuration until the eventual dismantling and decontamination activities occur.
- Entombment: This option entails placing the facility into a condition that will allow the remaining on-site radioactive material to remain on-site without the requirement of ever removing it totally. This option usually involves reducing the size of the area where the radioactive material is located and then encasing the facility in a long lived structure such as concrete, that will last for a period of time to ensure the remaining radioactivity is no longer of concern. (Source: Reisenweaver and Laraia, 2000; NEI, 2007)

There is no right or wrong method for decommissioning. Table 6.8 shows some of the approaches used to decommission nuclear stations. This list is by no means definitive but what it does show is that considerable experience is being gained in the actual process of closing and dismantling old nuclear stations.

Decommissioning accounts for some 2 per cent of the overall costs of a nuclear station. Although this may be a small proportion of the overall cost, it does represent a considerable sum of money and has to be factored into the overall cost model of a nuclear station. Typically the cost of decommissioning is the responsibility of the owner or operator, although there is no single mechanism for funding decommissioning. Typical mechanisms are:

- Prepayment, where money is deposited in a separate account to cover decommissioning costs even before the plant begins operation. This may be done in a number of ways but the funds cannot be withdrawn other than for decommissioning purposes.
- External sinking fund (Nuclear Power Levy): This is built up over the years from a percentage of the electricity rates charged to consumers. Proceeds are placed in a trust fund outside the utility's control. This is the main

US system, where sufficient funds are set aside during the reactor's operating lifetime to cover the cost of decommissioning. In the US utilities collect 0.1–0.2 cents/kWh to fund decommissioning. By 2001, $23.7 billion of the total estimated cost of $35.1 billion for all US nuclear power plants had been collected.
- Surety fund, letter of credit, or insurance purchased by the utility to guarantee that decommissioning costs will be covered even if the utility defaults.

In the US, they must then report regularly to the NRC on the status of their decommissioning funds.

An OECD survey published in 2003 reported US dollar (2001) costs by reactor type. For western PWRs, most were $200–500/kWe, for VVERs costs were around $330/kWe, for BWRs $300–550/kWe, for CANDU $270–430/kWe. For gas cooled reactors the costs were much higher due to the greater amount of radioactive materials involved, reaching $2600/kWe for some UK Magnox reactors (OECD/NEA, 2003).

## Sources of nuclear fuel

Nuclear is not classified as a renewable resource as it uses a material that is mined from the Earth's surface. Nuclear fuel can also be obtained from the reprocessing of spent nuclear fuel. The raw material used to make nuclear fuel is uranium. Uranium is a slightly radioactive metal that occurs throughout the Earth's crust. It is about 500 times more abundant than gold and about as common as tin. It is present in most rocks and soils as well as in many rivers and in seawater. It is, for example, found in concentrations of about four parts per million (ppm) in granite, which makes up 60 per cent of the Earth's crust. Rock is taken from mainly open-cast mines all over the world. The ore that is taken contains around 1.5 per cent uranium. After the uranium has been mined it is milled. The uranium ore is crushed into fine slurry, which is then leached with sulphuric acid to produce concentrated $U_3O_8$, commonly known as yellowcake.

The yellowcake is first refined to produce uranium dioxide. This can be used as the fuel for those types of reactors that do not require enriched uranium, such as the CANDU and

**Table 6.8** Decommissioning strategies

| Country | Reactor types | Method | Comment |
|---|---|---|---|
| France | 3 gas-cooled reactors | Partial dismantling | Postponed final dismantling and demolition for 50 years |
| UK | 25 Magnox reactors | Extended period of care and maintenance in the Safestore phase | Ultimately they will be dismantled – Berkeley is first site to be decommissioned and other will follows same pattern |
| Spain | 1 gas-graphite reactor | Dismantle (allows much of the site to be released) and 30 years Safestor | The cost of project was €93 million. |
| Japan | 1 UK designed Magnox Reactor | Closed 1998. After 5–10 years storage, unit dismantled and the site released for other uses about 2018. | The total cost will be 93 billion yen – 35 billion for dismantling and 58 billion for waste treatment |
| US | A total of 31 reactors have been closed and decommissioned | 14 power reactors are using the Safestor approach, while 10 are using, or have used, Decon. | US experience is varied. Procedures are set by the Nuclear Regulatory Commission (NRC), and considerable experience has now been gained. |

*Source:* Adapted from WNA, 2007a

Magnox reactors. Most is then converted into uranium hexafluoride, ready for the enrichment plant. Natural uranium contains two isotopes of which a small fraction (0.7 per cent) is the fissile uranium 235 (U-235) which is capable of undergoing fission, the process by which energy is produced in a nuclear reactor. The remainder is uranium 238 (U-238). Enrichment produces a higher concentration, typically between 3.5 per cent and 5 per cent of U-235, by removing over 85 per cent of the U-238. There are two enrichment processes in large scale commercial use, each of which uses uranium hexafluoride as feed: gaseous diffusion and gas centrifuge. They both use the physical properties of molecules, specifically the 1 per cent mass difference, to separate the isotopes. The product of this stage of the nuclear fuel cycle is enriched uranium hexafluoride, which is reconverted to produce enriched uranium oxide.

The uranium oxide is processed in various ways to produce the fuel pellets for the reactors. The type of fuel varies depending on the type of reactor in which it is used. For example, the LWRs, used throughout the world, use a type of oxide fuel, uranium dioxide, in the form of pellets. The uranium dioxide fuel pellets are then stacked inside zirconium alloy fuel tubes. The tubes are grouped together to form a fuel assembly. For example, the fuel assembly for Sizewell B in the UK is made up of 264 zirconium alloy tubes, each containing about 300 pellets.

## Reprocessing

Spent fuel is about 95 per cent U-238 but it also contains about 1 per cent U-235 that has not fissioned, about 1 per cent plutonium and 3 per cent fission products, which are highly radioactive, with other transuranic elements formed in the reactor. Reprocessing separates the spent fuel into its three components: uranium, plutonium and waste, containing fission products. Reprocessing enables recycling of the uranium and plutonium into fresh fuel and produces a significantly reduced amount of waste (compared with treating all used fuel as waste). The uranium from reprocessing, which typically contains a slightly higher concentration of U-235 than occurs in nature, can be reused as fuel after conversion and enrichment, if necessary. The plutonium can be directly made into mixed oxide (MOX) fuel, in which uranium and plutonium oxides are combined. In reactors that use MOX fuel, plutonium substitutes for the U-235 in normal uranium oxide fuel (WNA, 2008a; BNFL, 2003/4).

## Nuclear resources

Climate and energy security concerns have generated fresh interest in nuclear power. In addition the very steep increases in the price of fossil fuels impacts the costs of electricity to the consumer. For example, a doubling of international prices translates into generation cost increases of about 35–45 per cent for coal fired electricity and 70–80 per cent for natural gas. In contrast, a doubling of uranium prices increases nuclear generating costs by only about 5 per cent. This interest in nuclear generation has impacted the uranium market where prices have risen significantly in recent years as shown in Figure 6.8.

The world uranium market has had to adjust rapidly to this change in expectations which has led to a position where the market has had to catch up with demand, as shown in Figure 6.9.

This has led to an expansion in exploration and mine development with expenditure on exploration increasing fourfold between 2001 and 2006. According to the World Energy Council uranium resources are plentiful and pose no constraint on future nuclear power development (WEC, 2007b). Table 6.9 shows global resources for uranium.

Table 6.9 is based on proven or reasonably assured resources and inferred resources. These are defined as:

- *Proved reserves*: correspond to the NEA category 'Reasonably Assured Resources' (RAR), and refer to recoverable uranium that occurs in known mineral deposits of delineated size,

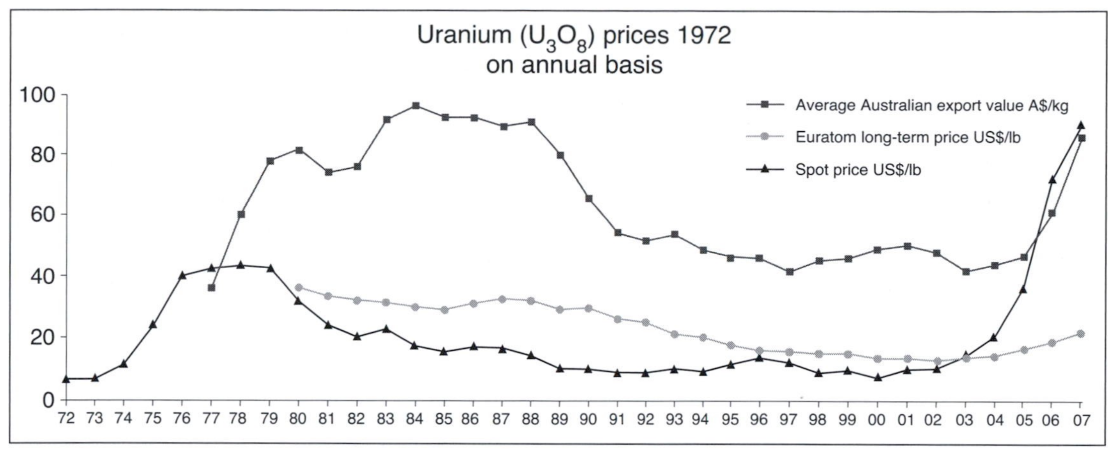


*Source:* WNA, 2008b

**Figure 6.8** Evolution of uranium prices

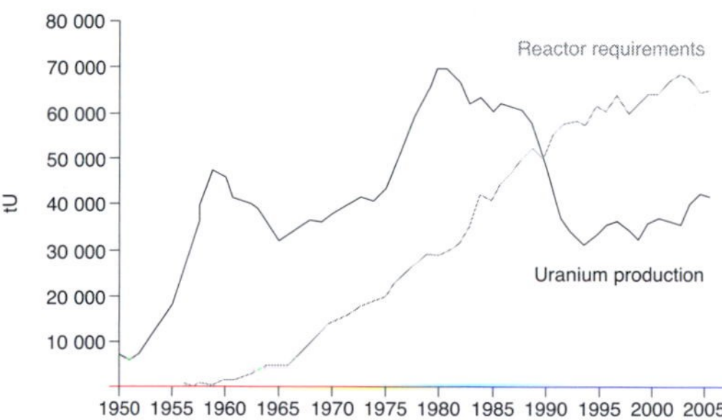


*Source:* WEC, 2007b:197

**Figure 6.9** Global annual uranium production and reactor requirements

grade and configuration such that the quantities which could be recovered within the given production cost ranges with currently proven mining and processing technology can be specified. Estimates of tonnage and grade are based on specific sample data and measurements of the deposits and on knowledge of deposit characteristics. Proved reserves have a high assurance of existence.

- *Inferred resources*: refers to recoverable uranium (in addition to proved reserves) that is inferred to occur, based on direct geological evidence, in extensions of well-explored deposits and in deposits in which geological continuity has been established, but where specific data and measurements of the deposits and knowledge of their characteristics are considered to be inadequate to classify the resource as a proven reserve (WEC, 2007b, p208).

**Table 6.9** Known recoverable resources of uranium (tonnes U, per cent of world)

| Country | Tonnes U | Per cent of world |
|---|---|---|
| Australia | 1,143,000 | 24 |
| Kazakhstan | 816,000 | 17 |
| Canada | 444,000 | 9 |
| US | 342,000 | 7 |
| South Africa | 341,000 | 7 |
| Namibia | 282,000 | 6 |
| Brazil | 279,000 | 6 |
| Niger | 225,000 | 5 |
| Russian Fed. | 172,000 | 4 |
| Uzbekistan | 116,000 | 2 |
| Ukraine | 90,000 | 2 |
| Jordan | 79,000 | 2 |
| India | 67,000 | 1 |
| China | 60,000 | 1 |
| Other | 287,000 | 6 |
| World total | 4,743,000 | |

*Note:* Reasonably Assured Resources plus Inferred Resources, to US$130/kg U, 1/1/05, from OECD, NEA and IAEA, Uranium 2005: Resources, Production and Demand.

*Source:* WNA, 2008c

Mining produces about 55 per cent of the uranium used for power generation, some 42,000 tonnes. The balance is made from stockpiles (these have run down recently), ex-military material and from reprocessing spent fuel. Annual discharges of spent fuel from the world's reactors total about 10,500 metric tonnes of heavy metal (t HM) per year. Two different management strategies are being implemented for spent nuclear fuel. In the first strategy, spent fuel is reprocessed to extract usable material (uranium and plutonium) for new mixed oxide (MOX) fuel (or stored for future reprocessing). Approximately one-third of the world's discharged spent fuel has been reprocessed. In the second strategy, spent fuel is considered as waste and is stored pending disposal.

Major commercial reprocessing plants are operating in France and the UK, with capacity of over 4000 tonnes of spent fuel per year. The product from these re-enters the fuel cycle and is fabricated into fresh MOX fuel elements. About 200 tonnes of MOX is used each year which is equivalent to about 2000 tonnes from mines. Military uranium for weapons is enriched to a higher level than civil fuel. Weapons-grade is about 97 per cent U-235. This can be diluted at approximately 25:1 with depleted uranium (or 30:1 with enriched depleted uranium) to reduce it to about 4 per cent, suitable for reactor use. From 1999 the dilution of 30 tonnes of weapons grade material displaced about 10,600 tonnes per year of mine production. The US and Russia have agreed to dispose of 34 tonnes each of military plutonium by 2014. Most of it is likely to be used as feed for MOX plants, to make about 1500 tonnes of MOX fuel which will progressively be burned in civil reactors.

## Thorium as a nuclear fuel

Today uranium is the only fuel supplied for nuclear reactors. However, thorium can also be utilized as a fuel for CANDU reactors or in reactors specially designed for this purpose. Neutron efficient reactors, such as CANDU, are capable of operating on a thorium fuel cycle, once they are started using a fissile material such as U-235 or Pu-239. Then the thorium (Th-232) atom captures a neutron in the reactor to become fissile uranium (U-233), which continues the reaction. Some advanced reactor designs are likely to be able to make use of thorium on a substantial scale.

The thorium fuel cycle has some attractive features, although it is not yet in commercial use. Thorium is reported to be about three times as abundant in the Earth's crust as uranium. The 2005 IAEA-NEA 'Red Book' gives a figure of 4.5 million tonnes of reserves and additional resources, but points out that this excludes data from much of the world (WNA, 2008c).

## Summary

Despite the controversy surrounding nuclear power there is considerable interest in developing further capacity, driven primarily by climate and energy security concerns. This is particularly true in the OECD where typically countries are resource poor in terms of fossil fuels. In the EU public opinion does appear to be shifting in favour of nuclear power, although globally there are mixed views. Nuclear technology has made progress since Chernobyl in terms of safety, but it should be recognized that with any complex technology an accident or failure is likely at some time. If sufficient safeguards and containment measures are built into nuclear systems, then the likelihood of a repeat of Chernobyl is small. But there are some difficult issues that remain. Finding a long-term and safe method for dealing with waste, either from de-commissioning or spent fuel, is problematic simply because of the time scales. Our technological experience to date does not allow thinking beyond a relatively short timescale. Although nuclear power is a reliable power source, it is likely to remain controversial. The future of fusion is uncertain.

## References


American Nuclear Society (2005) Fast Reactor Technology: A Path to Long-Term Energy Sustainability: Position Statement November 2005. Available at: www.ans.org/pi/ps/docs/ps74.pdf

BNFL (2003/4) Manufacturing Nuclear Fuel: A Briefing Note. Available at: www.bnfl.co.uk/UserFiles/File/150_1.pdf

The Chernobyl Forum (2005) *Chernobyl's Legacy: Health, Environmental and Socio-Economic Impacts*. Second revised version. Available at: http://chernobyl.undp.org/english/docs/chernobyl.pdf

ENS (2005) 'The EPR Becomes Reality at Finland's Olkiluoto 3', ENS News, Issue 10. Available at: www.euronuclear.org/e-news/e-news-10/Olkiluoto-3.htm

Eurobarometer (2008) 'Attitudes towards radioactive waste', Special Eurobarometer 297. Available at: http://ec.europa.eu/public_opinion/archives/ebs/ebs_297_en.pdf

Fairlie, I. and Sumner, D. (2006) *The Other Report on Chernobyl (TORCH)*. Available at: www.greens-efa.org/cms/topics/dokbin/118/118559.torch_executive_summary@en.pdf

GIF (2002) 'A Technology Roadmap for Generation IV Nuclear Energy Systems'. U.S. DOE Nuclear Energy Research Advisory Committee and the Generation IV International Forum. Available at: www.gen-4.org/PDFs/GenIVRoadmap.pdf

GIF ( 2007) Generation IV International Forum, Annual Report. Available at: www.gen 4.org/PDFs/annual_report2007.pdf

Greenpeace ( 2007) 'The impacts of climate change on nuclear power station sites: a review of four proposed new-buildsites on the UK coastline'. Study by Middlesex University Flood Hazard Research Centre. Available at: www.greenpeace.org.uk/files/pdfs/nuclear/8176.pdf

Harding, J. (2007) 'Economics of New Nuclear Power and Proliferation Risks in a Carbon-Constrained World', Nonproliferation Policy Education Center, US. Available at: www.npec-web.org/Essays/20070600-Harding-EconomicsNewNuclearPower.pdf

Hill, R. O'Keefe, P. and Snape, C. (1995) *The Future of Energy Use*, London, Earthscan

IAEA (1997) International Atomic Energy Agency Information Circular INFCIRC/56, 24 December 1997

IAEA (2002) Radioactive Waste Management: Status and Trends-Issue #2 (Vienna, Austria: Sept. 2002) 24. Available at: www-pub.iaea.org/MTCD/publications/PDF/rwmst2/IAEA-WMDB-ST-2-Part-1.pdf

IAEA (2003) *Radioactive Waste Management Glossary – 2003 Edition*, publication STI/PUB/1155 (2003). Available at: www-pub.iaea.org/MTCD/publications/PDF/Pub1155_web.pdf

IAEA (2005), Global Public Opinion on Nuclear Issues and the IAEA: Final Report from 18 Countries, Vienna. Available at: www.iaea.org/Publications/Reports/gponi_report2005.pdf

IAEA (2007) Spent Fuel and High Level Waste: Chemical Durability and Performance under Simulated Repository Conditions: Results of a Coordinated Research Project 1998–2004. EA-TECDOC-1563. Available at: www-pub.iaea.org/MTCD/publications/PDF/te_1563_web.pdf

The Keystone Center (2007) Nuclear Power Joint Fact Finding. Available at: www.state.nv.us/nucwaste/news2007/pdf/njff07jun.pdf

Lean, G. and Owen, J. (2008) 'Defects found in nuclear reactor the French want to build in Britain', *The Independent*, Sunday, 13 April 2008. Available at: www.independent.co.uk/news/uk/home-news/defects-found-in-nuclear-reactor-the-french-want-to-build-in-britain-808461.html

NEI (2007) 'Decommissioning of Nuclear Power Plants', NEI Factsheet. Available at: www.nei.org/filefolder/decommissioning_of_nuclear_power_plants_0807.pdf

*Nuclear Engineering International Handbook* (2007) London: Nuclear Engineering International, Progressive Media Markets

OECD/IEA NEA (2005) *Projected Costs of Generating Electricity – update*

OECD/NEA (2003) *Decommissioning Nuclear Power Plants – policies, strategies and costs*

Postnote (2003) The nuclear energy option in the UK, Parliamentary Office of Science and Technology, Crown Copyright. Available at: www.parliament.uk/documents/upload/postpn208.pdf

Reisenwaever, D. and Laraia, M. (2000) Preparing for the End of the Line – Radioactive Residues from Nuclear Decommissioning, IAEA (2000) IAEA Bulletin 42/3/2000, IAEA. Available at: www.iaea.org/Publications/Magazines/Bulletin/Bull423/42305085154.pdf

Schneider, M. and Froggatt, A. (2007) *The World Nuclear Industry Status Report 2007* (Updated to 31 December 2007). Report Commissioned by the Greens-EFA Group in the European Parliament. Available at: www.greens-efa.org/cms/topics/dokbin/206/206749.pdf

UNEP (2003) 'Impacts of summer 2003 heat wave in Europe', Environment Alert Bulletin. Available at: www.grid.unep.ch/product/publication/download/ew_heat_wave.en.pdf

USNRC (2003) *Biological Effects of Radiation*, USNRC Technical Training Center, Reactor Concepts manual. Available at: www.nrc.gov/reading-rm/basic-ref/teachers/09.pdf

USNRC (2004) 'Biological Effects of radiation', Fact Sheet, United States Nuclear Regulatory Commission. Available at: www.nrc.gov/reading-rm/doc-collections/fact-sheets/bio-effects-radiation.pdf

WEC (2007a) Performance of Generating Plant: Managing the Changes – Executive Summary. World Energy Council. Available at: www.worldenergy.org/documents/pgp_es_final_cmyk_print.pdf

WEC (2007b) Survey of Energy Resources, World Energy Council, UK. Available at: www.worldenergy.org/documents/ser2007_final_online_version_1.pdf

WEC (2007c) *The Role of Nuclear Power in Europe*, London: World Energy Council

WEC (2008) Focus: Nuclear Waste Management. Available at: www.worldenergy.org/focus/nuclear_waste_management/387.asp

WNA (2006a) What is uranium? How does it work? World Nuclear Association. Available at: www.world-nuclear.org/education/uran.htm

WNA (2006b) International Nuclear Waste Disposal Concepts, World Nuclear Association. Available at: www.world-nuclear.org/info/inf21.html

WNA (2007a) Decommissioning Nuclear Facilities. World Nuclear Association. Available at: www.world-nuclear.org/info/inf19.html

WNA (2007b) Radiation and Nuclear Energy, World Nuclear Association. Available at: www.world-nuclear.org/info/inf05.htm

WNA (2008a) The Nuclear Fuel Cycle, World Nuclear Association. Available at: www.world-nuclear.org/info/inf03.html

WNA (2008b) Uranium Markets, World Nuclear Association. Available at: www.world-nuclear.org/info/inf22.html

WNA (2008c) Supply of Uranium, World Nuclear Association. Available at: www.world-nuclear.org/info/inf75.html

WNA (2008d) Fast Neutron Reactors, World Nuclear Association. Available at: www.world-nuclear.org/info/inf98.html

WNA (2008e) Advanced Nuclear Power Reactors, World Nuclear Association. Available at: www.world-nuclear.org/info/inf08.html

# 7

# Renewable Energy Resources

## Introduction

Energy technologies based on renewable sources have seen rapid change in recent years, especially in terms of the scale of implementation and the public and commercial attitude to their use. Indeed, it is now widely accepted that renewable energy technologies have a major part to play in our future energy generation, not least because of the environmental imperative to move away from conventional fossil fuel-based energy sources. There is an immediate potential for the exploitation of renewable energy technologies such as wind and biomass and a very large potential for the use of solar and marine technologies in the next few years and into the future.

Technologies exist to address all of the energy sectors, for example, electricity production from water, wind or photovoltaics, heating and cooling from solar or biomass and fuels from biomass or renewable derived hydrogen. In terms of numerical potential, we could supply all our energy needs from renewable sources, especially if energy efficiency measures are also fully implemented. However, creating the necessary infrastructure and building enough generating plant to exploit our renewable resources fully could take many years unless governments choose to support a more rapid changeover than the market would adopt if left to its own devices. Many countries are now setting targets for the percentage of energy generation from renewable technologies. For example, in 2007, the European Union (EU) adopted a binding target of 20 per cent of energy from renewable sources on average across Europe by 2020, with each member state having an individual target related to their energy and economic circumstances (COM, 2008). This target is linked to an energy efficiency target of 20 per cent reduction in demand, since this is another important aspect of the effort to address climate change (see Chapter 4 for a discussion of energy efficiency). It is interesting to note that, across much of Europe, the electricity generation and distribution network is due for renewal and upgrading and this would provide an excellent opportunity to incorporate the features required for the introduction of a significant percentage of renewable energy generation on that network. In countries with less developed electricity distribution networks, of course, the potential exists to include significant amounts of renewable generation as the network is installed and improved.

In this chapter we will describe, briefly, those renewable energy technologies that could make a major contribution to the energy supplies of the world. The basic principles of energy conversion for each technology remain constant, although the techniques required to obtain high conversion efficiency evolve through ongoing research and development. Perhaps the most rapidly developing aspects are, however, the political and commercial attitudes and the status of the implementation. Therefore, this chapter can only give a snapshot of the renewable energy situation as it stands now, bearing in mind that substantial growth in the sector is both technically possible and environmentally important.

There are several sources of statistics on renewable energy, from national records installations to reviews carried out by organizations such as the International Energy Agency (IEA) and the Worldwatch Institute. These tend to arrive at slightly different numbers depending on the method of counting, but there is general agreement on the growth and balance between technologies. Figures 7.1 to 7.3 show data from a 2008 global status report from the REN21 Global Policy Network as an example (Martinot, 2008).

Figure 7.1 shows the contribution of renewable energy technologies to global final energy consumption in 2006. It can be seen that the majority of the contribution comes from traditional biomass (essentially the collection of firewood, animal waste, etc. for heating and cooking), with a significant contribution from large hydropower and a much smaller contribution from technologies such as wind and photovoltaics for power conversion, solar heating and biofuels. However, we would expect the contribution from these latter technologies to grow rapidly over the coming years, whilst there is less scope for an increase in generation from either traditional biomass or large hydropower due to resource issues. Some care should be taken when considering the measurement of the contribution of different sources, since the energy generated from each source must be converted to a common unit (at present, usually, million tons of oil equivalent) in order to make a comparison. The detailed numbers depend on how that conversion has been carried out and this varies between different studies and can lead to some distortion of the numbers.

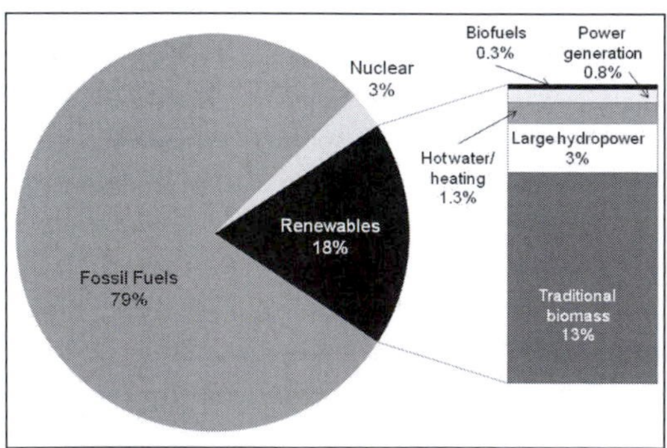


*Source:* Martinot, 2008

**Figure 7.1** Renewable energy share of global final energy consumption, 2006

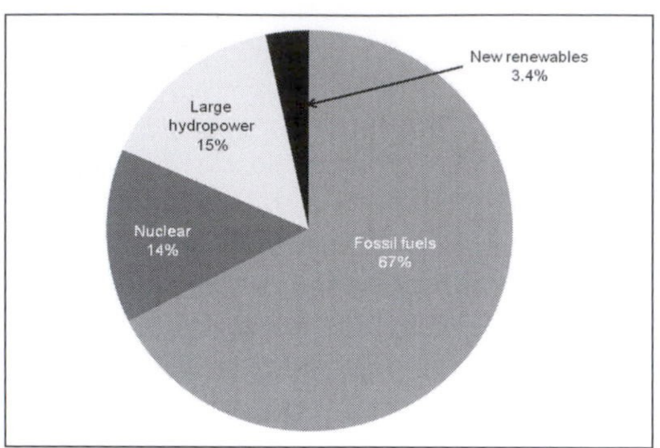


*Source:* Martinot, 2008

**Figure 7.2** Share of global electricity from renewable energy, 2006

Figure 7.2 shows a similar comparison, but now concentrating on the contribution of renewable technology to demand for electricity in 2006. Large hydropower dominates but, as before, we would expect the category of new renewables to gain an increasing share over the next few years, especially in countries where firm political targets have been set. Finally, Figure 7.3 illustrates growth in the implementation of renewable technologies since 2002 in percentage terms. It should be noted that those which show the highest average annual growth rates (photovoltaics and biofuels) start from a relatively low base, but nevertheless it can be seen that the use of all the new renewable technologies (except marine, which is not yet commercialized) is growing rapidly.

Renewable energy technologies, whilst differing in their energy conversion techniques, share some general characteristics that influence the way in which they are exploited and how they can be incorporated into the energy supply. Of course, there are specific exceptions and these will be discussed as we consider each technology

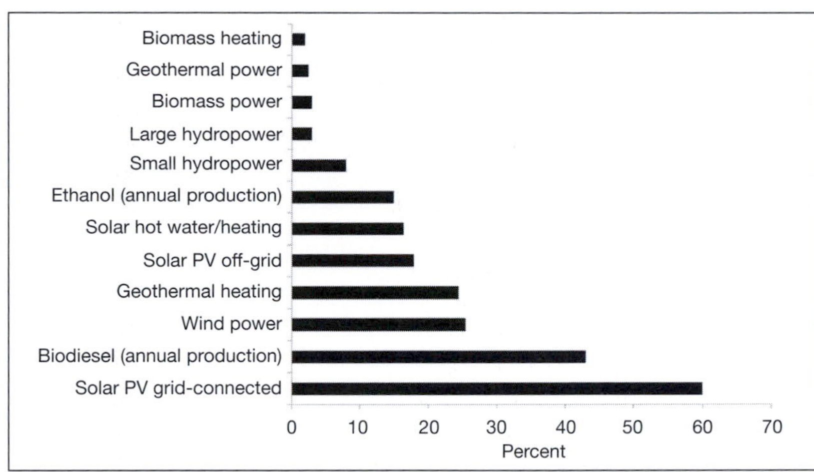


*Source:* Martinot, 2008

**Figure 7.3** Average annual growth rates of renewable energy capacity, 2002–2006

in turn. The most obvious exception is biomass, which is a stored form of energy and therefore has some characteristics in common with fossil fuels. However, in general, the output from a renewable energy generation system is climate, and therefore location, dependent and intermittent in nature. Whilst having relatively low running costs (due to no fuel costs), the major expenditure is the initial capital cost. Both these aspects are in contrast to the conventional fossil fuel-based energy system and therefore some changes are required in both the technical and financial treatment of energy generation.

Figure 7.4 is a flow diagram of the energy balance of the Earth. The major sources are the sun and tides, and the energy from nuclear, thermal and gravitational forces within the Earth itself. More than 99 per cent of incoming energy is solar radiation. Tidal energy and geothermal energy inputs are much lower than solar.

Almost all renewable resources derive their power, directly or indirectly, from the sun, so we will first consider solar radiation and the technologies for converting it into socially useful forms of energy. Solar radiation is the light and heat received by the Earth from the sun. The sun emits radiation because its surface is hot – just as an electric fire emits light and heat when the element is hot. Radiant energy is emitted in a range of wavelengths (corresponding to colours in the visible spectrum) that depends on the temperature of the radiating object. The sun is very large and very hot (see Table 7.1 for details), and so it emits enormous amounts of energy in the visible spectrum.

Only a tiny fraction of the sun's energy (about two parts in one billion) arrives on Earth; most misses the Earth and other planets and disappears into space. Of the energy received by the Earth, nearly one-third is reflected back into space by the clouds, ice and the oceans, etc. The other two-thirds keeps the Earth warm, drives the weather, makes crops grow and powers most of the world's natural processes.

The average temperature of the Earth's surface, night and day throughout the year, remains remarkably constant, albeit with some recent concerns about rising trends in that average. If the temperature were to drop by only a few degrees centigrade, then we should have a new ice age, and if it were to rise by a similar amount then the polar ice caps would melt and large areas of the world would be flooded. This is one of the major concerns of the current warming trend. To keep the temperature at its present average, the Earth must radiate into space as much energy as it

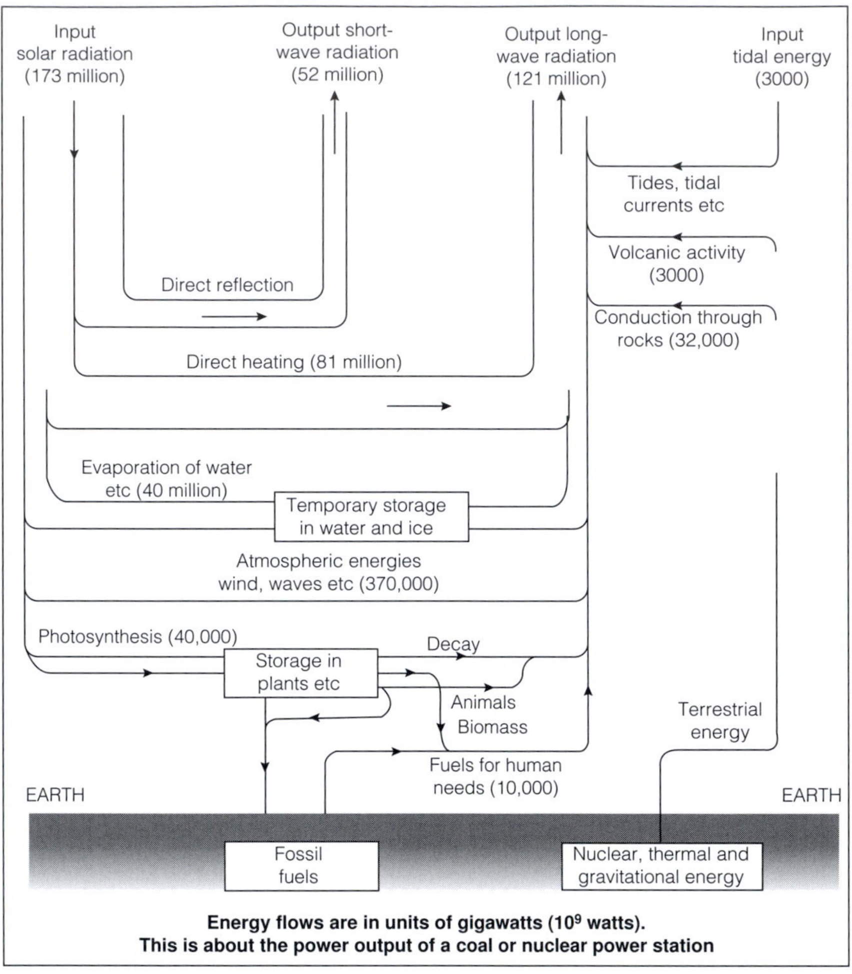


*Source:* Adapted from Hill et al, 1995

**Figure 7.4** Energy flows of planet Earth (units: gigawatts)

gets from the sun. During the day the Earth gets more energy from the sun than it can radiate and thus the temperature rises, but at night it radiates more than it gets and so the temperature falls again. As clouds can reflect some of this energy back to the Earth, cloudy nights are not as cold as clear nights. The concern over global warming is that so-called greenhouse gases trapped in the atmosphere reduce the overall radiation to space and thus the average temperature of the Earth increases, with effects on crops, sea level, etc.

Solar radiation reaching the Earth varies in a complex way, but one which has some highly predictable aspects. In any given location, it is at its maximum during the day but is zero at night and it is generally less in winter than in summer. Winter and summer seasons are defined in relation to the position of the sun in the sky, which is highest in the northern hemisphere around the summer solstice (21 June or close to this date) and lowest around the winter solstice (21 December or close to this date). In the southern hemisphere, the opposite is true, with high solar elevation in December and low solar elevation in June, hence the seasons are opposite. Day lengths are also reduced in the winter and increased in

**Table 7.1** Selected statistics of the sun

| Distance from Earth | 150,000,000km (93,000,000mi) | Diameter | 1,392,000km (864,000mi) |
|---|---|---|---|
| Rotation period | Equator 26 d<br>Poles 38 d | Angular diameter | 32 minutes of arc |
| Composition (per cent mass) | | Temperature | |
| Hydrogen | ~ 75 | Surface | 5700°C |
| Helium | ~ 23 | Central | 16,000,000°C |
| Oxygen | 1 | Sunspots | 4200°C |
| Carbon | 0.4 | Corona | 1,000,000°C |
| Iron | 0.16 | Energy source – fusion of hydrogen nuclei into helium | $4H \rightarrow HE + 2e^{+} + 2v + y$ |
| Silicon | 0.1 | | |
| Nitrogen | 0.1 | | |
| Magnesium | 0.09 | | |
| Neon | 0.07 | | |
| Other elements | Traces | | |
| Mass | $2 \times 10^{33}$g ($10^{27}$t) | Power output | $3.8 \times 10^{2}$MW |
| Density | 1.41g/cm$^3$ | Solar constant | 1.353kW/m$^2$ |
| Average | 150g/cm$^3$ | Rate of mass loss through conversion to energy | 4,500.00t/sec |
| Central | (13 × density of lead) | | |
| Surface | $10^{-7}$g/cm$^3$<br>(0.0001 × density of air) | | |
| Life expectancy | 10,000,000,000 y | | |
| Present age | 5,000,000,000 y | | |

the summer (for anywhere other than directly on the equator). When the sun is lower in the sky, the solar radiation has to travel further through the atmosphere and so the power (energy/second) is reduced. Sunshine also varies from day to day because of cloud cover and variation of absorption and scattering within the atmosphere (due to water vapour, pollutants or other particulates). Thus, we can predict the position of the sun at any time at any location with some accuracy but the climatic conditions then cause variation in the amount of sunlight received. If the best use is to be made of solar energy we need to know the average amount available and how variable it is and this is discussed in the next section.

## Solar radiation

Solar radiation (sometimes called irradiation or insolation) is a measure of the energy received on a specified surface over a specified period, usually 1m$^2$ of horizontal surface in one day. It differs in different places and varies with the seasons. Figure 7.5 shows the annual average daily irradiance in Europe. It can be seen that the values only differ by a factor of about 2.5 between southern Europe (around 5kWh/m$^2$ per day) and northern Europe (around 2.2kWh/m$^2$ per day). The locations with the highest daily averages tend to be in the desert regions of North Africa and Australia where levels between 6 and 7kWh/day can be found (Figure 7.6). Clearly, there are some seasonal variations and these tend to be higher in higher latitudes because of the greater variation in the length of the day. Table 7.2 gives some examples of annual and seasonal variation in daily irradiation values for some capital cities around the world. Note that, although the total irradiation generally increases as latitude decreases, it is also influenced by climatic variations. Thus, Johannesburg gets more sunlight than Rio de

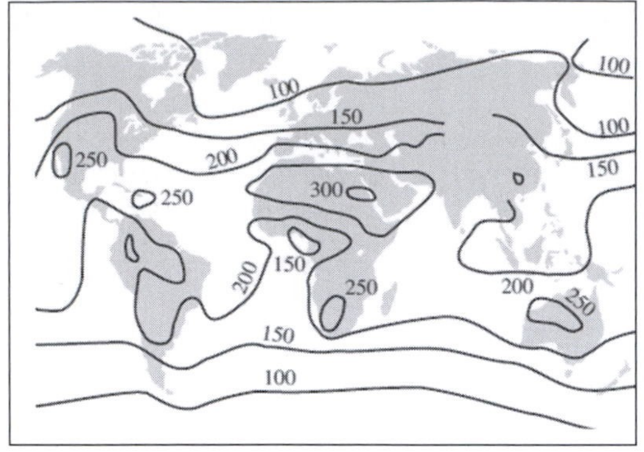


*Source:* Adapted from Hill et al, 1995

**Figure 7.5** Average annual irradiance in Europe (kWh m$^{-2}$/day)

Janeiro despite being at a higher latitude. It is also common for locations around the equator to get reduced solar radiation in certain seasons because of, for example, heavy rains. Also, the seasonal variation tends to increase with latitude, with the difference in June and December values being around a factor of 10 in London compared to around a factor of 2 in Tokyo.

For locations at latitudes between the Tropic of Cancer (23.45°N) and the Tropic of Capricorn (23.45°S), the sun is directly overhead at noon, twice during the year. The tropics are so named because that is where the sun appears to turn and move in the opposite direction at the solstices. For all locations at higher latitudes, the sun is never directly overhead. Therefore, in almost all cases, sunlight strikes a horizontal surface at an angle. The power density (watts/m$^2$) received by this surface is always less than the density of the sunlight on a plane normal to the direction to the sun (see the account of the laws of radiation, below). Therefore, it is advisable to mount solar collectors at an angle to the horizontal which maximizes the amount of irradiation falling on them. The best angle varies from equal to the latitude angle to around (latitude – 15 degrees) as the latitude of the site increases. So, in the UK, we would usually mount solar collectors at an angle of about 40° to the horizontal to get the maximum irradiation over the year. Of course,

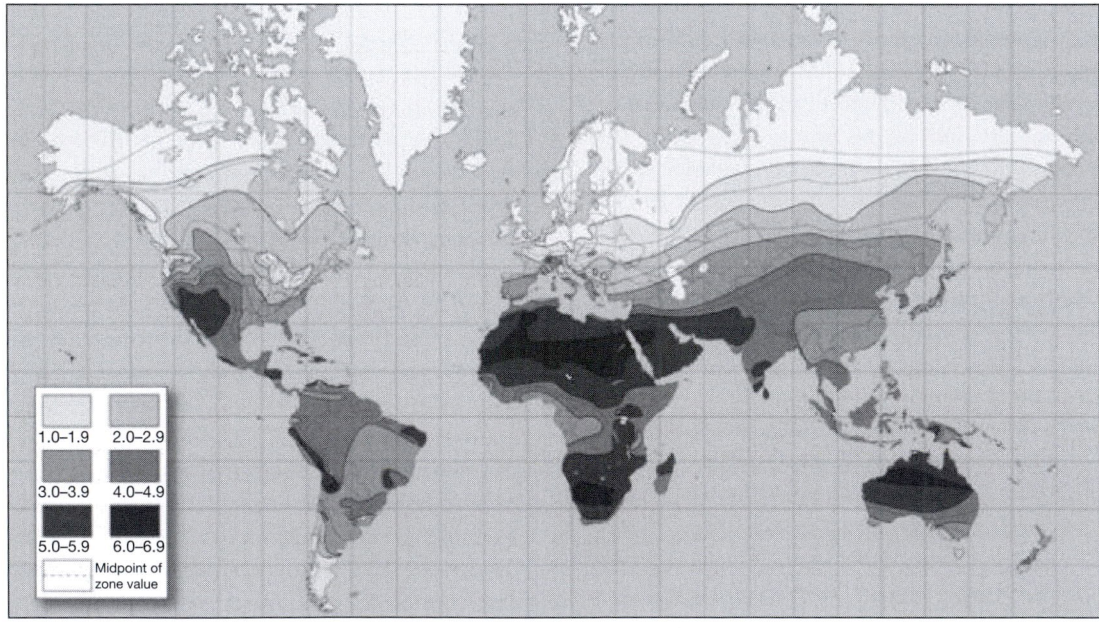


*Source:* OkSolar.com

**Figure 7.6** Annual average irradiance for the world (kWh m$^{-2}$/day)

**Table 7.2** Average daily irradiation values for various cities around the world for a horizontal surface, in kWh/m$^2$

| City, country and latitude | Annual average daily irradiation | Average daily irradiation, March | Average daily irradiation, June | Average daily irradiation, September | Average daily irradiation, December |
|---|---|---|---|---|---|
| London, UK 51.4°N | 2.62 | 2.26 | 4.87 | 2.93 | 0.48 |
| Madrid, Spain 40.4°N | 4.55 | 4.55 | 7.43 | 5.00 | 1.58 |
| Washington DC, US 39.1°N | 4.07 | 3.90 | 6.20 | 4.43 | 1.77 |
| Moscow, Russia 55.6°N | 2.65 | 2.48 | 5.20 | 2.37 | 0.32 |
| Tokyo, Japan 35.3°N | 3.49 | 3.71 | 4.20 | 3.23 | 2.23 |
| Sydney, Australia 33.5°S | 4.42 | 4.22 | 2.33 | 4.63 | 6.12 |
| Beijing, China 39.5°N | 3.68 | 3.71 | 5.47 | 3.90 | 1.81 |
| Rio de Janeiro, Brazil 22.5°S | 4.63 | 5.22 | 3.20 | 4.23 | 5.61 |
| Johannesburg, South Africa 27.5°S | 5.68 | 5.90 | 3.97 | 6.13 | 7.03 |

*Source:* Meteonorm v4.0 (solar data software)

if we choose to use a higher mounting angle (e.g. the vertical façade of a building), then we can reduce the variability of the solar radiation received for higher latitude locations.

## Variability of solar irradiation

We are all aware that solar irradiation varies from season to season and this needs to be taken into account when considering how best to design a system to harness it. The data shown in irradiation maps, including those in Figure 7.5, are for all days in the year averaged over many years. But the energy received in any given month in any given year will be different from that received in the same month in any other year and it is important to remember this when dealing with the prediction of outputs from renewable energy systems. So it is usual to take the maximum and minimum radiation expected in 1 year in 10 as the extremes. As an example, Figures 7.7 (a) and (b) show these maximum and minimum rates of irradiation for June and Figures 7.7 (c) and (d) for December, both for the UK. In June, the range runs from 8kWh/m$^2$ to 2kWh/m$^2$ and in December that range is reduced by a factor of about ten. However, maxima and minima taken from 1 year in 10 will conceal the greater variations that would be experienced by someone living for, say, 70 years.

To deal with these greater variations it is necessary to consider the number of occasions in 100 years when the average energy received will be less than a certain value. This is known as the 'cumulative frequency distribution' and, in Figure 7.8, it is given for the months of June and December. Not only is the average energy much lower in December than in June, but the variability is much greater. However, it is important to remember that months vary from fortnight to fortnight, week to week, even day to day. It is

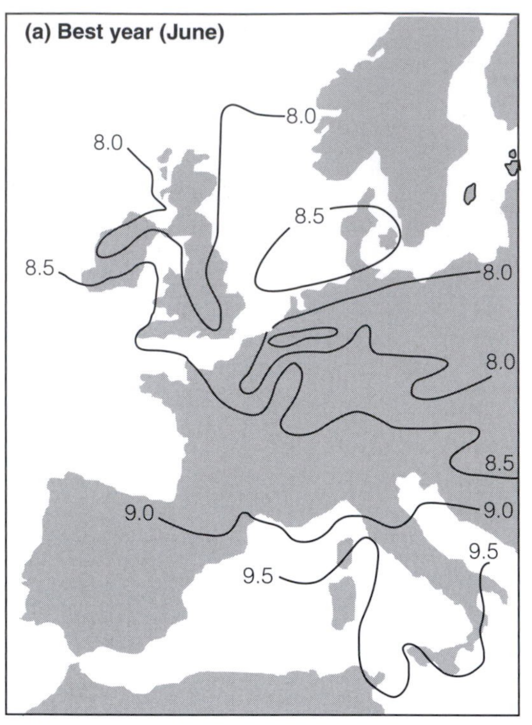


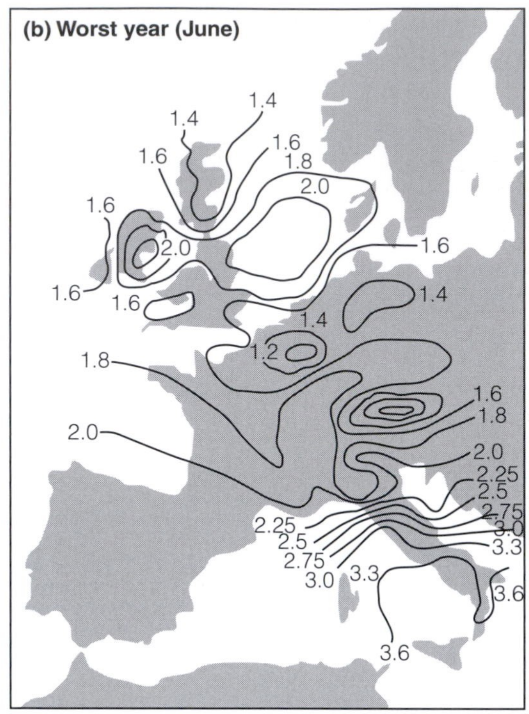


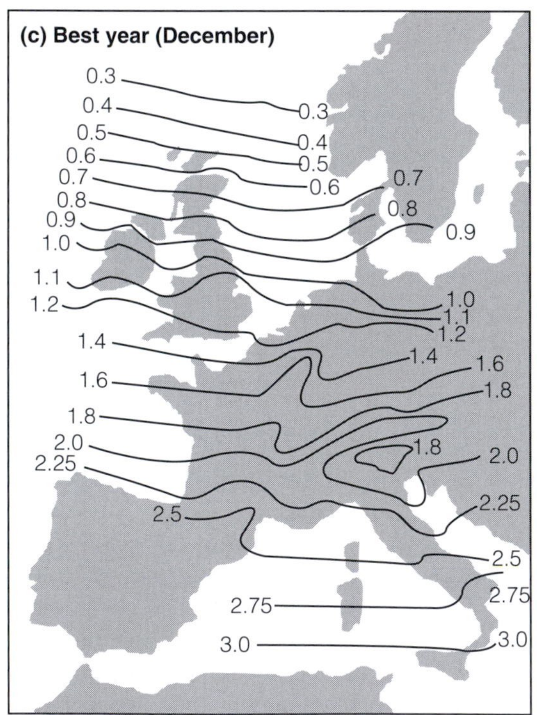


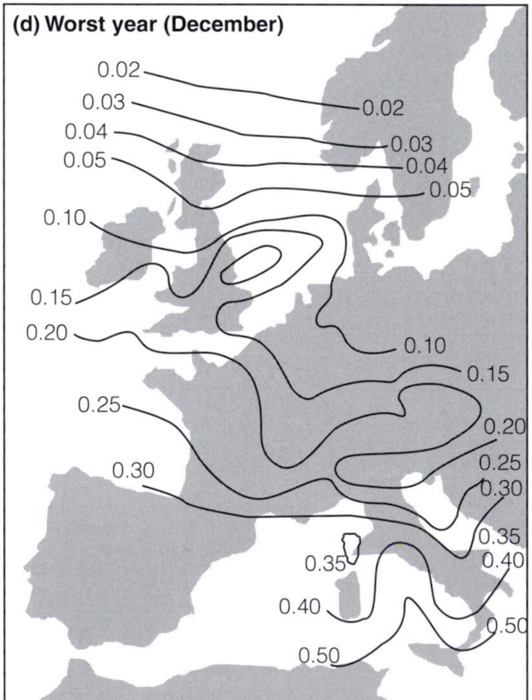


*Source:* Adapted from Hill et al, 1995

**Figure 7.7** Solar irradiation for the best/worst year in ten (kWh m$^{-2}$/day)

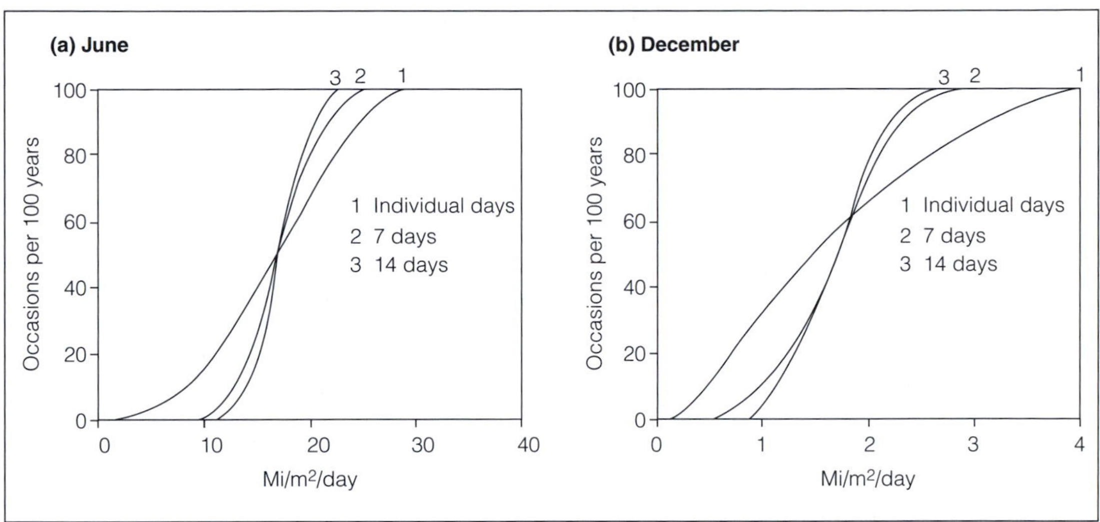


**Figure 7.8** Solar irradiation probability distribution

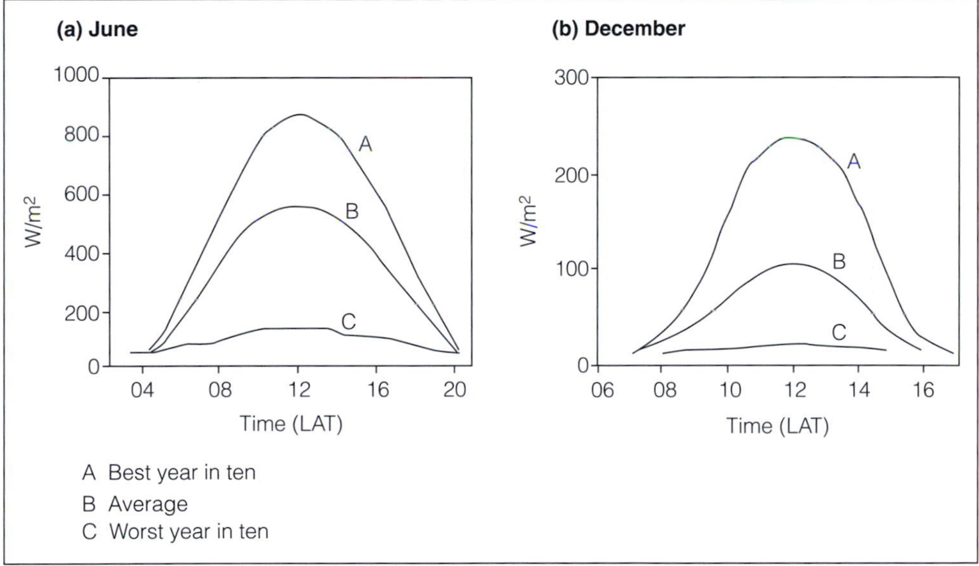


**Figure 7.9** Variation of solar incidence with time

also possible to plot a century long, cumulative frequency distribution for fortnights and weeks, which could be a useful guide to the chances of a sunny or a wet summer holiday. Solar energy varies in intensity during the day and these variations, measured by the amount of energy falling on a horizontal surface of $1m^2$, is shown in Figure 7.9 for daytimes in June and December. The graphs give averages for the best and worst days in every 50 and the absolute average.

## Spectral dependence of solar irradiation

Light, from the sun, ranges in colour from ultraviolet, through the visible spectrum to infrared. Extreme ultraviolet burns the skin and can cause skin cancer, near ultraviolet tans the skin and infrared simply makes us feel hot. The energy in sunlight is at its strongest around the yellow part of the visible spectrum, the part to which the eyes of living creatures are most sensitive. Outside the Earth's atmosphere the spectral distribution of sunlight is very like that from a black body with a surface temperature of 5800° Kelvin (see 'Laws of radiation' below). The carbon dioxide and water vapour in the Earth's atmosphere absorb some wavelengths of sunlight more than others. The distance through the atmosphere that sunlight must travel is determined by the height of the sun in the sky. If it is directly overhead then the light travels vertically through the atmosphere. If the sun's elevation is less than 90°, the light travels through a longer path, so more absorption occurs. The length of this path is described by the Air Mass Number (AM), which is defined as the secant of the zenith angle (the angle between the vertical and the line joining the observer and the sun). The sunlight in space is termed AM0, because the light does not travel through any of the atmosphere. For describing the output of a photovoltaic module, the standard spectrum is defined as AM1.5 global, which is typical of good sunlight conditions at moderate latitudes (e.g. southern US, northern Africa). The AM0 and AM1.5 standard spectra are compared in Figure 7.10 and the effect of water vapour absorption can be seen in the AM1.5 spectrum at wavelengths around 940, 1130, 1380 and 1850nm.

Much of the sunlight that we receive, especially at higher latitudes like the UK, is not direct but diffused, scattered by water droplets and dust particles in the atmosphere. Scattered sunlight is less intense than direct sunlight and also has more irradiation in the blue part of the spectrum. This can be seen on a clear day because blue light is scattered eight times more efficiently than red and so makes the sky look blue. On the other hand when the sun is low in the sky, either at dawn or at dusk, it looks very red because the blue light has been completely scattered. Most solar energy equipment responds well to both direct and diffuse radiation, and so will still work in cloudy conditions.

## Laws of radiation

### *Kirchoff's Laws*

An object that is perfectly black absorbs all radiation that falls on it, so no real object can absorb more radiation than a 'black body' of the same size. The absorptance ($\alpha$) is the ratio of the radia-

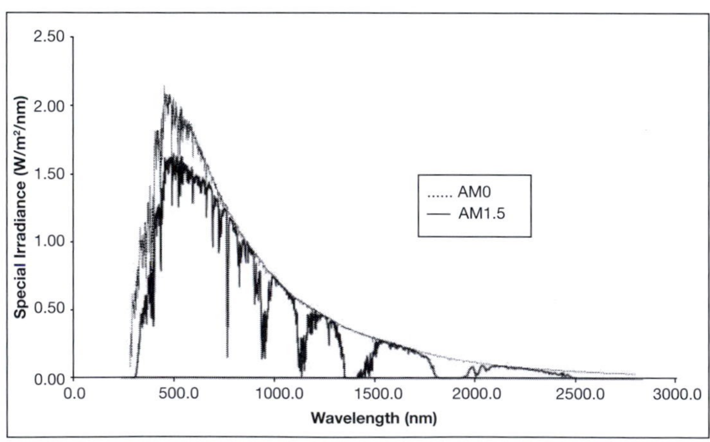


**Figure 7.10** Solar spectrum for space (AM0) and terrestrial (AM1.5) irradiation

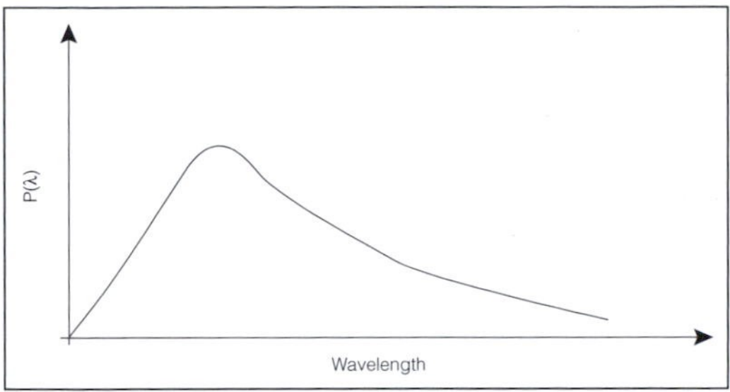


**Figure 7.11** Power output per unit wavelength versus wavelength for black body

tion absorbed by a surface to the radiation falling on the surface, and for a black body $\alpha = 1$. For any other object $\alpha$ is between 0 and 1.

Kirchoff also showed that no real object can emit more radiation than a similar black body at the same temperature. The emittance ($\varepsilon$) of a surface is the ratio of the radiation emitted from the surface and that emitted by a similar black body at the same temperature. For a black body $\varepsilon = 1$, while for any other object $\varepsilon$ is between 0 and 1. For radiation of any given wavelength ($\lambda$), the absorptance of a surface for that monochromatic radiation is equal to its emittance. This result is true for all surfaces when the emittance and absorptance are measured at the same surface temperature.

## *Planck's Radiation Law*

The power per unit wavelength radiated by the unit area of a black body at temperature T (K) is given by:

$$P = C_1 / (\lambda^5 [\exp (C_2/T) - 1])$$

where $C_1 = 3.74 \times 10^{-16}$ Wm$^2$ and $C_2 = 0.0144$ mK

If we plot the power density as a function of wavelength, we see a very characteristic curve (Figure 7.11). In deriving his radiation law, Planck had to make the assumption that radiant energy occurred in tiny discrete chunks called 'quanta'. He thus founded quantum theory, which underlies all of modern electronic technology.

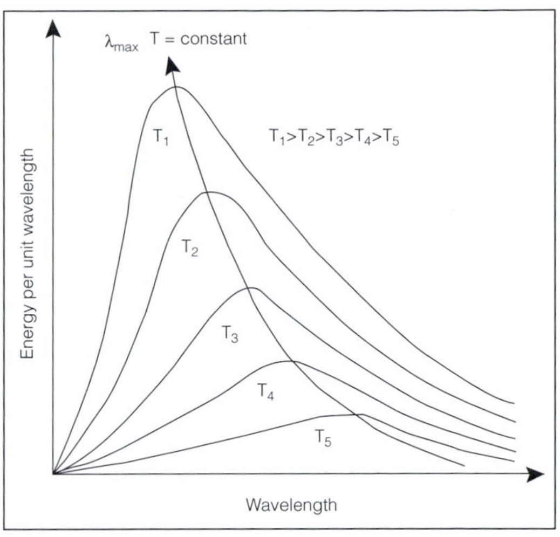


**Figure 7.12** Variation in black body radiation with temperature

## *Wien's Law*

Wien discovered a relationship between the temperature of a surface and the wavelength ($\lambda_{max}$) (Figure 7.12) at which the power per unit wavelength is a maximum. Wien's Law states that:

$$\lambda_{max} T = \text{constant} = 2898 \times 10^{-6} \text{ mK}$$

For the sun, where T = 5800K, $\lambda_{max}$ = 0.5 μm yellow light

For the Earth, where T~280K, $\lambda_{max}$ = 10 μm infrared light

The infrared radiation which the Earth radiates into space can be absorbed by carbon dioxide and

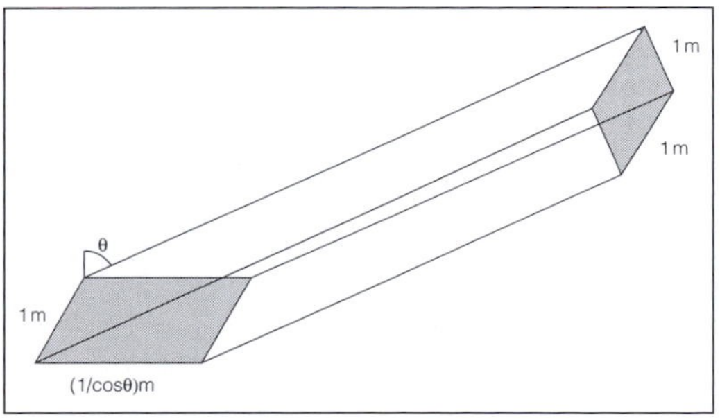


**Figure 7.13** Representation of decreases in irradiance with incidence angle

other greenhouse gases in the atmosphere. This is the mechanism by which the temperature of the Earth rises and why we wish to reduce the level of greenhouse gases in the atmosphere.

### Cosine Law of Radiation (Lambert's Law)

If a rectangular beam of light (1m × 1m, say) falls at an angle on to a surface, the beam will cover an area of that surface which is greater than $1m^2$ and therefore have a lower power density than for the surface that is normal to the beam. If the beam falls at an angle $\theta$ to the normal to the surface, then the beam will cover an area of surface equal to $(1 \times 1/\cos\theta)m^2$. If the power density of the beam is P watts/$m^2$, then this power P is now spread over $(1/\cos\theta)m^2$ of the surface (Figure 7.13). The power density received by the surface is therefore $P/(1/\cos\theta) = P\cos\theta$ watts/$m^2$.

Let us consider an example. We assume a clear day in the UK with a power density in sunlight of 900 watts/$m^2$ for a surface which is positioned normal to the sun's beam. In June at midday the sun is (latitude −22°) from the vertical, i.e. about 30°, so the power density falling on the ground (horizontal) is 900 cos 30 = 900 × 0.866 = 780watts/$m^2$. If we consider December at midday and assume the same power density normal to the sun, the sun is (latitude +22°) from the vertical, i.e. about 75°, so the power density falling on flat ground is 900 cos 75 = 900 × 0.259 = 233watts/$m^2$.

### Stefan's Law

The power per unit area of radiation emitted from a black surface at temperature T (K) is given by Stefan's Law as:

$$P = \sigma T^4$$

where $\sigma$ = Stefan's constant
$= 5.67 \times 10^{-8} Wm^{-2}K^{-4}$

For a surface with an emittance $\varepsilon$

$$P = \varepsilon\sigma T^4$$

The sun behaves rather like a black body with surface temperature of 5800K so:

$$P = 5.67 \times 10^{-8} (5800)^4 = 64MW\ m^{-2}$$

The Earth's surface has many different colours and therefore many different values of emittance for different areas: sand, sea, forest, etc. Taking an average emittance of 0.7 and an average surface temperature of 280K, $P = (0.7)\ 5.67 \times 10^{-8} (280)^4 = 240Wm^{-2}$

## Photovoltaics

### Solar cells

A solar cell converts light to electricity. The cells produce both electric current and voltage by

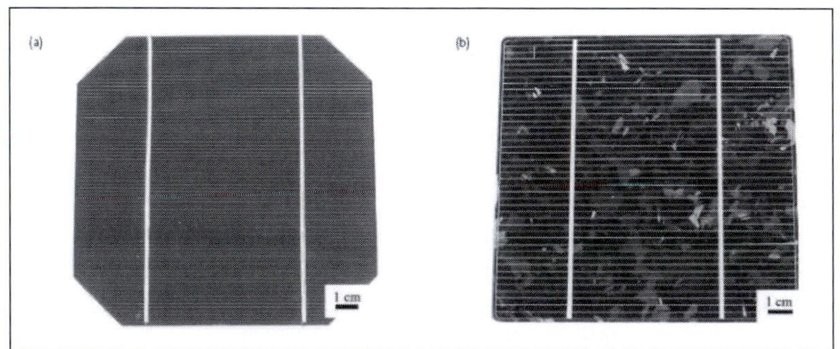


*Source:* Miles, 2007

**Figure 7.14** Typical single crystal and multicrystalline silicon solar cells

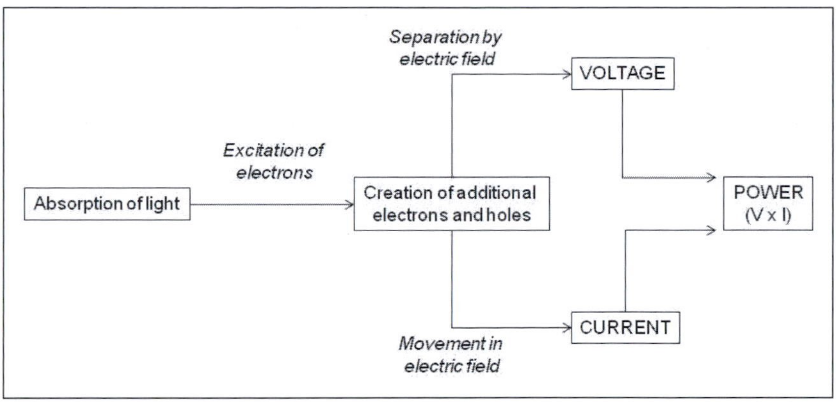


**Figure 7.15** Conversion of light to electricity by the photovoltaic effect

the 'photovoltaic effect', and the technology is often given the name 'photovoltaics' (sometimes solar photovoltaics to clearly identify the energy source). Solar cells are electronic devices made from semiconductor materials such as silicon.

Cells fabricated from crystalline silicon have traditionally dominated the market, accounting for typically 93–95 per cent of the market over the last few years. Whilst thin film solar cells based on other semiconductor materials are now making a growing impact on the market, experts agree that crystalline silicon will continue to be a major part of the market in the medium term. Silicon cells are usually in the form of thin slices (known as wafers) about 0.25mm thick. The positive contact is a layer of metal on the back of the wafer, while the negative contact on top of the cell must collect the current and also allow as much light as possible to enter the device. Therefore, the top contact is usually made in the form of a grid, as shown in Figure 7.14.

The process by which the absorption of light in a solar cell can produce DC (direct current) electrical power is represented by the schematic diagram in Figure 7.15. Note that a cell must produce both current and voltage to generate power, since power = current × voltage. In bright sunlight, a typical single crystal silicon cell of dimensions 15 × 15cm ($225cm^2$) would have an output of about 0.5 volts and 7 amps, i.e. about 3.5 watts of power. Manufacturers quote the output of their cells for a sunlight intensity of $1kW/m^2$, at a cell temperature of 25°C and for a defined solar spectrum (AM1.5 Global). The output under these Standard Test Conditions is often labelled 'peak watts' or 'Wp'.

The current generated by a solar cell varies linearly with the intensity of light except at very low light levels. So if the light is halved, then the current output will also halve. The current is also proportional to the cell area. The voltage is dependent on the material and design of the cell and has a small dependence on light level (actually a logarithmic dependency). The voltage also depends on the temperature of the cell and, for crystalline silicon, decreases by about 0.5 per cent for every one degree rise above 25°C.

In most cases, the structure of the solar cell is the same as a diode and the current–voltage

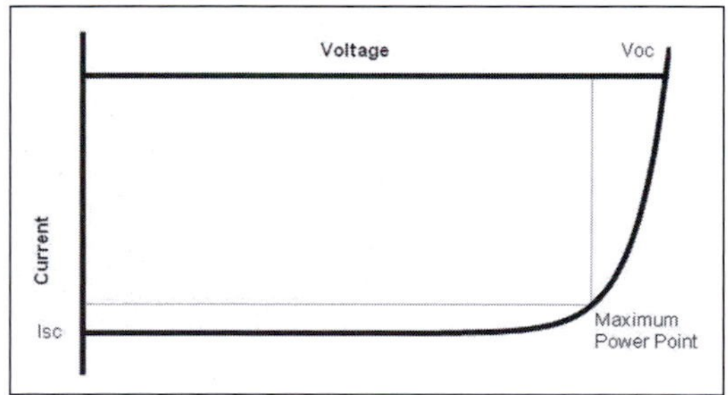


**Figure 7.16** The current–voltage characteristic of a solar cell

characteristic has the same shape as a conventional diode curve except that the current flow is in the reverse direction. Figure 7.16 shows the shape of the characteristic curve, together with the main parameters: short circuit current, open circuit voltage and maximum power. For any given resistive load across the cell terminals, the cell will operate at the point on the characteristic curve where Ohm's law, $V = I R$, is met (where V is the voltage, I is the current and R is the resistance of the load). We can extract most power from the solar cell if we can load the cell with a resistance that corresponds to the point on the curve at which the product of I and V is largest, designated the Maximum Power Point.

## Production of crystalline silicon solar cells

The majority of today's commercial solar cells are made from silicon, which is a plentiful natural resource making up more than a fifth of the Earth's crust and the chief component of ordinary sand. The silicon used in solar cells must be purified to a high degree and for this the silicon oxide (sand) is first heated to the point where the oxygen is driven off, leaving impure silicon. This is reacted with hydrogen chloride to give a liquid silicon compound which, in turn, is purified by fractional distillation. The resulting ultra-pure trichlorosilane is then heated until it dissociates, leaving pieces of silicon which have a purity of about one part per billion. These silicon pieces are then melted in a furnace and, using a small seed crystal, grown into a large crystal or 'boule' which can be over 1m in length and usually 200–300mm in diameter. If a small amount of boron is added to the molten silicon, this makes the silicon electrically conducting through positive mobile charges. The boron-doped silicon is referred to as 'P' type silicon and is used to form the base of the cell.

To make solar cells, or any other electronic device such as silicon chips, the boule is cut, usually by a diamond impregnated wire, into very thin slices called 'wafers'. These are then polished to about 0.25mm thick and, with one face covered, are put into a furnace containing a vaporized phosphorus compound which diffuses into the exposed surface of the wafer to a depth of about 1/1000mm. This phosphorus doping gives mobile negative charges and the resulting surface is called 'N' type silicon; the wafer is now a semiconductor diode. The junction between 'P' and

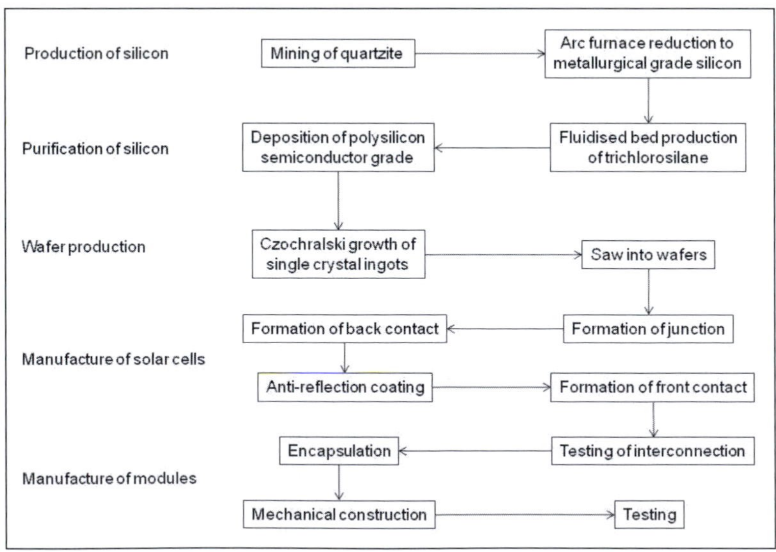


**Figure 7.17** The basic stages in the manufacture of crystalline silicon PV modules

'N' silicon creates an electric field. Contacts are screen printed on to the back of the cell, covering the whole area, and in a grid formation on the front, allowing in as much light as possible. The top silicon surface is then coated with a transparent layer to reduce the amount of light reflected from the surface of the cell. All the cells are tested to see that they reach their specifications and are then sorted into groups depending on their quality. Figure 7.17 provides a flow chart of the production stages of a photovoltaic module of silicon cells from the raw materials.

Many of today's silicon cells are produced from multicrystalline silicon, that is material made up of a number of crystals not just a single crystal. To produce this material, molten silicon is poured into a container and allowed to cool under controlled conditions, resulting in ingots of silicon with large columnar crystals (or grains) growing from the bottom of the container upwards. The ingots are then sliced into wafers and processed in a similar way to single crystal wafers. There is some reduction in performance due to the boundaries between crystals, but the cells still usually exhibit power outputs of over 80 per cent of the equivalent single crystal silicon cell. The advantages of using multicrystalline material in comparison with single crystal are lower capital costs, lower energy and processing costs, higher throughput and lower sensitivity to the silicon quality. Commercial silicon solar cells have efficiencies in the range of 12–16.5 per cent, whilst the best cell efficiency in the laboratory has reached 24.7 per cent (Green et al, 2008).

## Photovoltaic modules

Because single solar cells give only small amounts of power, they are commonly assembled into photovoltaic (PV) modules. The top contact of each cell in the module is connected to the back contact of the one which precedes it (a 'series' connection) and this results in the voltage output of the module being the sum of the voltages from each cell. The module current is equivalent to that of a single cell and is governed by the cell size. Traditional modules, originally designed for remote systems, comprise 30–36 silicon cells and this ensures that the output will exceed 12 volts even in moderate sunlight and, hence, will charge a 12 volt battery. This is one consequence of the logarithmic dependence of voltage on light intensity, since it means that the voltage remains quite high even at rather low light levels. The current–voltage characteristic of the PV module is the same shape as that of the cell, now with modified parameters according to the number of cells connected together.

New applications of modules in buildings and in large power plants, together with the use of new cell materials, have led to the development of a wide range of modules with different numbers of cells to reflect the voltage requirement of the application. It is also possible to increase the module size to accommodate 70–100 cells, but to have several series strings incorporated into the module to keep the voltage at a reasonably low level. These strings would then be connected in parallel, so increasing the current from the module. Many new designs of module are now available on the market.

The module must be strong enough to withstand the elements, and to protect the cells and their electrical contacts from attacks by moisture and atmospheric pollutants throughout their lifetime of 25–30 years or more. Cell temperatures can vary from –20°C on a cold night to over 60°C on a hot day, so the module design must allow for thermal contraction and expansion of the cells and other materials. For the crystalline silicon module, a schematic of the module construction is shown in Figure 7.18. A typical sequence of manufacture is as follows:

- the individual cells are first connected to give the correct electrical configuration;
- they are then arranged in the physical configuration required, for example, 4 adjacent rows of 9 cells in each row;
- the module front sheet is toughened glass, usually around 4mm thick – it is made from glass with a low iron content to ensure high transmittance in the blue region of the spectrum;
- the cells are laid out on the glass with a thin sheet of encapsulant material (e.g. ethyl vinyl

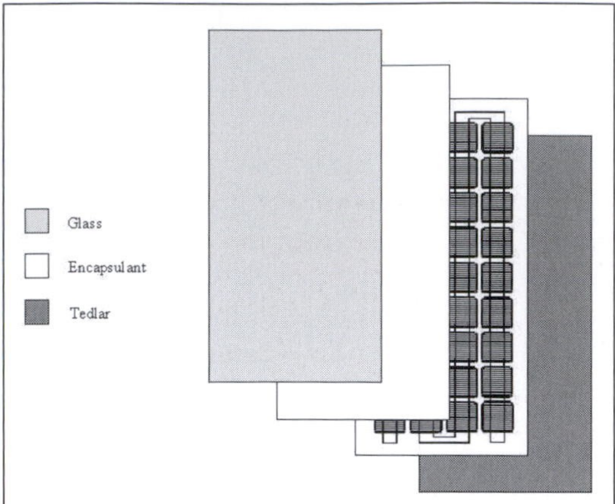


**Figure 7.18** Schematic of construction of a crystalline silicon PV module

acetate (EVA)) below and above; additional encapsulation material is placed at the edges of the module to ensure a complete barrier to moisture after processing;

- the back sheet of a polymer material, such as tedlar, is placed on top and the whole sandwich clamped together to prevent lateral movement;
- the module assembly is laminated, i.e. heated under pressure, as follows:
  1. the structure is placed in the laminator, glass side down;
  2. air is pumped out to ensure that there are no air bubbles in the module and then reintroduced above a flexible membrane to provide pressure on the top of the module structure;
  3. the module is heated at a temperature (typically 150–180°C) for a length of time (typically 20–30 minutes) depending on the encapsulant material – in this period, the material flows around the cells and crosslinking of the polymer provides a strong physical bond and produces a transparent material;
  4. the module is allowed to cool before removing from the laminator;
- the junction box is added to the rear of the module;
- an optional metal frame round the edges gives added strength and a means of attaching the module to a structure.

The power output of the module varies with the number and efficiency of the cells used, but is typically between 50 and 120Wp. If a semi-transparent module is required, often for architectural applications such as an atrium roof, then the back sheet is usually also glass and the lamination process must allow for the thermal conductivity of a thick glass sheet rather than the thinner tedlar layer.

## Thin film PV cells and modules

So far, we have considered crystalline silicon cells but there are several other semiconductor materials that have good properties for making solar cells. Over the last few years, there has been a growing production capability of thin film cells and modules based on three main materials, amorphous silicon (a-Si), cadmium telluride (CdTe) and copper indium gallium diselenide (CIGS). Because these materials absorb visible light over much shorter distances than crystalline silicon, the cells can be made from very thin layers of material only a few thousandths of a millimeter in thickness. Coupled with processing techniques that also use less energy, this means that we can reduce the cost of the cells once the volume of production is high enough.

The amorphous silicon (a-Si) cell is made by depositing thin films of silicon onto a glass or metal substrate. It usually has a slightly different structure to the other cells, since an intrinsic (no doping) layer is introduced to make what is termed a p-i-n junction. Nevertheless, the operating principle and the shape of the current–voltage curve are similar to the other devices discussed here. Even though the material is silicon, the film is amorphous (disordered) in structure and one of the results is that it absorbs light more strongly than the crystalline material. One of the disadvantages of the a-Si cell is that the output degrades in bright sunlight conditions over the first few months of operation. It has been found

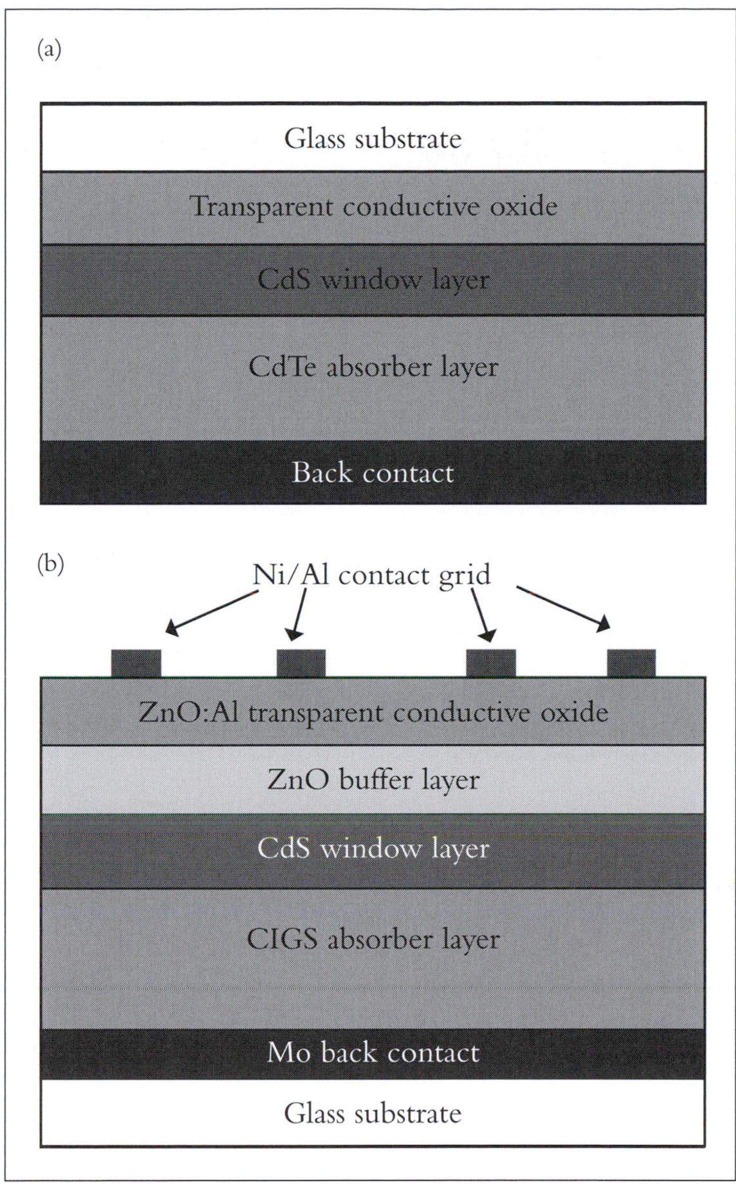


*Note:* layers not drawn to scale.

*Source:* Miles, 2007

**Figure 7.19** Cross sectional schematic of (a) CdTe and (b) CIGS solar cells

that thinner i-layers are more stable, so the best cells now have two or three junctions, each with a thin i-layer but with the overall thickness of material being sufficient to absorb most of the light.

Figure 7.19 shows schematic diagrams of the structures of a CdTe cell and a CIGS cell. These are both heterojunction structures with the CdTe or CIGS forming the p-layer and cadmium sulphide (CdS) forming the n-layer of the diode. Most of the light is transmitted through the CdS layer and is absorbed by either the CdTe or CIGS layer. The top contact in both cases is a transparent conductive oxide (i.e. a film which conducts electricity but also transmits light) rather than a metal grid.

The layers for each cell can be formed by a variety of techniques including thermal evaporation, sputtering, electrodeposition and plasma enhanced processes. All thin film cells consist of a number of deposited layers that are then patterned to produce discrete cells on a single large substrate. These can be considered as single cells in terms of connection, as for the crystalline silicon cell, even though they are physically positioned on the same substrate. Thus, the module is made directly by the sequential deposition and patterning of the layers. This removes the step where the cells need to be connected together and placed in the module and so thin film module production can be more automated and less costly. However, it is necessary to be able to carefully control the deposition of the layers over large areas, so that the cells are as uniform as possible.

Because the cells are all made on a large substrate, we cannot consider individual cell efficiencies at the commercial product level, but only the module efficiency. This is typically around 10 per cent for CIGS, 9.5 per cent for CdTe and between 4 and 7 per cent for a-Si depending on the design. The highest thin film cell efficiency reached in the laboratory is for CIGS at 19.2 per cent, whereas CdTe has reached 16.5 per cent and a-Si 9.5 per cent (stabilized under 1 sun for 800 hours) (Green et al, 2008).

## Other photovoltaic materials

We have discussed the PV materials that are commercially available for power modules, but there are several other interesting materials and concepts which we will mention here although we will not discuss them in detail.

Cells based on materials such as gallium arsenide, indium gallium arsenide and others from the same family (so-called III-V materials) are used for space applications and in systems using concentrated sunlight. Epitaxial growth methods are used to produce cells with several p-n junctions and very high efficiencies. Several cell designs have yielded efficiencies well over 30 per cent under concentrated sunlight (Green et al, 2008). These materials are also the basis of some advanced cell structures which promise even higher efficiencies if they can be fabricated successfully in volume.

There has been a considerable amount of recent research and development on organic- and polymer-based cells. Perhaps the most well-known of these is the dye-sensitized cell, which uses titanium dioxide particles coated in a photosensitive dye and immersed in an electrolyte. The promise of this family of devices is low cost due to the materials and processing techniques, although they also currently have lower efficiencies than inorganic cells. Green et al (2008) report a best organic cell efficiency of 5.15 per cent but this has yet to be translated to an equivalent module efficiency. It is likely to require a few more years of development before organic cells can compete on efficiency and stability with the current PV products, but they represent an interesting route to lower cost.

## Photovoltaic arrays

As with cells, modules can be connected together in series (positive to negative) to increase the voltage or in parallel (negative to negative, positive to positive) to increase the current. The photovoltaic array consists of a number of electrically connected modules, fastened to a secure structure, which can either be fixed in the best position to receive the greatest amount of sunlight or they can be driven so that they follow the changing position of the sun (known as sun tracking). These arrays can vary in size from just a few modules, for purposes like telecommunications, to hundreds of thousands for large supplies to grid-connected utilities. The power capacity of the array is calculated as the sum of the rated outputs of the constituent modules and the current–voltage characteristic is the same shape as for the cells and modules, but now reflects the number of modules connected together.

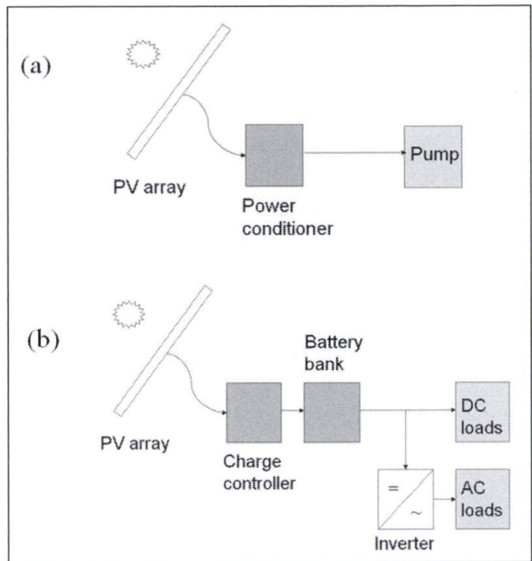


**Figure 7.20** Block diagram of (a) direct coupled PV system and (b) stand-alone PV system with battery storage

## Photovoltaic systems

A PV module, or array of modules, generates DC electricity. To provide a useful service it must be incorporated into a system and these vary in complexity depending on their purpose. The stand-alone or autonomous system provides the sole power supply for a specific load. These are sometimes also called an off-grid system, since by definition the load is not connected to the grid supply. A water pump working only in daylight could adequately be powered by a direct coupled PV system (see Figure 7.20a) in which the DC load is the motor of the pump. On the other hand, lights are mostly used at night so a lighting system would have to include battery storage which would be charged during the day and would power the lights at night (see Figure 7.20b). For systems incorporating batteries, it is usual to include a charge controller, which selects the optimum charging current and will also prevent the batteries from discharging to a level that could cause damage. The system can incorporate multiple loads, such as those in a house, and, for instances when either the load or the solar resource is very variable, it is possible to include a second independent power supply (e.g. wind, diesel) to form a hybrid system.

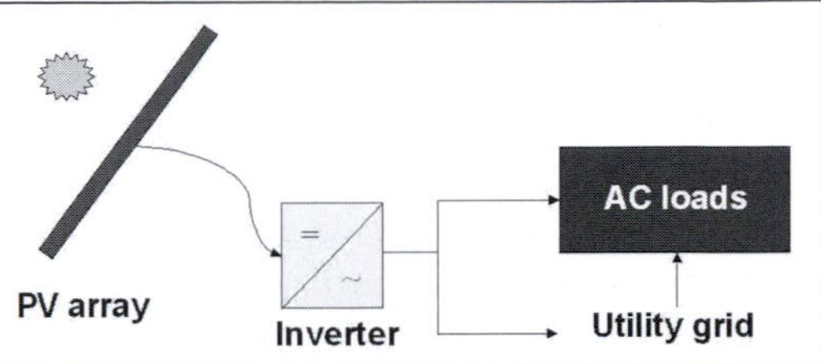


**Figure 7.21** Block diagram of a grid-connected PV system

In concept, the grid connected system is very simple (Figure 7.21). The PV system includes an inverter that will convert the DC output of the PV array into AC (alternating current) output at the correct voltage and frequency to match with the electricity grid. The inverter incorporates a maximum power point tracker to ensure that the system delivers as much power as possible as conditions change. Some PV arrays are integrated into the fabric of a building (roof and/or façade) and feed power to local loads within the building. These are generally known as building integrated PV or BIPV systems. The output from the PV system is connected in parallel with the input from the grid, such that they are both capable of supplying the loads without switching between the two. When the PV system is generating less than required by the load, the shortfall is supplied by the grid, and when the PV system generates more than required by the load, the excess is fed into the grid. This happens automatically with no required action from the users.

The other main type of grid connected PV system is designed to feed electricity directly into the grid. Typically, these systems range in size from a few hundred kW to several MW. In this case, all the electricity produced is simply exported to the grid. This kind of system can also contribute to strengthening the grid by boosting voltage at the end of long feeder lines, especially where the main load on the grid coincides with high output of the PV system, for example, air conditioning loads.

## Concentrating PV systems

Most PV systems in use today do not involve concentration of the sunlight, but there is increasing interest in concentrator systems for the grid feed-in applications discussed above. This type of system involves the use of lenses or mirrors to focus the sunlight onto the solar cell. They allow the use of smaller cell areas, which in turn means that the cells themselves can be more expensive without increasing system costs. Because they use high efficiency cells, the concentrator module can also have higher conversion efficiency than the flat plate module. However, it is only possible to concentrate the direct irradiation, so these systems work best in locations where clear weather conditions predominate. As such, systems have been installed in the US, northern Africa and parts of southern Europe but are much less attractive for more northerly locations.

The concentration ratio is usually expressed by the factor X. For example, 100X means that the light falling on the cell is 100 times more intense than the light falling on the outer surface of the collector (the lens or mirror). The maximum concentration ratio achievable depends on the optical system used. The concentrator system needs to track the sun's position and the higher the concentration ratio, the more accurate that tracking needs to be. Also, due to the higher irradiance levels, there will be more heating of the cell and the module design needs to address the dissipation of that heat in order to maintain efficiency and prevent damage to the cell.

There are two main designs of module. The first uses a Fresnel lens to concentrate the sunlight onto a cell positioned beneath it. The Fresnel lens is basically a collection of small prisms and because of its construction can produce higher uniformity than a conventional lens, whilst also being both thinner and cheaper to manufacture. A square Fresnel lens can achieve a maximum concentration ratio of about 70X, which can be improved by the use of secondary optics to give an added concentration stage.

The second design uses a mirror system to reflect the light onto the cell target. A circular reflective parabolic dish has a high maximum concentration ratio of about 800X and is consequently often used for solar thermal systems. The parabolic trough reflector has a much lower concentration ratio (about 30X) because it only concentrates light in one axis.

The sun tracking system needs to follow the sun to the required accuracy, with the ability to either continue to track in cloudy conditions or to be able to re-acquire the sun rapidly once the cloud has cleared. It also needs to return the array to the starting position either at the end of the day or at the beginning of the following day. Tracking systems are now used on some of the large flat plate (i.e. no concentration) PV systems to increase the yield. Because they increase the initial cost of the system, the gain in output needs to offset this extra cost and this also favours locations with clear sky conditions.

## Photovoltaic applications and markets

Photovoltaic cells have no moving parts and no fuel costs, and they can be designed to supply power ranging from less than 1W to many megawatts. As already discussed, they can be used for generating electricity to feed into the grid, be integrated into buildings to provide power for local loads or be designed to meet a specific load in a stand-alone system. There are very many examples of these applications in practice, from water pumping in India to covering the roofs of major sports stadia in Europe, from providing power for a house in Liverpool to facilitating the operation of a clinic in Botswana.

Energy, including electricity, is crucial for social and economic development and the IEA has estimated that 1.6 billion people still do not have access to electricity, mostly in Africa and the Indian subcontinent (IEA, 2006). At the same time, poor distribution, lack of maintenance services and increasing oil prices make diesel generation both expensive and unreliable. Photovoltaic generators not only produce electricity more cheaply, but are also more reliable. It can often be cheaper to install a photovoltaic generator in a small village than to connect it to a grid. There are several uses for which photovoltaic systems have, in many circumstances, proved to be the best choice in developing countries on both engineering and economic grounds. In telecommunications their reliability and low maintenance needs not only reduce costs but also increase revenue because people can be sure that telephones will work when they want them. PV lights homes, shops, clinics, hospitals, communal buildings or camps not connected to the grid, reliably and cost effectively, using high efficiency DC lamps. Many thousands of PV powered water pumps, both for drinking water and for irrigation, are already in use. At the end of the cold chain small but reliable refrigerators are needed for keeping vaccines. PV powered refrigerators cost more than those powered by kerosene or similar fuels, but because they are so reliable the cost per effective dose is significantly lower.

The main obstacle to the widespread use of photovoltaic systems for purposes like these is financial. Although the lifetime costs are lower than for diesel systems, the initial capital cost is beyond the reach of villagers in developing countries. Solutions in terms of agricultural banks and local loan systems are now being established, but progress remains slower than it should be in addressing these needs.

Notwithstanding the need for solar systems in developing countries, by far the fastest growing sector is that of grid-connected PV systems and, indeed, the overall PV market has been the fastest growing of all the renewable technologies in the last five years albeit from a relatively low base. The availability of financial support through capital grants or enhanced feed-in tariffs has promoted the use of PV, especially in Europe. At the end of 2007, it is estimated that Germany had almost 4GW of PV capacity installed, this being around half of the worldwide installed capacity, as a result of their long-term market development programme and, not unconnected, they were also the strongest country in terms of PV industry when both cell manufacture and systems level expertise are taken into account (IEA-PVPS, 2008). Japan has the second largest total, with almost 2GW at the end of 2007, and the US was third with around 830MW. One of the challenges of the PV market is to develop the capacity in all the other countries that can use substantial amounts of solar energy to match that in the three countries named above.

The exact balance between grid connected and off-grid systems is quite hard to quantify since the latter are not always fully reported, but we can estimate that the former accounted for around 90 per cent of the cumulative installed capacity in 2007 with an annual growth rate of almost twice that of the off-grid market. In recent years, there has been a trend towards ground mounted systems in the multi-megawatt range due to the favourable investment conditions brought about by feed-in tariff schemes. However, many schemes are now looking at balancing the support provided to encourage the installation of distributed systems where PV is particularly well suited to providing power in the urban environment. The largest building integrated systems are around 1–5MW in size, compared to up to 60MW for ground mounted plants, but a major market for PV is the millions of buildings that could benefit from systems of a few kW to a few hundred kW. Architects are finding new ways of integrating PV arrays to add multi-functionality, including using them for passive solar shading, assistance with natural ventilation and visual features. As we move forward, PV will be incorporated into some of the world's most iconic buildings and be a common site on housing all over the world.

PV systems have no emissions in use and so the main environmental impacts relate to manufacture and disposal (in common with most other renewable technologies). Fthenakis and Alsema (2005) have shown that the energy payback time for multicrystalline silicon modules (2004 production, European, 13.2 per cent efficiency) in a rooftop installation in an average southern European location with 1700kWh/m$^2$ irradiation is around 2.2 years. Therefore, this should allow over 25 years of operation with net energy gain. For thin film cadmium telluride modules at 8 per cent efficiency, the energy payback time for the same system is around half this (about 1 year). Thin film modules have lower energy payback times because they use less material and lower energy processes. For the UK, where the irradiation levels are a little lower, these payback times would be increased by 50–60 per cent but this would still mean that the energy of manufacture is less than 20 per cent of the expected energy generation. In the same study, Fthenakis and Alsema showed that the greenhouse gas emissions from PV systems, assuming the same operating conditions, are comparable with those from nuclear power stations (in the range of 20–40 $gCO_2$-eq./kWh) and about one-tenth of those from gas fired power stations. Since the greenhouse gas emissions are almost entirely associated with the use of conventional energy for cell and material processing, these values will reduce as more renewable technologies are introduced into the energy generation system. Several European PV companies have formed an association to develop recycling techniques for PV modules and a number of them already offer this service. The challenge, for a large number of distributed systems and some of them in remote locations, is to collect all components for recycling or controlled disposal.

The barrier to widespread use of photovoltaic systems remains the capital cost, but the rapid development of the market and the industry will help to reduce manufacturing costs, which are still affected by the relatively small scale of production. It is expected that PV systems will be competitive with conventional electricity generation technologies in southern Europe and similar climates in the next decade and throughout Europe soon after 2020, at least for distributed systems. Coupled with the suitability of PV for off-grid applications and the simplicity of use, solar electricity from photovoltaics can make a major contribution to the future of energy supply.

## Solar thermal technologies

When an object absorbs sunlight it gets hot. This heat energy is usable in various ways, to provide space heating or cooling, to provide domestic or industrial hot water, to boil water or other fluids for industrial processes, or to drive engines. The solar heating or cooling of housing or working spaces can be accomplished simply by the appropriate design of the buildings and without any machines or moving parts. Such buildings first appeared in Greece over 2000 years ago, and were common throughout the last millennium in Islamic architecture. The advent of cheap and abundant fossil fuels led to the abandonment of these traditions, but they are now being re-established on a firm scientific basis and termed passive solar technologies.

**Table 7.3** Thermal conductivities of some common building materials

| Material | Thermal conductivity (W/mK) |
|---|---|
| Aluminium | 204 |
| Steel, Iron | 52 |
| Brick | 0.6–0.7 |
| Concrete (varies with density) | 0.12–2.0 |
| Glass | 0.8 |
| Tiles | 1.2 |
| Hardwood | 0.17 |
| Polyurethane foam (PUR) | 0.025–0.035 |
| Cavity wall isolation | 0.05 |
| Air | 0.023 |

### Passive solar heating

All rooms with a window facing the sun are heated when the sunlight shines in. These unplanned solar gains contribute 10–20 per cent of the annual space heating of a typical house in the UK. Some houses have conservatories or greenhouses which are designed to make use of solar heating, and there are a growing number of houses and other buildings which are designed to minimize the total energy needed for space heating and to maximize the contribution which solar energy can make.

When the temperature inside a building is higher than that outside, heat can be lost by conduction, convection and radiation. Conduction of heat depends on temperature gradients across, and the thermal conductivity of, a given material. For a given difference, ΔT, between inside and outside temperatures the thicker the materials, for example in the house walls, the lower the temperature gradient and hence the lower the heat losses. Also, the lower the thermal conductivity, k, of the material, the lower is the conduction. In general, the rate of energy loss through a wall of area, A, and thickness, L, is $P = kA\Delta T/L = UA\Delta T$, where U is the thermal conductance (the 'U value') of the construction. The thermal conductivities of some common building materials are shown in Table 7.3. A window pane of 3mm glass has a U-value of 350. A 5cm thick layer of still air between two sheets of glass reduces this to 0.52, showing the benefits of double glazing. A brick wall 10cm thick would have a U-value of 6, and two such walls a U-value of 3. The cavity between the walls is wide enough to allow convection currents in the air, so heat is transported readily from the inner to the outer wall. If the cavity (say 10cm wide) is filled with polyurethane foam, the U-value of the wall is reduced to 0.25. The U-value of a typical plastered plaster board ceiling is around 20. Laying 10cm of mineral wool over the ceiling reduces the U-value to around 0.35. A well designed and well insulated house would have an overall U-value of around $0.2–0.3 Wm^{-2}K^{-1}$.

Hot air rises while cold air sinks, thus producing convection currents. These are very efficient at transporting heat and are the main mechanism by which outside walls lose heat. Inside older houses, convection currents are called draughts and result in substantial heat loss. As we remarked in the section on solar radiation, all objects radiate energy. The rate at which they do it and the wavelength of the radiation depends on the temperature of the object.

The sun radiates mainly in the visible spectrum, while any surface at room temperature radiates in the infrared at a wavelength of around 10μm. Rooms with large windows present a large area through which infrared radiation from the room can escape. On sunny days a south-facing window gains more energy from the sunlight than it loses through the infrared radiation from the surfaces in the room, and the larger the window the warmer the room. If the sun is not shining, however, the larger the window the more infrared radiation escapes and the cooler the room. At night this effect can be mitigated by drawing curtains or blinds, but this is not sensible during the day. The solution is to use windows which transmit visible radiation but reflect infrared radiation back into the room. Glass itself reflects infrared more than visible light, but it is possible to enhance this effect considerably by special coatings on the glass. This then minimizes the loss of heat by infrared radiation while allowing the use of large south facing windows to maximize the solar gain. In hot countries, other coatings can reflect much of the solar radiation while transmitting most of the infrared and so minimize the cooling needed to keep the room at a comfortable temperature.

From all this we can define the main requirements for a passive solar building. The south-facing glazed area should be large, to maximize solar gain, while the north-facing glazed area should be small to minimize radiation losses, and the construction should have a low overall U-value. (Note: this section assumes a building in the northern hemisphere where the sun is primarily to the south. Transposing south and north provides the rules for a passive solar building in the southern hemisphere, but east and west directions remain the same). To avoid claustrophobia in the rooms on the north side of the building, the windows should be tall but narrow, so that the sky

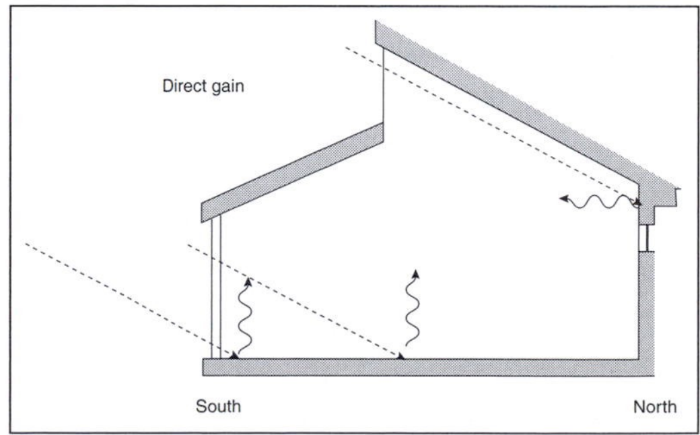


**Figure 7.22** Passive solar gain

and ground are both visible, while keeping the window area small. The simplest type of passive solar building is shown in outline in Figure 7.22 and such structures are said to provide 'direct solar gain'. Additional solar gain can be provided in the morning by east-facing windows but west windows may need to be shaded to avoid unwanted solar gain on late summer afternoons. Excessive heating can be prevented by designing the roof structures that overhang the windows to provide shade against the high summer sun, while admitting the low winter sun. Alternatively, a plantation of deciduous trees in front of the south façade would provide considerable shading in the summer but little in winter when the leaves have been shed.

The solar collection area can be expanded to the entire south façade if 'indirect solar gain' techniques are used. The most common of these is a conservatory on the south wall with a means of circulating the warm air throughout the house (see Figure 7.23). A more effective, but expensive method is to glaze all or much of the south-facing wall (see Figure 7.24). A black wall absorbs the sunlight, and the air between the glazing and the wall rises as it is warmed and is distributed throughout the building. This type of structure was developed by Felix Trombe in the 1950s and is often called a 'Trombe Wall'. In some cases, transparent insulation is used as the glazing, to reduce the heat otherwise lost by conduction and radiation. Transparent insulation was first noticed in the coats of polar bears. Each strand of fur acts like an optical fibre, transmitting sunlight to their skin. Plastic fibres laid side by side can be glued together into sheets or other shapes. Light is transmitted down the fibres, so the sheet is transparent, but if the fibres are, say, 20cm long, the U-value of the sheet would be about 0.5 or so. The great advantage of transparent insulation is that it can be made in standard, self-supporting shapes, like bricks, from which walls can be built.

The economics of passive solar buildings are attractive. The additional cost of passive solar and thermal insulation over and above that of a conventional new building is usually no more than 5–10 per cent. Nevertheless, passive solar features are not commonly included in new domestic properties in the UK, although this may change as building regulations require lower energy consumption targets to be met. Many commercial buildings incorporate atria and other passive solar features for the benefits that they bring to space heating and day lighting.

## Passive solar cooling

Buildings can be designed so that the heat of the sun induces convection currents which draw cool air into the building and so reduce the inside temperature. Islamic architecture has used this principle for centuries and many of its buildings have a 'chimney' which draws up hot

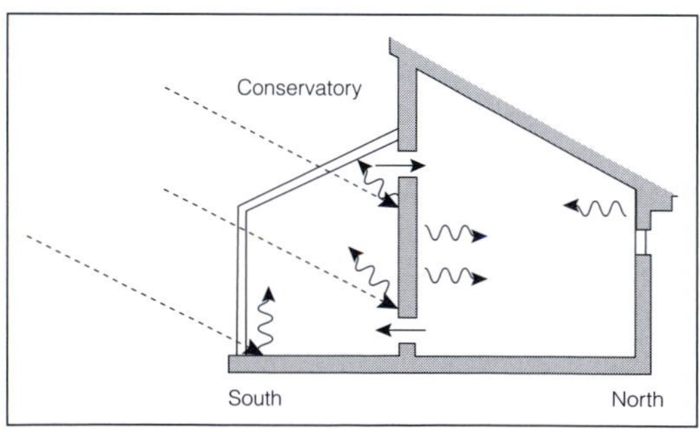


**Figure 7.23** Use of a conservatory for passive solar gain

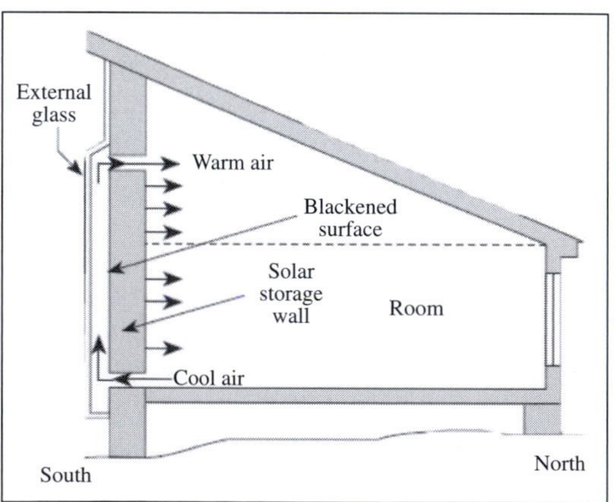


**Figure 7.24** Solar wall system

air and brings air into the building past north facing surfaces which remain cool throughout the day. A modern variant of this uses a Trombe Wall to create the air movement. Instead of being circulated in the building, the hot air is vented to the atmosphere while the incoming air is cooled by underground, or north-facing, heavy masonry surfaces.

## Active solar heating

Sunshine can be used to raise the temperature of a working fluid and the increases can vary from a few degrees to over 2000°C, depending on the type of system used.

### *Low temperature water heaters*

If water is run through a black hose-pipe exposed to sunlight it will come out warmer than it went in. If it is filled with water and both ends are sealed, and it is then coiled flat, the water will heat up on a hot sunny day to as much as 60°C. The water temperature will rise until the rate at which the hose loses heat to its surroundings is equal to the rate at which it gains heat from the sun. This equilibrium (also called the stagnation temperature) depends on the rate of heat loss as well as the rate of heat gain. The rate of heat gain depends on the intensity of the sunlight and the efficiency with which it is absorbed by the surface of the hosepipe. The rate of heat loss depends on the conduction, convection and radiation of energy from the surface of the hose.

If water is flowing through the hose, the useful energy is carried away by the heated water (e.g. to heat a swimming pool). This useful energy is delivered at a rate equal to the difference between the rate of energy input from the sun and the rate of energy loss to the surroundings. The rate of energy input from the sun depends on the intensity of solar radiation I ($Wm^{-2}$), the absorbance capacity ($\alpha$) of the surface of the hose and a factor F (between 0 and 1), which takes account of any other influence on absorption of sunlight, such as reflections, surface roughness and surface geometry. The rate of energy loss to the surroundings depends mainly on conduction at low water temperatures, and so varies with the difference between water temperature Tw and ambient temperature Ta. It can be written as U(Tw – Ta), where U is the 'U-value' of the solar collector system. The rate of delivery of useful energy Q (Watts) can then be written Q = $\alpha$FI – U (Tw – Ta). This is known as the Hotel–Whillier–Bliss equation, after the three people who first derived the equation and used it to study the performance of solar heat collectors.

The efficiency of a collector is the ratio of useful heat delivered to the incident solar radiation, i.e. Q/I, so the efficiency $\eta$c = $\alpha$F – U (Tw – Ta)/I. To produce collectors with the highest efficiency, we should increase the absorption efficiency $\alpha$F, reduce the heat loss factors and operate at the highest solar intensity. Figure 7.25 shows a plot of collector efficiency against (Tw – Ta)/I, for different types of collectors. If the U-value of the collector was independent of temperature, the relationships would be straight lines. In fact at higher temperatures, radiation begins to play a larger role in heat loss and the efficiency falls off more rapidly.

The standard solar water heater consists of a flat metal plate with a black upper surface and pipes, in good thermal contact, attached to the back. The black surface is exposed to the sunlight and so is heated. Water flowing through the pipes

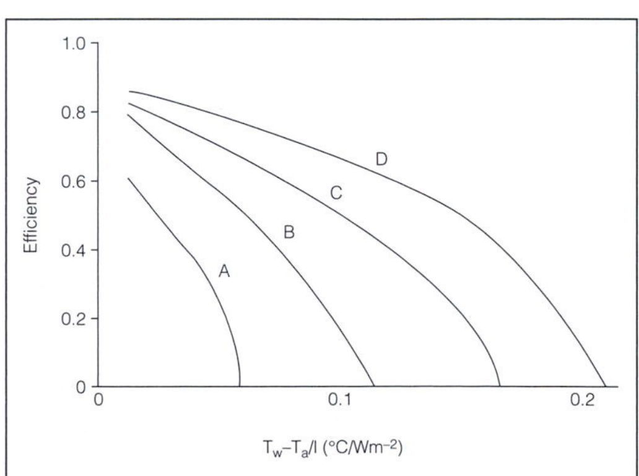


**Figure 7.25** Efficiency curves for solar water heaters

is warmed by heat conducted from the plate to the pipes. A great deal of heat would be lost if the plate and the pipes were exposed to winds or convection currents, so they are enclosed in a well-insulated box, with a glass or plastic front cover. For the highest efficiency the front may be double glazed, although this adds to the cost. The upper surface of the plate is coated with a black paint or other coating which absorbs sunlight efficiently. Most black surfaces are good absorbers of visible light and also efficient radiators of infrared. They therefore radiate heat through the glass front. Selective surfaces are good absorbers of sunlight but poor radiators of infrared and so reduce the radiation heat losses by factors of five or more. Such surfaces give rather higher efficiencies at low temperature, but, more importantly, they maintain efficiency up to much higher temperatures, as shown in curve c of Figure 7.25.

The evacuated tube collector has the lowest heat loss of any of these devices. It consists of a glass tube, sealed at both ends and evacuated of all air, with a heat collector tube running down the middle. The bottom half of the outer tube is silvered on the inside, so sunlight is focused on to the inner collector tube. The collector tube may be hollow, with water flowing through it, or it may be a heat pipe (a device which transmits heat very efficiently along its length). Since there is no air between the collector tube and the outer tube, there can be no convection or conduction loss, except at the ends where the collector tube passes through the seals. Radiation losses are reduced by using selective black surfaces on the collector tube.

A solar collector will consist of 10–20 of those evacuated tubes side by side in a rectangular box, having a common water inlet and outlet. Although they are more expensive than the standard flat-plate collector, they are much more efficient at higher temperatures and can easily produce low pressure steam in high solar intensities, and produce very hot water even in winter sunshine in countries like the UK. In use the solar collector must be plumbed in to a building's water supply and heating system. The fluid in the solar collector is a mixture of water and anti-freeze, so it must be kept separate from the domestic water supply. When the fluid in the solar collector is hotter than the water in the pre-heat tank, the circulating pump is switched on. The hot fluid from the collector is sent through the heat exchanger to heat the water in the pre-heat tank, from where it is either used directly or stored in the hot water tank.

Nearly 50 million households had solar collectors to provide hot water by the end of 2006 (Martinot, 2008).

If very large areas of collector are required, it can be more cost effective in some circumstances to use a solar pond. Any area of shallow water exposed to sunlight is heated by the solar energy absorbed by the water and the bed of the pond. Much of this heat is lost in normal ponds by convection currents which bring the hot water to the surface, where it loses heat to the atmosphere. The density of the water may be increased by adding salt. The concentration of salt is made to be higher at the bottom of the pond, and decreased gradually toward the surface. Even if the water at the bottom of the pond is hotter than at the surface, it remains more dense as it has more salt dissolved in it. In this way convection currents are suppressed. In these ponds the bottom layer of brine can become quite hot, and the upper layers of water act as a good insulator, making an efficient solar collector. The concentrated brine can be pumped through a heat exchanger to heat water or other fluids and, if the temperature is high enough, they can be used to drive a turbine and generate electricity.

## Concentrating solar power systems

If higher temperatures are required than can be achieved by flat-plate collectors, usually for generating electricity, then sunlight can be concentrated by mirrors or lenses. Concentrating collectors can achieve temperatures up to 1000°C depending on the configuration of the system and the concentration ratio (as described in the section on concentrating PV systems). There are three main configurations of concentrating solar power (CSP) collector which have been commercially demonstrated:

**Table 7.4** Characteristics of different configurations for CSP systems

| | Capacity (MW) | Concentration | Peak solar efficiency | Annual solar efficiency | Thermal cycle efficiency |
|---|---|---|---|---|---|
| Parabolic trough | 10–200 | 70–80 | 21% (d) | 10–15% (d)<br>17–18% (p) | 30–40% ST |
| Power tower | 10–150 | 300–1000 | 20% (d)<br>35% (p) | 8–10% (d)<br>15–25% (p) | 30–40% ST<br>45–55% CC |
| Parabolic dish | 0.01–0.4 | 1000–3000 | 29 percent (d) | 16–18% (d)<br>18–23% (p) | 30–40% Stirling<br>20–30% GT |

*Note:* d – demonstrated, p – projected, ST – steam turbine, CC – combined cycle, ST – steam turbine, solar efficiency = net power generation/incident beam radiation

*Source:* Adapted from DLR, 2005)

- parabolic trough collectors;
- parabolic dish collectors;
- solar power tower systems (a collection of mirrors reflecting the light to a single point on a central tower).

Table 7.4 compares the main characteristics of the three options.

For low concentration ratios, mirrors in the shape of a long trough whose sides form compound parabolas can be used to collect sunlight wherever the sun is in the sky without having to move the mirrors. The troughs are oriented east–west (Figure 7.26), so that the sunlight enters the mirrors throughout the day. The mirrors can collect sunlight from both high and low solar elevations, that is in both summer and winter. The shape of these compound parabolic reflectors is quite complex and not so easy to manufacture. Furthermore, although the mirrors accept sunlight at low solar elevations, the energy density of the solar irradiance is reduced in proportion to the cosine of the angle of the sun. Higher concentration levels are achieved by using one-axis tracking of long parabolic trough-shaped mirrors (see Figure 7.27). The absorber is usually a metal tube with a 'super black' surface coating enclosed in an evacuated glass tube. The working fluid is usually an oil with good chemical stability at high temperatures. The heated fluid is then used in a heat exchanger to generate steam, which in turn is used in a conventional Rankine cycle steam power plant to generate electricity. It is also possible to heat generate steam directly in the collector if the concentration ratio, and therefore temperature, is sufficiently high.

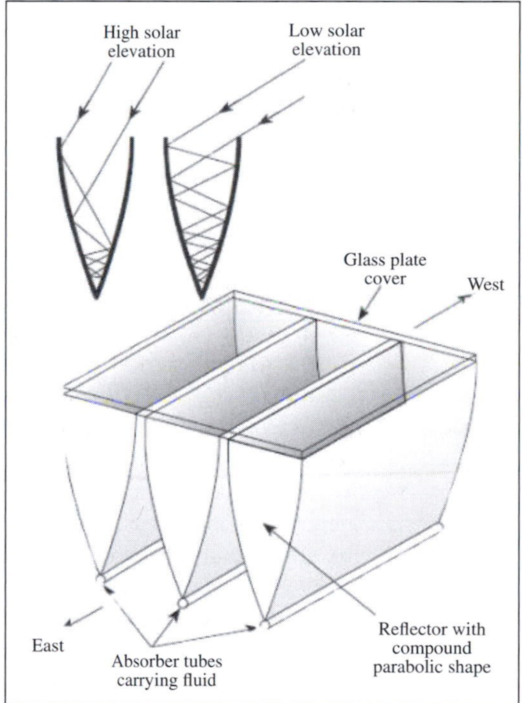


**Figure 7.26** Compound parabolic reflector concentrating sunlight

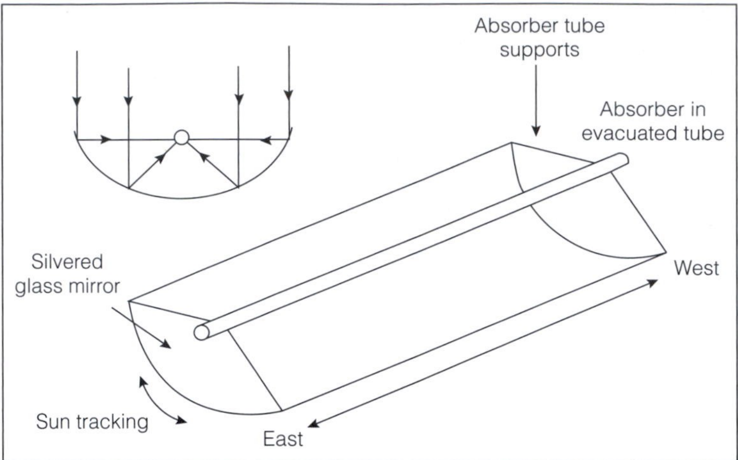


**Figure 7.27** Tracking parabolic reflector

By contrast, parabolic dishes tend to use air as the working fluid. A Stirling engine, which is a high efficiency, closed cycle hot air engine, is positioned at the focal point of the collector. Since this requires the generator to be part of the collector assembly, these systems are usually smaller than the parabolic trough systems and currently compete with photovoltaics or diesel engines for remote power supplies. The parabolic dish can achieve concentration ratios of over 1000 and is one of the most efficient configurations of CSP system.

Very high temperatures can be reached in solar furnaces, where many mirrors are controlled to reflect sunlight onto a single absorber mounted high up on a tower – the 'power tower'. The mirrors are mounted on a two-axis tracking system and each is controlled by computer to reflect and focus an image of the sun onto an absorber. As an example, a solar power tower plant inaugurated in Seville in 2007 is rated at 11MW; it has 625 heliostats (movable mirrors), each 120 $m^2$, focusing light onto a tower 115m in height. As with the parabolic dish collectors, these systems use a working fluid, often a molten metallic salt, which is then used to generate steam to feed into a conventional generator. A typical salt might be a mixture of sodium nitrate and potassium nitrate, melting around 600°C.

The use of a molten salt as the working fluid also allows some storage of the heat in a suitable storage chamber, and this will allow the generation of electricity in cloudy periods or at night with the result that the power output from a large CSP plant is not as variable as that from a photovoltaic system of the same capacity. It has also been suggested that CSP plants could be used for the generation of hydrogen as a fossil fuel replacement.

Clearly, because the systems can only concentrate direct sunlight, CSP systems are suited to locations with predominantly clear sky conditions, but there are a number of areas that meet these conditions, including Mediterranean countries, northern Africa, the Middle East, the south western US, China and Australia. Although CSP systems were first installed commercially in the 1990s, a combination of high capital costs and resistance to large scale solar plants meant that there was then no activity for a decade or so. Interest has renewed in the last few years and by the end of 2007, the total of installed and contracted CSP plants had reached 2GW in capacity (Martinot, 2008). It now looks likely that CSP will make a significant contribution to renewable energy generation over the next 20 years or so.

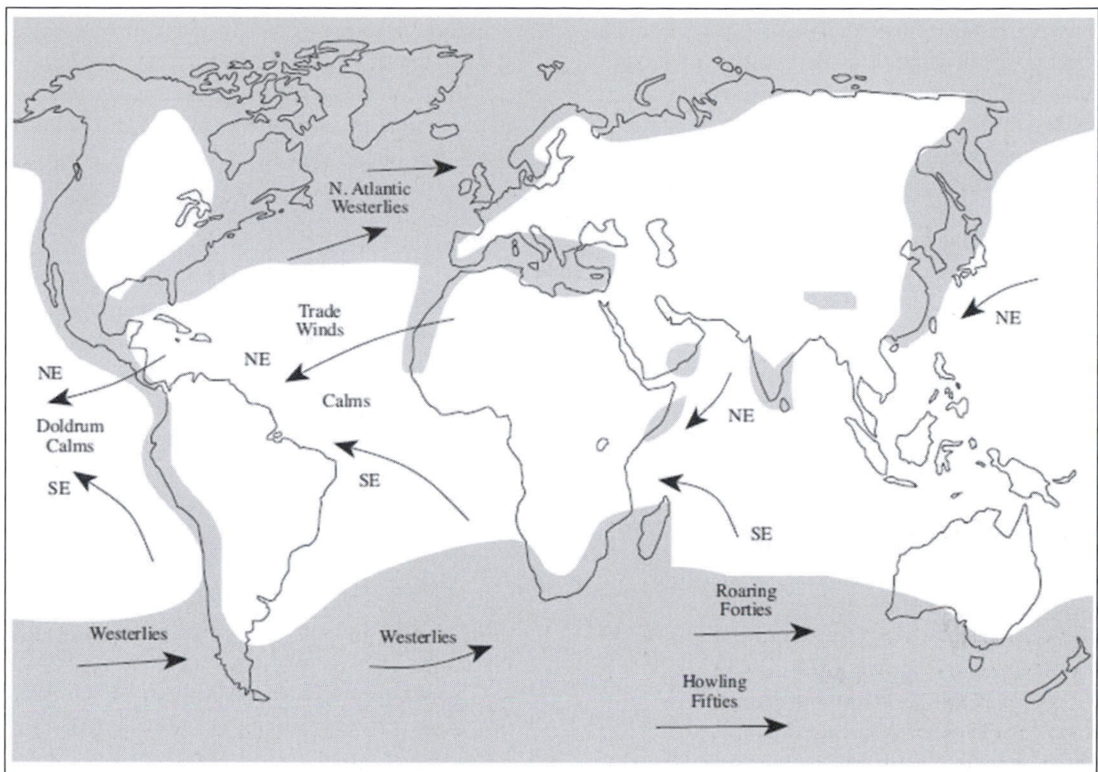


The map shows the prevailing strong winds. The shaded areas indicate regions where the wind energy is attractive for power generation with average speeds over 5ms$^{-1}$, and average generation over 33% of rated power. Note the importance of marine situations, and beware of non-site-related generalizations

**Figure 7.28** Prevailing strong winds

# Wind energy

Solar radiation over the equator heats up the air which rises as the cold polar air sinks. This establishes the basic global pattern of wind circulation (Figure 7.28). There is an enormous resource of wind power worldwide, although, since the power in the wind depends on the wind speed and that varies with height above the ground (see the section on siting a wind turbine), the potential amount of energy that can be extracted from the wind is not a fixed quantity but depends on the average size of turbines. The estimate also varies depending on the inclusion of onshore and offshore resources and how these are defined. Grubb and Meyer (1993) estimated the worldwide potential resource to be just under 500,000TWh per year, based on a hub height of 50m and a conversion efficiency of 26 per cent (see Grubb and Meyer (1993) for more detail on the assumptions). However, this should only taken as an approximate figure, since the resource changes with the development of technology and with the other assumptions made. The UK has some of the best wind conditions in the world, due to its location, and the highest estimated onshore and offshore resource in Europe (although estimates for the latter vary considerably depending on the sites considered).

Wind power is pollution free in operation and onshore turbines need less maintenance than conventional power stations, all of which makes it an attractive form of power production. Offshore wind turbines are more challenging from a technical point of view, since they need to withstand sea conditions and the power has to be transmitted back to shore, but the wind speeds are higher and more consistent. A combination of onshore and offshore wind farms is expected to make a major contribution to meeting the target of 20 per cent of Europe's energy demand from renewable sources by 2020.

## Wind energy assessment

The energy in the wind is the kinetic energy of a moving mass of air.

$$\text{Kinetic energy} = \tfrac{1}{2}mv^2$$

where $m$ = mass of the moving air and $v$ = velocity (see Figure 7.29)

$$\text{Mass of the air} = \text{volume} \times \text{density}$$

If the air is moving with velocity $v\text{ms}^{-1}$, the volume of the air passing through $1\text{m}^2$ of area in one second = $v\text{m}^3$.

$$\text{The mass of this air} = 1 \times v \times \rho,$$

where $\rho$ is the density of air.

The energy in the air which passes in one second is therefore:

$$\tfrac{1}{2}mv^2 = \tfrac{1}{2}\rho v^3 \text{ watts/m}^2$$

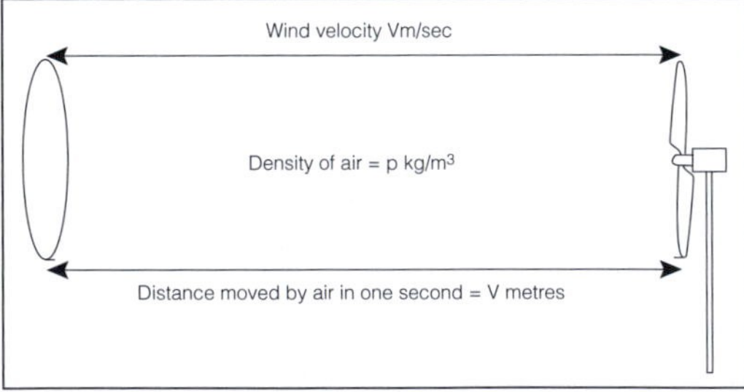


**Figure 7.29** Representation of a column of air moving past a rotor

Thus, the power output of a wind turbine varies as the cube of the wind speed.

This means that the power varies substantially as the wind speed changes and careful siting is needed to obtain the maximum power output.

## Siting a wind turbine

Windmills have existed for between one and two millennia, and although modern turbines have the advantage of aerodynamic design based on an understanding of air flow and turbulence, all designs have in common that the amount of energy they produce is heavily dependent on where they are sited, since this affects the wind speed and variability. Placing a wind generator at the foot of a hill would be pointless, but if it is placed at the top where the wind has been forced to accelerate by the rise in the ground, it will work very successfully. Ground drag also makes a large difference, as the less there is, the faster the wind. A wind generator sited on a sheet of ice or concrete will produce its expected output, but put behind trees or large bushes it will not. We may notice that open areas, even in our cities, tend to be windy while forests are relatively calm. By the same token, wind generators are quite tall and around 50m above the ground is usually high enough to escape the drag factor. Note that issues of drag can strongly influence the output of small wind turbines placed on buildings in an urban environment, due to the effect of the surrounding buildings.

In order to determine the output potential for a given site, we need to know the average wind speed but this varies with height above the ground. We will discuss the design of wind turbines in the next section, but as turbine size increases, the height of the turbine also increases. Often the measurement of wind speed is made at a lower elevation (frequently about 10m) than that of the hub of the turbine which is eventually installed and a correction must be made based on an extrapolation from the measurement height.

It turns out that a power law can often give a good approximation of the variation in wind speed, although this is more useful for open sites than for complex terrain. For a given average wind speed $v_1$ measured at a height $h_1$, we can derive the wind speed $v_2$ at height $h_2$ from the following equation:

$$v_2 = v_1[h_2/h_1]^x$$

where $x$ is a coefficient that varies with local terrain. The value of $x$ is around 0.14 for open sites but rises with increasing complexity of the terrain and may be around 0.3 for obstructed urban sites. This is easy to understand, since the wind at high levels is unaffected by the terrain, but at lower levels it is slowed by obstructions and therefore there is a greater differential due to height in these cases. Of course, the coefficient will also vary depending on the actual values of $h_1$ and $h_2$ and how they relate to the site and more complex mathematical relationships exist to express this.

## Wind turbines

Wind turbines convert the energy of the wind into socially useful energy, usually electricity, and they come in many shapes and kinds. Two main categories can be defined, horizontal and vertical axis machines, and there are different designs in each category.

For machines with a horizontal axis (Figure 7.30) the dominant force is lift and the rotor blades may be in front of (upwind) or behind the tower (downwind). Upwind turbines need a tail or some other mechanism to point them into the wind. Downwind turbines may be quite seriously affected by the tower, which can produce a wind shadow and turbulence in the path of the blades. Both kinds of machine with a capacity greater than about 50kW are turned into the wind by an electric motor. Multi-bladed rotors which have a high torque in light winds are used for pumping water and other tasks calling for low frequency mechanical power. For generating electricity, turbines having one, two or three blades have all been considered over the last 20 years or so. Single bladed machines are the most structurally efficient and the blade can be stowed in line with the tower under high wind conditions to minimize storm damage. However, they have reduced aerodynamic efficiency due to higher tip losses

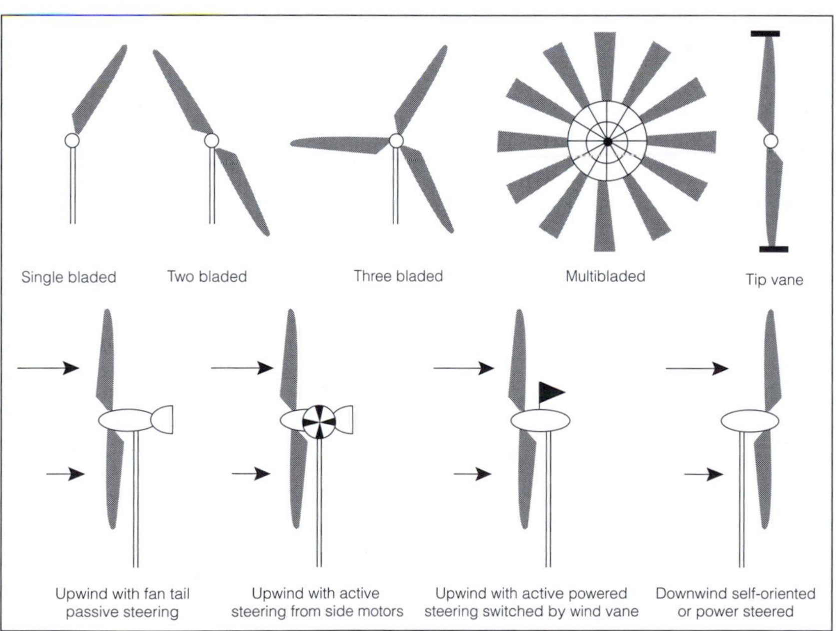


**Figure 7.30** Horizontal axis wind turbines of different designs

and require a counterweight to balance the rotor statically. The higher tip speed (the speed of rotation) also leads to higher noise levels.

The two bladed rotor is a little less aerodynamically efficient than the three bladed version but this is partially offset by the simpler structure. However, although not technically necessary, two bladed designs have also tended to operate at higher tip speed with consequently more noise issues. The visual aspect of both the single and two bladed designs is also not judged to be as pleasing as for the three bladed design. In commercial applications, the three bladed horizontal axis turbine makes up the majority of the market.

Wind turbines with vertical axes have the advantage that they may, without adjustment, be driven by wind from any direction, but the torque from wind variation during each turn of the blades can produce unwanted vibrations. Because the angle of attack of the wind on the turbine blades changes as the turbine rotates, the aerodynamic torque changes and so the vertical axis machine is inherently less efficient than the horizontal axis machine. As a consequence, it has made little impact on the commercial wind farm market. Nevertheless, there has been renewed interest in small vertical axis machines for use in urban environments because of easier installation, lower noise levels and lower sensitivity to the varying wind regime resulting from the complex terrain. It is not yet clear what design the market will ultimately favour for these small systems, since assessment of the use and performance level of wind turbines in an urban environment is still at a relatively early stage.

We do not have the space to consider all the variations of vertical axis wind turbines that have been developed, so will mention just two of the basic designs. The Savonius rotor (Figure 7.31) consists of two (or sometimes three) scoops and rotates because of differential drag between the two scoops. Essentially the drag is higher when moving against the wind than it is when moving with the wind. A Savonius rotor can even be homemade from two halves of an oil drum for simple water pumping applications. However, a machine working on the principle of drag is much less efficient than one working on the principle of lift.

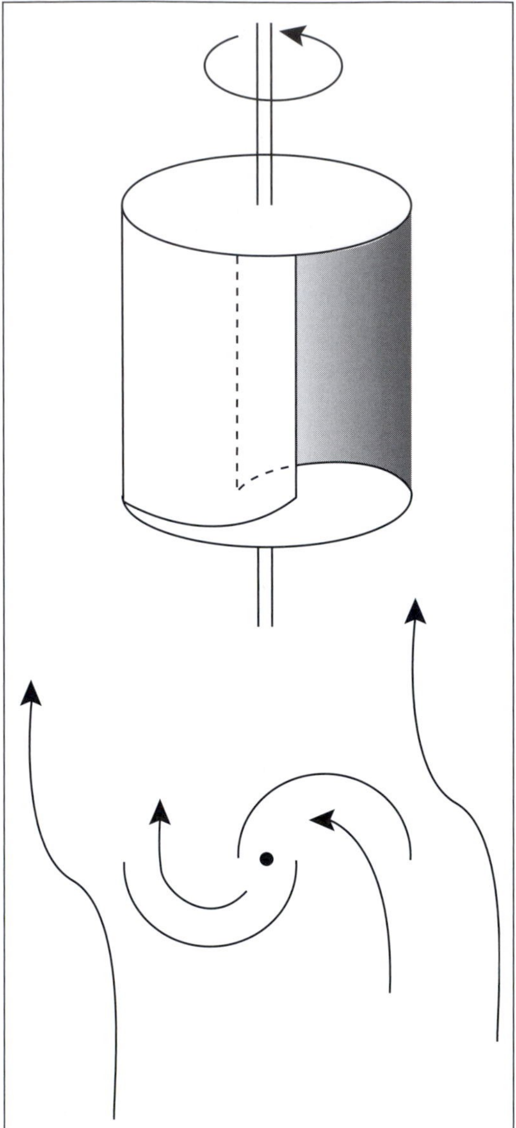

**Figure 7.31** Savonius rotor

The Darrieus rotor (Figure 7.32) has two or three thin blades with an aerofoil section. The driving force is lift and maximum torque occurs when the blade is moving across the wind faster than the speed of the wind. But this rotor, which is used for generating electricity, is not, as a rule,

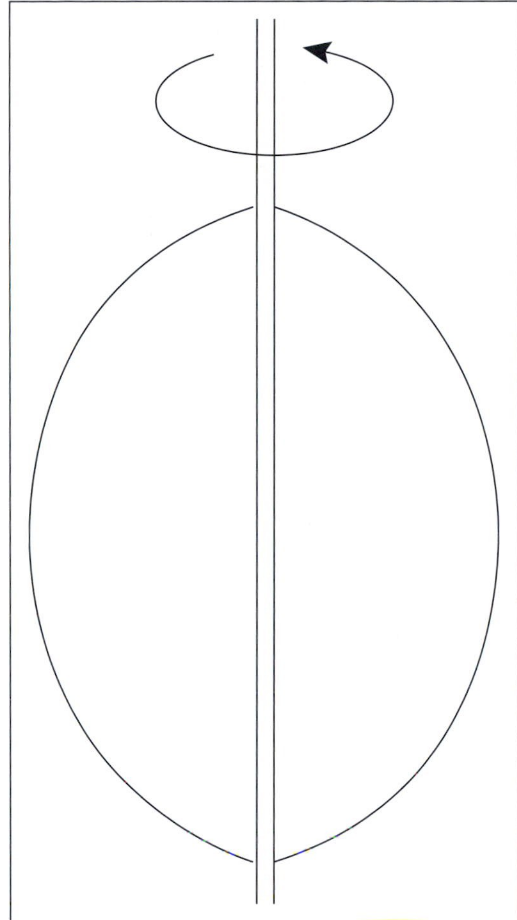

**Figure 7.32** Darrieus turbine

self-starting and has to be started by the generator itself which calls either for complicated controls or constant supervision by an operator. Also, a high proportion of the blade area is close to the axis and so the turbine rotates at a relatively low speed, resulting in reduced aerodynamic efficiency.

## Three bladed, horizontal axis wind turbines

We shall now consider the typical modern wind turbine, a horizontal axis machine with three blades, in a little more detail. The blades are usually made of composite materials, typically fibreglass and polyester or epoxy, but wood and carbon fibre are also used. They need to be lightweight and easily manufactured, but also strong enough to withstand the wind for a design lifetime of around 20 years. The rotating blades are connected via a gearbox and drive train to an electrical generator, with all this equipment usually being housed in a protective enclosure called a nacelle. The nacelle and the rotor are positioned on top of a tower, usually made from steel, and the nacelle/rotor assembly can move round in order to face the prevailing wind. Some rotors are directly connected to the generator, eliminating the need for a gearbox.

Turbines are designed either to operate at a fixed speed (i.e. the speed of rotation is maintained across a range of wind speeds) or at variable speed (i.e. the speed of rotation varies with the wind speed). Variable speed turbines are becoming more common even though they need additional power conditioning to ensure that the frequency of the power fed into the grid remains constant. All turbines also need some method of controlling the speed of rotation at high wind speeds, to ensure that the turbine is not damaged. The speed at which the blades will turn in a given wind speed depends on the shape of the blades and their attitude with respect to the wind direction. Pitch-controlled turbines rotate the blades to present a different profile as the wind speed increases, limiting the power output until the rated power is reached and a steady output is achieved. Stall-controlled turbines have fixed blades, which gradually go into stall conditions as the wind speed increases, also limiting the power but this time by passive rather than active means. Clearly this method does not require the means to control the blade pitch, but it is more difficult to achieve constant power conditions and the output tends to decrease again at wind speeds above the rated speed, so reducing the overall output from the turbine compared to the pitch-controlled option. Thus, pitch-controlled turbines are now a more popular design.

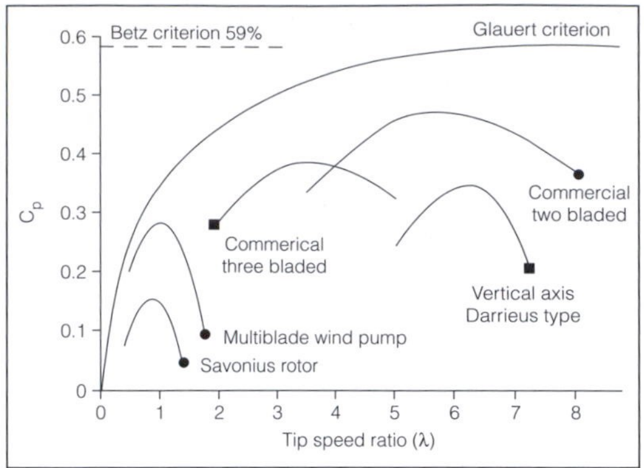


**Figure 7.33** Variation of efficiency versus tip speed ratio for various types of wind turbine

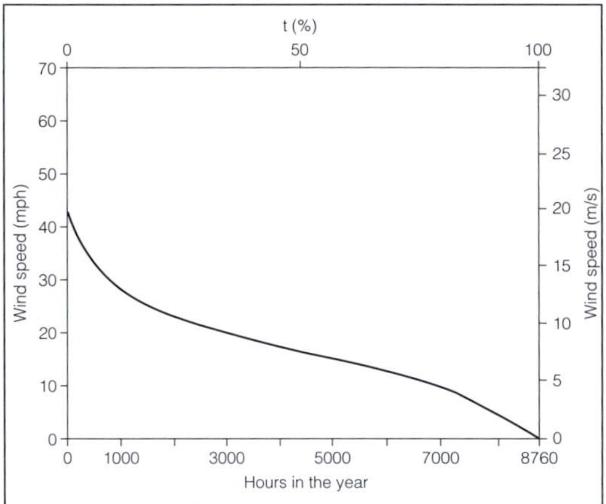


**Figure 7.34** Representative wind speed duration curve for the UK

## Betz Limit

Air flowing through a turbine cannot give all its energy to the rotors, otherwise the air velocity would be zero in front of the turbine and air could no longer flow through it. For a continuous stream of air passing through the turbine, the maximum power the air can deliver to the rotors is 59 per cent of its kinetic energy. This figure for the ideal efficiency was first derived by Betz and is known as the Betz Limit. All real wind turbines are less efficient than this. When a wind generator is rated at 70 per cent efficiency this means that it converts 70 per cent × 0.59 = 41 per cent of the wind energy into rotational energy.

## Tip speed ratio

The efficiency with which a wind turbine can use the wind energy varies according to the wind speed. At very low speeds the wind will not turn the rotors, while at very high speeds the rotors become more inefficient. Once the rotors start to turn they rotate more quickly as the wind speed increases and, to maintain efficiency, the ratio of the speed of the tips of the rotor compared to the wind speed should be kept constant. This 'tip speed ratio' is an important aspect of the design of a wind turbine, with the value of the 'tip speed ratio' which gives maximum efficiency being different for the different types of wind turbine (Figure 7.33).

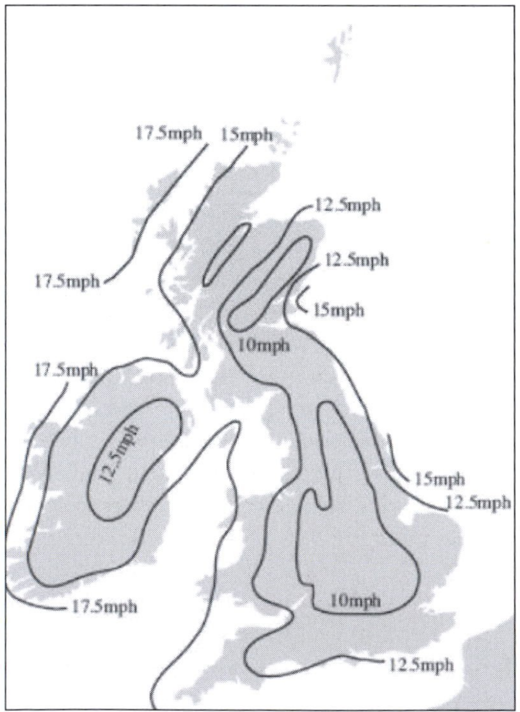


**Figure 7.35** Average wind speed data for the UK

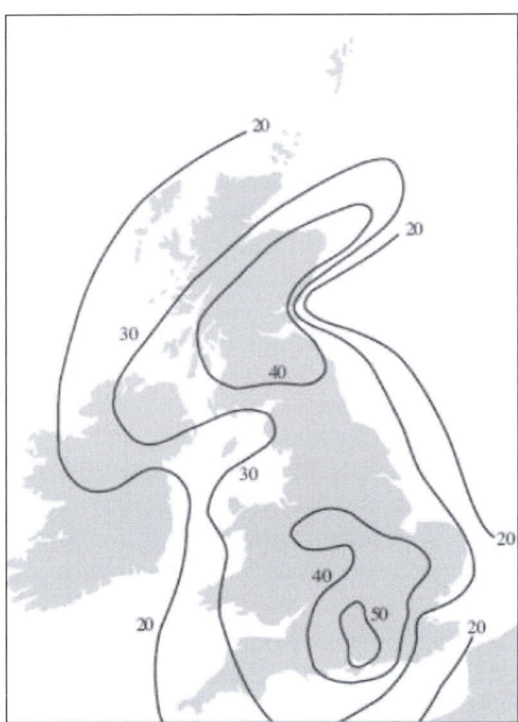


**Figure 7.36** Percentage of calm days in the British Isles

## Energy output from wind turbines

The energy generated by a wind turbine is the sum of the power it generates over a period, where that power is proportional to the cube of the wind speed (at least up to the rated wind speed of the turbine, where it may then be limited to that level). The watts generated in each second of the day are added up to give the energy output in that day. The outcome will depend on how often the wind blows and at what speed. In Figure 7.34 we give a wind velocity duration chart typical of a good site in the UK and it shows the number of hours in a year during which the speed is at or below a particular value. Figures 7.35 and 7.36 show the mean wind speed around the UK averaged over the year and it can be seen that there are large areas where wind speed exceeds 12.5mph, and where there are calm days for less than 40 per cent of the year.

## Development of wind turbines and the wind market

In recent years, there has been a remarkable growth in the wind turbine market, with both more installations and a move to much larger machines. The World Energy Council's 2007 report on energy resources estimates that the total installed capacity of wind turbines was just over 59GW at the end of 2005, yielding an annual total of around 105TWh of electricity (WEC, 2007). The countries with the most installed capacity were Germany, Spain and the US. The amount of power from a wind turbine depends on its cross-sectional area and there has been a move to larger blade sizes to increase the power output. Since the early 1990s the average rating of wind turbines has increased ten-fold from around 200kW (blade size of around 25m) to 2MW (blade size of around 80m), but much bigger turbines up to around 5MW are also available. In 2005, only about 750MW of wind power was installed offshore, but this sector is growing rapidly and pushes the market towards larger turbines, since the economics are improved for both installation and transmission of the power back to land.

Wind turbines can be installed singly or in wind farms with multiple turbines, depending on the location and output requirements. One interesting aspect of the growth in wind turbine size is the opportunity for 'repowering' existing sites by replacing the turbines with larger sizes (provided that the existing spacing and other location constraints allow).

There are many possible locations for wind farms, although care must be taken in terms of visual amenity, noise (installations are normally required to be at least 400m from dwellings) and interference with other functions (such as airport radar, bird migration routes, etc.). Nevertheless, there is the opportunity to obtain a significant proportion of our electricity supply from wind turbines, for which the technology is largely proven (with some development still needed for offshore systems) and the current costs are amongst the lowest of all renewable technologies.

# Power from water

Oceans, lakes, rivers and all bodies of water evaporate as they absorb sunlight. Water vapour joins the general circulation of the atmosphere subsequently to be released as rain, much of which falls on land and runs from higher ground back to its sources. During its path to the sea, water can be intercepted by dams and channelled through turbines, or it can be used to drive the contemporary equivalent of ancient waterwheels. Seas, and even large lakes, are not level surfaces and the differences can be used to generate power, either from waves which are caused by winds – and are, effectively, stores of wind power – or from tides which are caused by the gravitational pull of the moon and, to a lesser degree, the sun. The regularity of tides is useful because it makes the potential power output fairly predictable, although this is modified by the extent to which it is affected by wind.

## Energy output from water systems

Water power depends on local conditions. Most streams, rivers, lakes, tides or waves can be used to generate some power, but to use them cost effectively is another matter. Cost effectiveness is important not only to show that the resource is worth exploiting, but also to be sure that the devices used to convert water power to electricity are being used wisely. It would easily be possible to use more energy constructing the devices than they could produce throughout their lifetimes if they were employed in inappropriate conditions. In Chapter 2 we explored the ways in which cost effectiveness is critically dependent on the discount rate used in its calculation. This is particularly important in the case of water power, where the capital cost is usually high, but the lifetime of the plant can be very long. The price at which water power can produce electricity has also to be compared with the price from other sources. Any large scheme feeding power into the national grid must compete with other large base-load power stations. Small schemes designed to provide power to a farm or a single village should be compared to the retail price of electricity. In developing countries the price must be compared to alternative means of supply in any given locality, whether by the extension of the grid, the use of diesel powered generators or the use of other renewable resources.

In general, water based technologies give us low environmental impacts in operation, but there can be significant environmental impacts in construction. This is especially true of large hydropower and tidal schemes, and the impacts need to be fully assessed to make sure that the net result is positive.

Resource estimations are usually produced for the separate categories of water power: hydropower (generating electricity from the flow of water from high ground to low ground), tidal power (generating electricity from the flow of water due to the tides) and ocean power (generating electricity from waves, the flow of water due to marine currents or the thermal difference between surface and deep water). We will consider the resources as we look at each of these technologies in turn.

## Hydropower

Hydropower relies on the conversion of the potential energy that water loses when flowing from a higher to a lower level. This loss of potential energy each second is the power available and is given by:

$$P = MgH = \rho VgH$$

where $P$ is the power (watts), $M$ is the mass flow of water ($kgs^{-1}$), $g$ is the acceleration due to gravity, $H$ is the height in metres through which the water falls, $\rho$ is the density of water ($kgm^{-3}$) and $V$ is the volume flow of water ($m^3s^{-1}$). Taking $g$ as approximately $10ms^{-2}$, the power is then $P = 10MH$ watts or $P = 10\ VH$ kW, since $\rho = 1000kgm^{-3}$ for fresh water.

We can see that in order to get high power levels, we need $H$ (the difference between the water levels) to be as high as possible, within some constraints resulting from the turbines that we can use for the conversion, coupled with a high enough volume of water flowing through the turbine. Some hydropower systems make use of existing natural features such as weirs, but often we have to modify the water flow by diverting it through different channels and/or by building a dam.

Hydropower is one of the oldest major uses of renewable energy for electricity production and large hydropower schemes provided about 15 per cent of world electricity demand in 2006 (see Figure 7.2). Large hydropower installations include some of the largest artificial structures in the world and include famous sites such as the Grand Coulee Dam in the US (rated at about 6.5GW) and the Three Gorges Dam in China (planned to be18.2GW when fully operational in 2009). Such systems provide large amounts of energy over long periods. However, there are a limited number of sites where it is possible to build a large dam such as these and the energy generated has to be balanced against the disruption caused and the effects of change of land use. The construction of the Three Gorges Dam has reportedly required the movement of over 1 million people from their homes and the flooding of 1200 towns and villages. It has both environmental benefits, mainly in that it provides some control of the Yangtze river which may prevent devastating flooding downriver, and some environmental disadvantages, for example, the potential greenhouse gas emissions resulting from decaying vegetation in the flooded areas and effects on local fauna and flora. All these effects must be balanced when considering whether and where to build a large hydropower scheme.

Hydropower can also be used for systems of a wide range of sizes. Definitions vary, but large systems tend to be considered as being over 10MW in capacity, small systems being between 100kW and 10MW and those that are smaller than 100kW usually termed micro-hydro systems. They all work on the same principles but with differences in turbine choices, the voltage level at which the electricity is taken off and whether they are used to meet local needs or feed into an electricity distribution network.

As with other renewable technologies, the technical resource depends on the choices and assumptions regarding the technology and how much of the natural flow is considered to be converted. As an example of the potential of hydropower, the World Energy Council estimates that, at the end of 2005, the total worldwide technical potential of hydropower was at least 16,500TWh/year of which around 2800TWh or 17 per cent was being exploited (WEC, 2007).

## Turbine design

There are two main categories of turbine which are used to convert this water power into socially useful power, usually electricity. These are as follows:

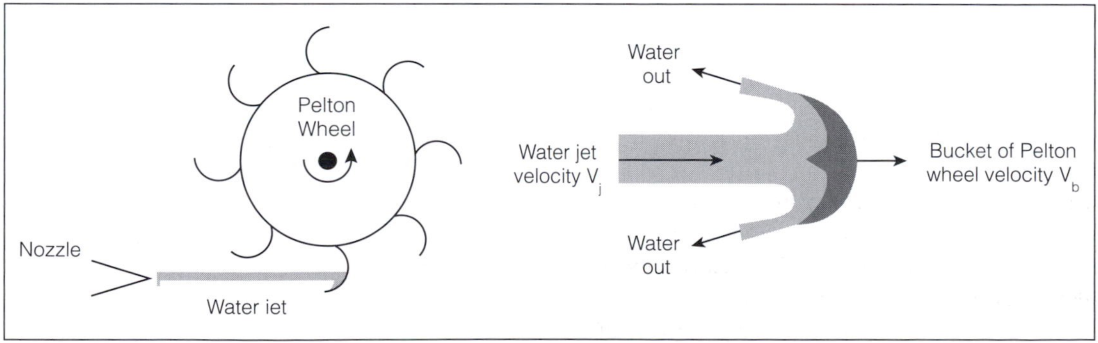


**Figure 7.37** Pelton wheel showing shape of bucket and splitting of outgoing water flow

1 *Impulse turbines*, where the flow of water hits open turbine blades in a jet and the power derives from the kinetic energy of the water.
2 *Reaction turbines*, where the turbine blades are fully immersed in the water and the power comes from the pressure drop across the turbine.

The choice of turbine for a particular system depends on head, flow rate and whether the turbine will be fully submerged.

## Pelton wheels

A Pelton wheel is an impulse turbine in which a jet of water hits a bucket attached to the wheel rim. The bucket is shaped as shown in Figure 7.37 so that the jet is divided into two equal streams and deflected from the incoming jet and out of the bucket. The direction of the water flow is reversed, giving the water a change of momentum equal to twice the momentum of the jet relative to the bucket.

The force on the bucket is

$$F = 2M(V_j - V_b)$$

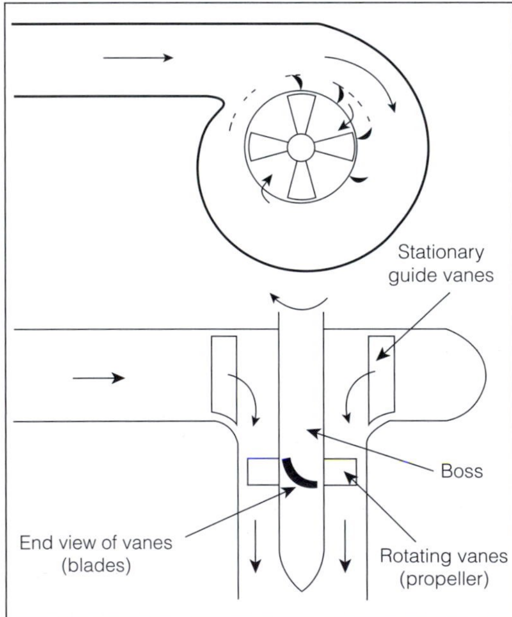


**Figure 7.38** The Francis turbine

where $M$ = mass flow rate. The power developed is

$$P = FV_b$$

$$\text{i.e. Power } P = 2M(V_j - V_b)V_b.$$

The power, in the ideal case, is equal to the total kinetic energy per second in the jet, that is 100 per cent efficiency. No real Pelton wheel can achieve this because of some friction of the water flowing round the bucket and the momentum of the outlet water not being quite equal to the momentum of the jet. However, efficiencies around 90 per cent are achieved in practice. It can be shown that the optimum speed of rotation of the wheel is half that of the speed of the water and Pelton wheels are used for high heads and high water speeds, but relatively low volumes of water.

## Francis turbine

Reaction turbines can be rather more efficient than a Pelton wheel, at the cost of greater mechanical and hydraulic complexity. The Francis turbine is shown in outline in Figure 7.38. Water flows into a casing around the working parts. The water flow is guided by fixed vanes onto the rotating vanes and the water leaves via the central outlet. To maximize the throughput of water, the machinery can be placed in a duct, of the same diameter as the rotating vanes, and the water then flows axially down the duct. The vanes are now similar to the propeller of a ship or aircraft and they rotate because of the stream of water passing over them. Francis turbines can cope with higher volumes of water than the Pelton wheel, but the rotation speed is equal to that of the water and so they are used at medium heads.

## Tidal energy

Tides are the result of the interaction of the gravitational pull of the moon, and to a lesser extent the sun, with the oceans. This results in a twice daily rise in sea levels at any given point and we can exploit the change in sea level to extract energy from the tidal flow, using similar technology to that described for hydropower. The tidal cycle is about 12 hours 25 minutes, since it depends on the lunar day which is 24 hours and 50 minutes long, and this is why the times of high and low tides change each day by a small amount. Although the sun is a much larger body than the moon, it is much further away so has less of an effect on the tides than the moon (a little less than half). When the sun, moon and Earth line up, the maximum forces are exerted on the oceans and the tide is highest (Spring Tide) and there is a correspondingly low tide (Neap Tide) when the forces from the sun and moon are in opposite directions. One of the most useful aspects of tidal energy is that it is predictable since we know when the tides will occur and their magnitude. However, there is some variation due to weather, particularly wind.

The difference in sea level is called the tidal range and depends on the gravitational forces and the topography of the land. It can be shown that the tidal range in deep water is around 0.5m and this is not sufficient for us to make use of it. However, where the flow of water is concentrated by the land, such as in a river inlet, it increases and can exceed 10m in suitably shaped estuaries. From a cost perspective a tidal range of at least 5m is generally required to make a site worthy of consideration.

The usual approach for a tidal energy plant is to construct a barrage (or dam) across the estuary mouth in an appropriate location. There are then two ways to control generation. First, as the tide is rising, water is allowed to flow in through the barrage and sluice gates are closed at high tide to trap the water behind the barrage. When the water level outside the barrage has dropped sufficiently, the water is allowed to flow back through the turbines, with the difference in height of water on each side of the barrage forming the required head. In the second case, during the rising tide water is kept on the outside of the barrage and when the difference in water level between inside and outside is sufficient, water is allowed to flow through the turbines into the basin behind. Generation of electricity only occurs when water is flowing through the turbines, so the profile is cyclic in nature. It is possible to combine the two approaches and generate electricity from water flow in both directions. This adds complexity and does not increase the overall energy extraction, but it is sometimes useful since there is a longer period of generation in each cycle.

It is also possible to generate power from turbines placed directly in the tidal current, although this technology is still at an early stage of development. We will consider current flow devices in the section on wave power. These devices would be placed where there is a sufficiently strong flow, such as in a channel between two islands.

Clearly tidal energy can only be exploited by those countries with a suitable coastline and then only in particular areas. The world's major tidal power sites are shown in Figure 7.39 and there has been little change in estimated resources since this figure was produced. The largest tidal power station is still La Rance in France which is rated at 240MW and was completed in 1966. However, there is renewed interest in a number of schemes around the world, including the Bay of Fundy in Canada, which has one of the highest tidal ranges at over 11m and where a demonstration project has recently been announced. The UK government is also reconsidering the Severn Barrage scheme, which has a 7m tidal range. With all barrage schemes, the consideration is whether the potential to generate energy outweighs the environmental effects of the plant on the local area and justifies the very large construction costs.

## Tidal turbines

For single direction flow, it is common to use similar turbines to those in hydropower systems, remembering that the volume of water is large and the head is relatively low. If we wish to have generation from either direction of water flow, we need a turbine that does not change its direction of rotation when the direction of water flow is altered. A Wells turbine has propeller blades which give the same rotation regardless of flow

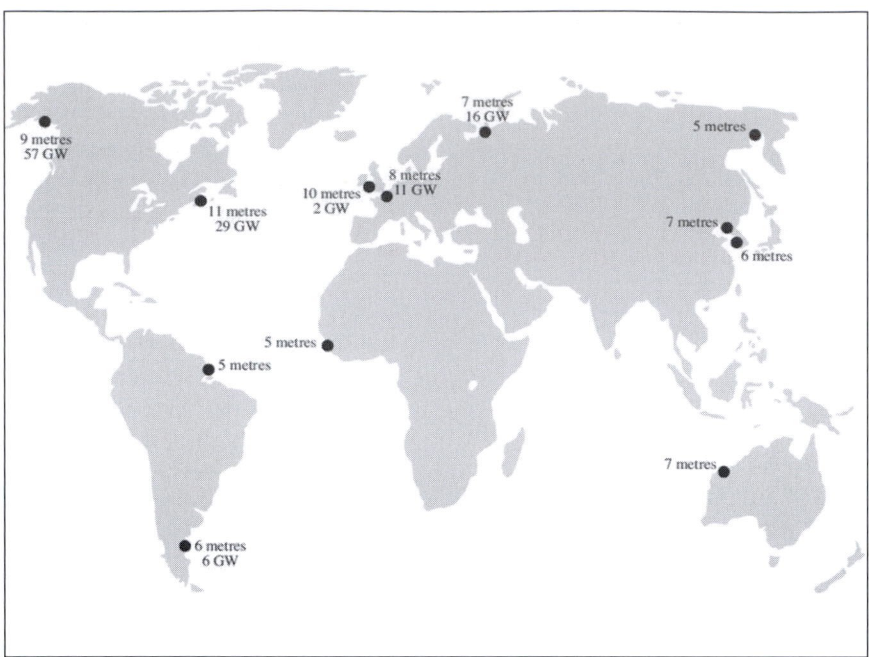


**Figure 7.39** Tidal ranges at selected sites around the world and estimated power output

direction, and can thus directly drive a generator from both incoming and outgoing tides.

A turbine to generate electricity from tidal currents can be considered as similar, at least superficially, to a wind turbine since both are designed to extract energy from a fluid flow. Several variations of turbine design have been developed and are now being tested, including both horizontal and vertical axis designs. The drag forces on a tidal turbine are more severe than those on a wind turbine, due to water having a higher density than air, and it is more of a challenge to fix the turbine securely in the tidal current. Turbines fixed to the sea bed are favoured in shallow water, but they are more often simply moored to the sea bed for deeper locations.

## Energy from tidal devices

For tidal barrage schemes, the energy output is calculated on the same principles as for hydropower systems, with the appropriate head and flow rate and remembering that the density of sea water is a little higher than that of fresh water (around $1025 kg/m^{-3}$).

In the case of tidal stream devices, as with wind turbines, the power output varies with the cube of the fluid velocity, so fast moving water would give a much higher output. However, unlike a wind turbine, the structure would not experience extreme speeds in gales or gusts and so would not need to be engineered to cope with excessive loads. Since water is much denser than air, $2ms^{-1}$ water current has the same power density as a $19ms^{-1}$ air current and many tidal and river sites could generate useful power. Tidal currents are variable, being zero when the tide turns twice a day so the load factor on a tidal marine turbine may only be 20 per cent, that is, its annual energy output is only 20 per cent of that which it could give if it operated at full power all the time. By contrast, a water turbine in a river whose flow varies little over the year might have a load factor of 80 per cent or so, giving four times the annual energy output of the same turbine in a tidal current and therefore the economics of a river turbine would be correspondingly more attractive. The tidal device also presents more challenges in terms of installation and maintenance due to its location. Nevertheless, there are several full-scale prototypes under test and tidal current devices are suited to providing power to coastal and island locations.

## Wave energy

Wave energy is extracted from ocean waves which are generated by the action of wind passing over a large stretch of water. Anyone who has been hit by a large wave knows how powerful they can be. The most powerful are those with a long period (the time taken for successive peaks to pass a given point) and great height, and they mainly occur in deep water because inshore waves lose much of their power through friction from the sea bed. Even so, coastal waves can have a very considerable average energy.

Figure 7.40 shows the average annual energy per metre of wave in various parts of the world. As with other renewable energy technologies, the calculation of the resource relies on technology assumptions and an estimate of the percentage of the energy in waves that could be extracted economically. Wavenet, a network of experts on wave energy, have suggested a technical resource of 5–20TWh/year for near shore devices and 140–750TWh/year for offshore devices, the wide range reflecting the fact that the energy extraction depends on both the wave regime and the nature of the device (Wavenet, 2003).

## Energy in waves

The energy in a wave consists of the kinetic energy of the moving water, and the potential

**Figure 7.40** Average annual wave energy per metre at selected sites (MWh)

energy associated with the peaks and troughs above and below mean sea level. For simplicity, let us consider a single wave with a wavelength $\lambda$ (the distance from crest to crest) and a wave period $T$ (the time between the passage of one crest and the following crest past a fixed point). The velocity at which the wave energy moves is given by $V = gT/2\pi$ and the wavelength is related to the wave period by $\lambda = VT = gT^2/2\pi$.

In both these relationships, $g$ is the acceleration due to gravity.

The total energy per unit surface area of the wave is given by

$$E = \tfrac{1}{2}\rho g a^2$$

where $\rho$ is the density of sea water and $a$ is the amplitude of the wave (the height of the crest from mean sea level, or half the height from trough to crest). The power per metre length of wavefront is

$$P = EV = (1/8\pi)\ \rho g^2 a^2\ T \text{ or}$$

$$P = 3.9a^2T\ (\text{kW})$$

when $a$ is in metres and $T$ in seconds.

For a wave of 1m amplitude and a wave period of 10 seconds, the power/metre is then 39kW. The length of such a wave would be about 150m, so these are, for example, the long Atlantic rollers which sweep into the west coast of Great Britain.

This analysis assumes that the waves are in deep water, where the sea bed has no influence on them. In shallow water, the velocity varies as the square root of the depth, so waves slow down in shallow water. This dependence of velocity on depth explains why waves usually arrive parallel to a shore and explains the breaking of waves when the trough is slowed so much that it is overtaken by the crest. Long before these effects occur, the wave is losing energy to the sea bed as it moves over the continental shelf into shallower waters, so waves close to shore are less powerful than waves in deep water.

The power per metre of a wave varies with the square of the amplitude. As waves are whipped up in a storm, the amplitude can increase significantly, increasing the power greatly. All wave energy devices must be engineered to withstand these large destructive waves. For about 1 per cent of waves, periods over 11 seconds are found. In these waves, the power/metre can exceed 1MW. The biggest waves, likely to be encountered once in a hundred years, may have an amplitude of 30m and power of 20MW $m^{-1}$ and few structures would withstand them.

Real seas, as opposed to the ideal single wave considered earlier, consist of a mixture of waves of different lengths and amplitudes. Away from the shore, waves may approach a device from different directions, so the devices must cope with a wide variety of waves at any given time and a much wider variety from calm to storm over their lifetime.

## Wave energy devices

Wave energy devices can be classified in a number of ways. The simplest classification is between active and passive devices. In active devices, some element moves with the wave and power is extracted from the relative movements of the various components. A passive device tries to capture as much energy as possible from the wave by presenting a large immovable structure in its path. At a more detailed level, wave power devices can be sub-divided into rectifiers, tuned oscillators and un-tuned oscillators or dampers. The rectifiers convert the energy of the wave into a head of water and the potential energy represented by this head is used to drive a water turbine. Tuned oscillators respond efficiently to a narrow range of wave periods with a fall-off in efficiency for waves of higher or lower periods. Un-tuned oscillators or dampers seek to absorb the energy from waves of all wavelengths efficiently, although for practical devices they will absorb some wavelengths more effectively than others.

The most detailed classification divides devices into ramps, floats, flaps, air bells and wave pumps. Ramps are passive devices which allow water to run up a sloping ramp into a reservoir and water from the reservoir runs back to sea through a turbine. Floats heave up and down on the waves, and the relative motion is used to drive a pump or generator. Flaps work by opening to allow a wave in and then closing to retain the head of water in a reservoir. Air bells usually float on the sea and have an open bottom below the water surface. The effect of a passing wave is to increase the pressure of air inside the bell as the crest passes and reduce the air pressure as the trough passes. The air escapes out of or into the air bell through a duct containing a turbine. A Wells turbine can be used to drive a generator from both the inward and outward motion of the air. Wave pumps exploit the pressure variations beneath the water surface to pump a fluid through a turbine.

An ideal wave device can convert waves of varying amplitude and direction into useful energy, whilst being robust enough to withstand storm conditions at its location. It needs to be able to be tethered in position and for the generated electricity to be taken off and transmitted to a distribution point. The best wave conditions are offshore, but this location presents challenges in terms of installation, mooring, power transmission and storm resistance.

There is a wide range of wave energy devices that have been proposed and it would be impossible to describe them all here. We will provide a brief summary of two examples, one usually operated near shore and one offshore, and which are amongst the closest to commercial deployment. However, this should not be taken to suggest that they are necessarily preferred over the many other options.

The concept of the Oscillating Water Column (OWC) is shown in Figure 7.41. It comprises a partially submerged chamber into which water is forced as the wave approaches, compressing the air trapped inside the chamber. The air is allowed to escape through the top (or sometimes side) of the chamber through a turbine, so generating electricity. As the wave recedes, the air space expands, the pressure drops and air is pulled back in to the chamber through the turbine. A Wells turbine is often used, providing rotation in the same direction for both the inward and outward movement of air. OWCs come in a variety of designs and can be used both at the shore line and near to shore.

Offshore devices are moored to the sea bed and need to respond to different wave heights and directions. The Pelamis wave device is just one example and three devices are currently on trial off the coast of Portugal. The Pelamis is named after the sea snake and consists of a series of cylindrical sections joined together by hinges (Figure 7.42). The sections can respond to the waves by oscillating in both the vertical and horizontal directions and the prototype versions are rated at 750kW. Each hinge section houses hydraulic rams which pump high pressure oil through hydraulic motors which in turn drive the electrical generators. The electricity is transmitted via a single

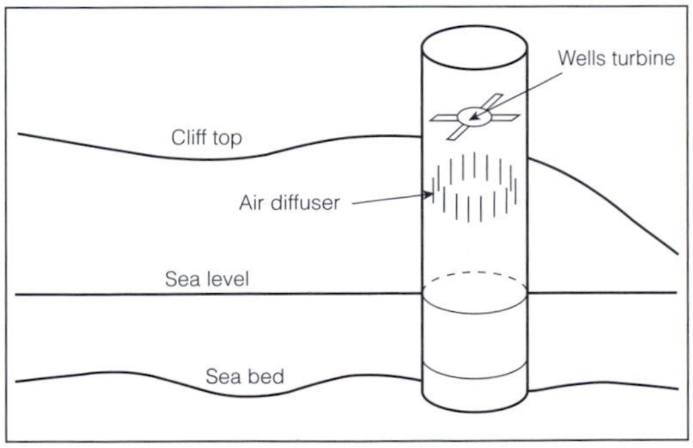

**Figure 7.41** Oscillating water column device

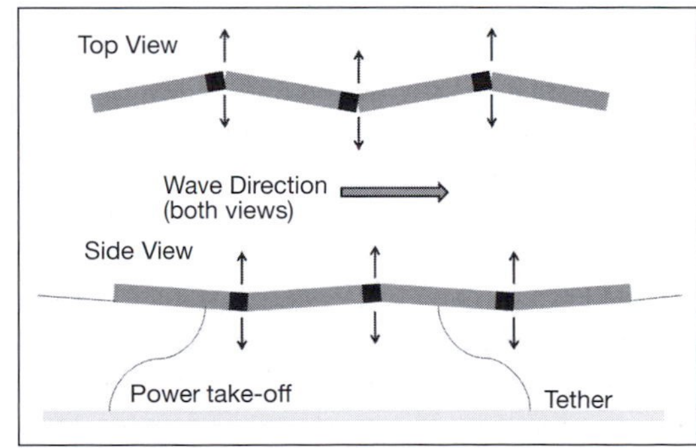


**Figure 7.42** Operation of Pelamis wave power device showing directions of movement

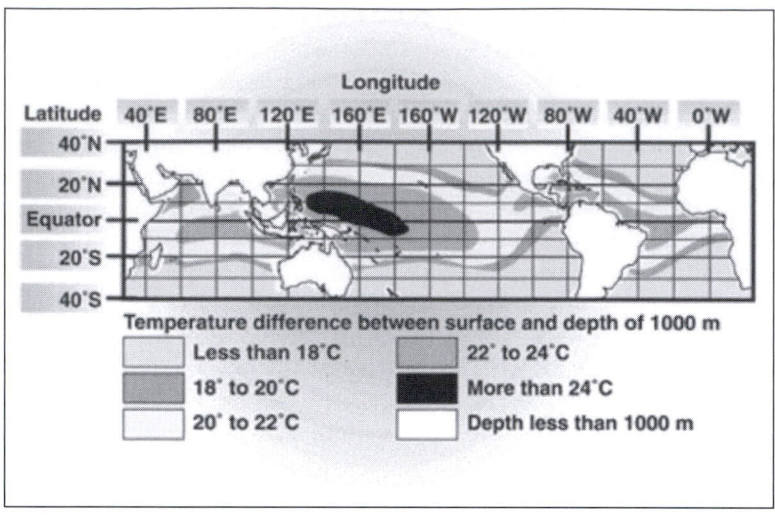


*Source:* NREL, 2008

**Figure 7.43** Average temperature differences between the surface of the ocean and a depth of 1000m

cable to a junction on the seabed and thence via cable back to shore. The device is designed to operate in depths of 50–70m and can be towed back to shore for maintenance when required. The device can be used singly or in wave farms (much like a wind farm).

Wave devices are still a relatively immature technology with a lot of technical and operational challenges, but a number of recent initiatives are allowing the gathering of valuable field data and wave energy is expected to provide a growing contribution in the coming years.

## Ocean thermal energy conversion

So far, we have discussed the conversion of the potential or kinetic energy of water, but it is also possible to consider the thermal energy in the oceans. Oceans make up over 70 per cent of the Earth's surface and they receive solar energy in the same way as the land does. As a result, the surface water is heated. Ocean thermal energy conversion (OTEC) exploits the temperature difference between water at the surface of the ocean and water at a depth of up to 1000m (this being a practical limit for extracting the cold water).

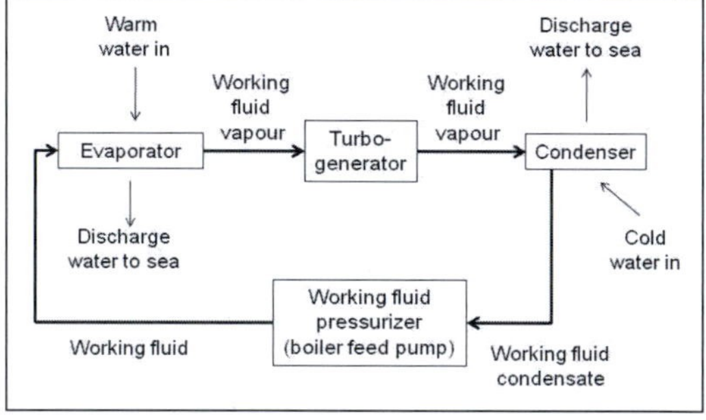


*Source:* NREL, 2008

**Figure 7.44** Schematic of closed cycle OTEC system

Figure 7.43 shows the average temperature difference for the world's oceans, where a value of 20°C is generally accepted as the minimum for OTEC systems to be able to extract viable amounts of energy. This restricts the technology to tropical or sub-tropical regions, between latitudes of around 20°N and 20°S. Interestingly, this includes many island nations, with relatively small economies and who are currently highly dependent on oil imports.

The system is basically a heat engine, as discussed in Chapter 1, and a schematic of the operation of a closed cycle OTEC system is shown in Figure 7.44. In this case, heat exchangers are used to transfer the heat from the surface water to a working fluid and, later in the cycle, from the working fluid to the cold ocean water pumped from deep water. OTEC systems can also use an open cycle where the warm sea water itself is used as the working fluid. Whilst the cold sea water is not generally a problem, some chlorination is needed to prevent biological fouling of the system from the warm water.

Similar to wave energy systems, OTEC plants can be land-based, near-shore-based or located offshore as either moored or floating platforms. Land or near-shore systems can be installed in relatively sheltered areas to protect them from storms and high seas and it is possible to operate them in conjunction with, in particular, other uses of the cold seawater. However, the water supply and discharge pipes have to cope with the stress of wave action in the surf zone at the shore and may need to be long to take the discharged seawater far enough offshore so that it is at the correct depth. By contrast, offshore systems have to cope with open ocean conditions and there is the challenge of power delivery from long undersea cables (not dissimilar to the challenges faced by offshore wave energy devices). It has also been suggested that large, floating OTEC systems could harvest the ocean thermal energy perhaps as self-propelled plantships. They would have to cope with storm conditions and transmitting the energy back to land would be very challenging. Therefore it has been suggested that these plants might be used to produce fuels such as hydrogen or methanol which could then be transported more easily.

OTEC power generation can also be coupled with other shore-based activities, either by using the power directly for applications such as desalination or by using the pumped cold water for refrigeration, cooling or mariculture (e.g. growth of phytoplankton and microalgae for use in fish and shellfish farms). This adds value to the OTEC installation.

Because the temperature difference between the warm and cold water is rather low, large quantities of water need to be pumped to get appreciable energy outputs and this, of course, also takes energy. Modern demonstration plants have provided acceptable net energy outputs, but the high capital cost of the systems has meant that there are no major commercial systems operating. It is likely that the first market will be for systems in the 5–10MW range, probably linked with some of the other applications described above, and these may be available in the next five years or so. Perhaps the most important advantage of OTEC systems is that they represent a major potential resource for small island states which are currently too dependent on oil imports, once the technical and financial aspects have been resolved.

## Bioenergy

Almost all life-forms need sunlight for their energy. Photosynthesis by green plants converts large amounts of sunlight into biological material – grass, trees, etc. – which are rich in energy and form the basis of food chains for other creatures. Human society derives products such as wood or alcohol from these basic photosynthetic materials and uses them to meet various human needs. The fossil fuels which we use at the moment are products of photosynthesis many millions of years ago and form an 'energy bank' on which our present society is drawing heavily.

Bioenergy refers to the extraction of energy from recently living plants. There are a number of different terms used to describe different aspects of bioenergy, with biomass generally used for solid sources and biofuels for the liquid or gaseous material, which is usually obtained from the processing of biomass. Bioenergy is used extensively throughout the world today (see Figure 7.1, where traditional biomass represents 13 per cent of primary energy demand, biofuels 0.3 per cent and there is also some contribution of bioenergy in the power generation category). The use of fuelwood, crop residue or cow dung for cooking is widespread in developing countries and, as it is not purchased, is often the only option for the rural poor. The use of bioenergy is, however, significant everywhere. For example, in 2006 the US obtained almost 5 per cent of its primary energy from bioenergy in all its forms, over half of the contribution from all renewable technologies (EIA, 2007).

Bioenergy encompasses a range of materials which are used in a variety of ways. Indeed, one of the main challenges is the diversity of types of bioenergy. We can distinguish the following categories:

- *wood fuels* – usually burnt to provide heating or lighting;
- *biomass for electricity generation* – energy crops or residues used, as the name implies, in electricity generation plants, often via co-firing with fossil fuels;
- *biofuels* – predominantly ethanol and biodiesel, used mainly for transportation and other liquid fuel applications.

We can also distinguish three main categories of biomass source material:

- *non-managed resources, such as existing forests* – this category includes much of the resource used for traditional biomass but causes some concern in regard to local deforestation and resource depletion;
- *energy crops* – trees and plants grown specifically for energy conversion;
- *waste and residues* – this category includes waste biomass from all kinds of industry (e.g. bagasse from the processing of sugar cane, forestry residue, waste from wood processing plants) together with human and animal waste and organic parts of municipal waste.

There are four major attractions in the use of bioenergy: it is an indigenous resource in most parts of the world; it provides stored energy unlike most renewable sources; it is very flexible in use; and, during its growth cycle, it captures $CO_2$. It also makes use of agricultural skills which are to be found throughout the world, and provides rural employment, and thus has the potential to address the drift from rural to urban areas.

## Wood fuels

The simplest use of biomass is to burn it and the most common biomass used this way is wood, although corn stalks, cow dung and many other agricultural residues are also burnt on open fires for cooking, heating and other social purposes. This method of conversion is rather inefficient and can involve problems in relation to local pollution, but may represent the only form of fuel available to a large number of the world's population. The next step in sophistication is to heat the wood in an enclosed space to make charcoal which is a very convenient fuel. It is much lighter than wood and so is much more easily transported. It burns without fumes and so is favoured in urban areas, but it is now a processed fuel, sold commercially to those who can afford it. The convenience of charcoal is at the expense of the energy lost in the conversion, since energy must be provided to dry out the wood and transform it. Overend (2007) provides a range of energy efficiencies for charcoal production from 25 per cent in Africa, using mainly artisanal methods, to around 48 per cent in Brazil, using industrial kilns. The other form of wood fuel is known as black liquor and is the waste product of the pulping industry.

In industrialized countries, there is a growing market for processed wood, which is dried and then made into briquettes or pellets for use in heating systems for both domestic and industrial purposes. These are burnt in purpose built furnaces that can deal with the higher volume of fuel required (biomass has a lower energy density than fossil fuels) and the ash content following combustion. In these cases, the wood usually comes from managed forests, from process residues (e.g. from the preparation of timber for the construction industry) or from energy crops such as poplar or willow.

Table 7.5 shows the 2005 consumption of wood fuels by continent. It can be seen that Africa and Asia use predominantly fuelwood, whereas, for example, North America has a higher proportional use of wood processing waste (black liquor) and a lower direct usage of fuelwood.

**Table 7.5** Consumption of wood fuels in 2005 (PJ)

| | Fuelwood | Charcoal | Black liquor | Total |
|---|---|---|---|---|
| Africa | 5633 | 688 | 33 | 6354 |
| North America | 852 | 40 | 1284 | 2176 |
| Latin America and Caribbean | 2378 | 485 | 288 | 3150 |
| Asia | 7795 | 135 | 463 | 8393 |
| Europe | 1173 | 14 | 644 | 1831 |
| Oceania | 90 | 1 | 22 | 113 |
| Total | 17,921 | 1361 | 2734 | 22,017 |

*Notes:* Original source of data was FAOSTAT. Fuelwood data expressed volumetrically and converted at 10GJ/tonne. Charcoal data expressed in terms of mass and converted at 30GJ/tonne. Black liquor converted at an average of 24GJ/tonne.

*Source:* WEC, 2007

More detailed information on a country basis can be obtained directly from FAOSTAT (Food and Agriculture Organisation of the United Nations).

## Biomass for electricity generation

The second largest use of biomass is for electricity generation. There has been a long history of using the residues from the sugar or wood processing industries to generate power locally, often in combined heat and power (CHP) systems. These are usually relatively low combustion temperature systems, frequently designed for a particular biomass source. In Brazil, this biomass is often bagasse, the residue from sugar cane processing, where about 90kg of bagasse is produced for every tonne of cane. In recent years, there has also been a growth in district CHP schemes in countries such as Sweden and Denmark, where significant woody biomass resources exist. CHP systems are typically designed to meet the required thermal load, with the electricity output as an additional output, but they are nevertheless an efficient way of utilizing biomass resources.

Incentives have recently been put in place in several countries to promote co-firing of biomass alongside conventional coal fired power stations. This allows some reduction of the $CO_2$ output of these stations, whilst still making use of the existing infrastructure. The biomass can be pre-mixed with the conventional fuel or mixed inside the boiler depending on the fuels used. A typical co-firing ratio is 5 per cent on an energy basis (remembering that this will be higher by volume since biomass has a lower specific energy content), although up to 15 per cent is generally considered to be technically possible without significant changes to the boiler system. Some attention needs to be paid to the ash from the combustion, since some biomass sources have inorganic content such as potassium which can lead to problems with boiler fouling, but many biomass sources produce a lower volume of ash than coal (NETBIOCOF, 2006). One interesting aspect of co-firing is in relation to the use of the fly ash, which has traditionally been used as a concrete additive. Many regulations only allow the use of coal generated fly ash and so preclude its use from co-firing plants. There appears to be no technical problem with the ash produced from wood combustion, but the alkaline residue from some other biomass sources may compromise its use in concrete production (IEA, undated).

The economics of co-firing of biomass vary with the cost of both biomass and coal sources, the local availability of biomass and any incentive programmes in place. It is generally more expensive to use biomass, but this is offset by the reduction in emissions from the plant. As regulations regarding $CO_2$ emissions are strengthened, this also has an increasing monetary value. Provided that the biomass proportion is not too large, its addition does not affect the efficiency of the plant by a significant amount. Co-firing is only really of interest in combination with coal, so it will also be necessary to continue to develop power generation plants fuelled only by biomass.

## Biofuels

The final category of usage comes under the general title of biofuels, where biomass is turned into a liquid fuel or into biogas to be used for the generation of heat, electricity (in some cases) or as a transportation fuel. This category encompasses a wide range of possible sources, processes and uses and so we will just concentrate on the main options.

In terms of publicity and promotional policies, the use of biofuels (ethanol or biodiesel) for transportation is the most widely known application. Many countries, including the US and most of Europe, have targets for the use of biofuels and, as with co-firing, one of the easiest routes to implementation is to produce a blend of ethanol and gasoline, rather than move directly to the use of ethanol alone (although some vehicles will run on ethanol only). Seventeen countries, including Brazil, India and the US, now have mandates for the blending of ethanol with gasoline (typically 10–15 per cent by volume) and biodiesel with diesel (typically 2–3 per cent) (Martinot, 2008). Brazil has operated a mandate for 30 years, with shares of around 20–25 per cent ethanol (generally produced from sugar) alongside other supporting policies. As part of the energy and environmental targets announced by the EU, the aim was to move to a 10 per cent share of biofuels for transportation by 2020. This has since been challenged in terms of sustainability and $CO_2$ reduction (see the discussion in the section on bioenergy and the environment), but nevertheless a substantial increase in biofuels is to be expected in the next few years.

Figure 7.45 shows a schematic of the biochemical route to the production of ethanol, which generally uses a fermentation process. Bio-conversion processing has been known for thousands of years, and is used to produce those

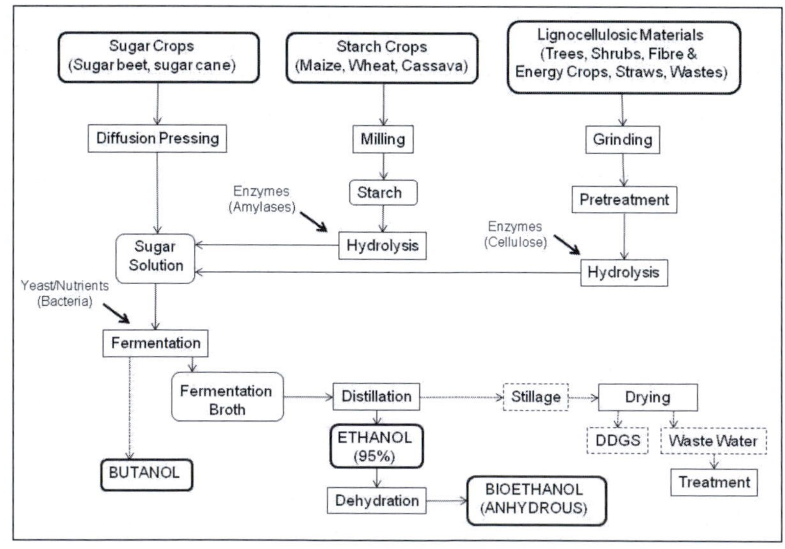


*Source:* European Biofuels Technology Platform, 2008

**Figure 7.45** The biochemical route to ethanol

**Table 7.6** World production of ethanol (hm³)

| Country | 2004 | 2005 | 2006 |
|---|---|---|---|
| Brazil | 15.10 | 16.00 | 17.00 |
| US | 13.40 | 16.20 | 18.40 |
| China | 3.65 | 3.80 | 3.85 |
| India | 1.75 | 1.70 | 1.90 |
| France | 0.83 | 0.91 | 0.95 |
| Russia | 0.75 | 0.75 | 0.75 |
| Germany | 0.27 | 0.43 | 0.77 |
| South Africa | 0.42 | 0.39 | 0.39 |
| Spain | 0.30 | 0.35 | 0.46 |
| UK | 0.40 | 0.35 | 0.28 |
| Thailand | 0.28 | 0.30 | 0.35 |
| Ukraine | 0.25 | 0.25 | 0.27 |
| Canada | 0.23 | 0.23 | 0.58 |
| Total of above | 37.60 | 41.60 | 45.90 |

*Source:* WEC, 2007

**Table 7.7** Production of biodiesel (thousand tonnes)

| Country | 2004 | 2005 | 2006 |
|---|---|---|---|
| Germany | 1035 | 1669 | 2681 |
| France | 348 | 492 | 775 |
| Italy | 320 | 396 | 857 |
| Malaysia | | 260 | 600 |
| US | 83 | 250 | 826 |
| Czech Republic | 60 | 133 | 203 |
| Poland | | 100 | 150 |
| Austria | 57 | 85 | 134 |
| Slovakia | 15 | 78 | 89 |
| Spain | 13 | 73 | 224 |
| Denmark | 70 | 71 | 81 |
| UK | 9 | 51 | 445 |
| Other EU | 6 | 36 | 430 |
| Total of above | 2016 | 3694 | 7495 |

*Source:* WEC, 2007

important solar products – beer, wine and spirits. In the Brazilian bioalcohol programme, a process similar to distilling spirits is used to ferment sugar cane and distil the ethanol; in the US grain is used for the same purpose. To ferment woody biomass, produced by coppicing energy plantations, its cell structure must first be broken down by hydrolysis, acids or enzymes to allow fermentation to proceed efficiently. Table 7.6 shows that bioethanol production has been increasing at about 10 per cent per year from 2004 to 2006, but a more rapid increase may be expected as more countries mandate its use.

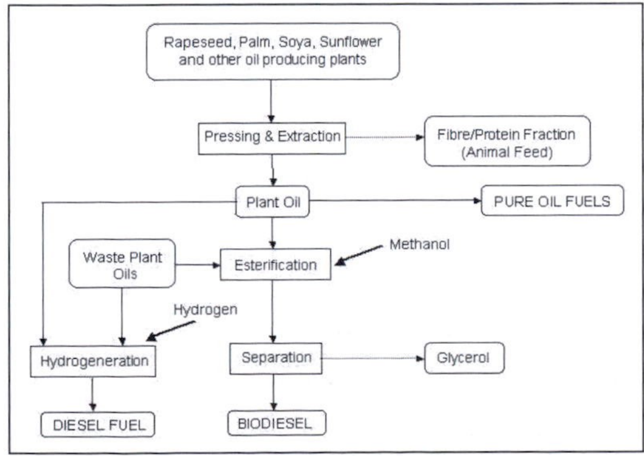


*Source:* European Biofuels Technology Platform, 2008

**Figure 7.46** Production route for biodiesel

Biodiesel is usually produced by esterification of animal fats or vegetable oils (see Figure 7.46). There are about 20 different species of crop that can be used to provide vegetable oils, including rapeseed, sunflower and soybean. The main product of the esterification process is FAME (fatty acid methyl ester) which is the precursor for biodiesel, with glycerol as a byproduct. The production of biodiesel has increased rapidly in recent years (Table 7.7), with palm oil becoming a major source because of its high energy ratio of around 8 (Overend, 2007). This has caused some concern in terms of the cutting down of forests to plant new palm oil plantations and how this affects the net $CO_2$ emissions relating to the biodiesel.

As well as liquid fuels, it is possible to produce biogas for use in heating systems by a process of gasification of a variety of source materials. In many parts of the world, animal and sometimes human wastes are used in biogas plants. The anaerobic digestion of these wastes produces a methane rich gas and leaves behind a benign residue which can be used as compost. There are millions of farm- or village-sized biogas plants in operation around the world, particularly in China and India. But collecting waste from animals which roam freely can be so time-consuming that biogas production is no longer worthwhile and it is only really practicable where animals are kept in pens. Nonetheless enough gas to satisfy the family's cooking and lighting needs can be produced from the waste of the livestock owned by a typical peasant farmer. This is particularly true where fairly rapid cooking is the norm, as with the Chinese wok. In general, the richer the farmer, the more animals the family will have, and hence the more likely it is that a biogas plant could supply the family with its cooking and lighting needs.

## Bioenergy and the environment

Photosynthesis results in the absorption of carbon dioxide and the emission of oxygen, so the growth of plants removes carbon dioxide from the atmosphere. This is re-emitted, possibly accompanied by methane, when they die and decay. The planting of trees is often regarded as a means of counteracting the increasing atmospheric concentrations of carbon dioxide and any reduction in the area of high productivity plants in the tropical forests and grasslands is a cause for concern in its effect on the natural carbon cycle.

The generation of bioenergy through combustion or other processes also releases the stored $CO_2$ in the material, so the simple view of bioenergy is that it is $CO_2$ neutral (i.e. what is absorbed during growth of the biomass is released during the energy generation process). However, this fails to take into account the energy required to manage the growth of the biomass (if fertilizers are used, for example), to harvest it, to process it (e.g. drying) and to transport it to the site of use. Therefore, in the best case, bioenergy is a small net emitter of $CO_2$, but the variability of the factors above mean that some care should be taken to ensure that the overall process is less carbon intensive than the energy source that is being replaced.

This is especially important when considering the transportation of the resource to the site where it is to be used to generate heat, electricity or motive power. Biomass has a lower energy density than fossil fuels and so requires a larger mass to produce the same power level. It is generally true that biomass resources should be used close to the location where they are grown in order to minimize the $CO_2$ (and cost) levels associated with this process step. This has some implications when considering the possibilities of poor growth conditions due to weather conditions in a particular area at certain times. Whilst the import of biomass from outside the affected region would allow continued generation, it may result in significant environmental impacts in terms of $CO_2$ and other emissions. There is also considerable variation with regard to the biomass source used. For example, ethanol made from corn is only marginally environmentally beneficial with only about 40 per cent more energy being produced than is used in manufacture and delivery, whereas in Brazil the production of ethanol from corn has an energy ratio of about 8 (eight times more energy produced than used in the supply chain) (Overend, 2007).

Of course, the process by which the energy is generated should be as efficient as possible. Because of the potential for emissions and pollutants, the combustion of biomass should be carefully controlled. Nevertheless, biomass contains little or no sulphur, so the acid emissions of sulphur compounds are avoided. The production of nitrogen oxides can also be reduced considerably in comparison with fossil fuels. Burning biomass as, for instance, on a garden bonfire can produce large quantities of noxious chemicals, some of which are carcinogenic, but if burning is properly controlled, in a furnace with flue gas cleaning, the overall acid emissions can be very small. The more advanced processing of raw biomass can be designed to emit very little into air, water or land.

Perhaps the most contentious issue with respect to the use of biomass (and, at present, biofuels in particular) relates to the competition for land between energy crops and food crops (both for human consumption and animal feedstock). A potential conflict arises because, as bioenergy implementation increases, more land is required for the growth of energy crops. If that land was previously in use for growing food, then that puts pressure on the supply of food with the potential for shortages and price rises. Some experts argue that the biofuel development policies, designed to promote the increased use of biofuels for environmental reasons, are having just that effect. It is clear to see that a farmer, especially where there are few profits to be made, will choose to grow the crop that provides the most income and if there are subsidies for energy crops but not for food crops, then he is likely to change his crop. In some cases, no change is required and it is only necessary to sell the corn or maize to a different buyer. What is not clear at the moment is how big an effect this is likely to have on the price and supply of food in the long term.

However, in the short term, a move from food to energy crops also has potential environmental implications. First, since different plants capture different amounts of $CO_2$ during growth, the change of land use can affect the $CO_2$ absorption potential of that land, resulting in the possibility of increasing the effective $CO_2$ burden of a particular energy crop depending upon what it has replaced. Second, considering only the commercial imperatives of the market has led to the transport of bioenergy resources over long distances and this can sometimes lead to the $CO_2$ footprint of that resource being higher than the fossil fuel that it replaces. There is ongoing assessment of the environmental impacts associated with the distortion of the market brought about by ambitious targets for the introduction of biofuels and some regions are stepping back from the targets whilst these issues are resolved. It may be that restrictions are placed on the source of the bioenergy to be used in particular countries so as to ensure positive gains in respect of $CO_2$ emissions.

### The potential of bioenergy

Quantification of the overall bioenergy resource is difficult, since it depends on evaluating a wide variety of sources and on assumptions relating to land usage in the face of growing demand for food. There is also the potential for the development of the productivity of energy crops in terms of output per unit area, by judicious choice and modification of crop characteristics. As the crops, land use and food requirements change over time, the annual resource will also change. Nevertheless, it is clearly a large resource, with a technical potential several times larger than the current world energy use, and bioenergy can make a substantial contribution to meeting our future energy needs, especially in sectors such as transportation which other renewable energy technologies do not specifically address.

Overend (2007) suggests that the 2006 usage of bioenergy amounts to around 24EJ from all sources (note that because of the wide range of applications, it is necessary to convert all outputs to a single energy unit and the assumptions in doing this have some impact on the total). This does not agree fully with the estimate of the IEA Bioenergy Programme, who estimate a bioenergy contribution of around 45–55EJ in 2004 (IEA, 2007). It is likely that the difference occurs in the way in which traditional biomass is counted, since this is a non-commercial sector and it is hard to derive the figures. This illustrates the difficulty of calculating an overall resource, although it is much easier to consider resources for specific countries or regions. For example, Overend describes the assessment of the US bioenergy resource and indicates that there is potential for some 20EJ of bioenergy per year using current technology. The IEA argues that worldwide bioenergy production could reach 200–400EJ per year by 2050.

## Geothermal energy

Geothermal energy is essentially the extraction of heat that is stored in rock, where that heat originates from the kinetic energy of the accreting particles that formed the Earth or, partially, from the decay of long-lived radioisotopes such as $uranium_{238}$. Geothermal energy has been used for many centuries, in terms of people using hot springs for washing or bathing, with electricity first being generated from geothermal sources in the early part of the 20th century. Unlike most of the other renewable sources discussed in this chapter, geothermal energy is not dependent on the climate but the ability to extract it does depend on the geological formation and so is location dependent.

Temperatures at the centre of the Earth are estimated to be around 4000–5000°C and heat flow is mainly by convection currents. This is a very efficient method of transferring heat and so the temperature difference with depth is relatively small. At the base of the continental crust (the solidified mass of rocks that forms the surface of the Earth, temperatures are in the range of 200–1000°C, depending on location. Heat transfer through the crust is by conduction and this leads to a significant temperature gradient with depth of around 25–30°C/km. Thus, assuming a mean annual temperature of around 15°C, a 3km well would have a bottom temperature of around 90–100°C. It is unusual to consider a well of greater depth than about 3km due to the practical difficulties of drilling such a well.

However, in certain locations, where there is volcanic plate activity or the rock formations are suitable, higher temperatures can be accessed at shallower depths. For electricity production from geothermal steam, we generally need well temperatures in excess of 150°C and so most large geothermal plants are sited in areas around the margins of the tectonic plates. In these regions, it is possible for wells at accessible depths to reach over 350°C. High temperature geothermal fields tend to be found in places where there is significant volcanic activity such as the so-called Ring of Fire around the Pacific or the East African Rift Valley. For well temperatures lower than 100°C, the geothermal energy is usually used directly for heating purposes and this can be done in a much wider range of locations. Nevertheless, the right characteristics of rock formations (porosity, hydraulic conductivity) are required if meaningful amounts of heat are to be extracted. Coarse grained volcanic ashes and some sandstones and limestones show high hydraulic conductivity.

To extract heat from the Earth, we need a heat source, a thermal fluid and an aquifer (or reservoir) of that fluid. The heat source can be the normal conduction of heat through the rocks or a high temperature source such as an intrusion of magma reaching to relatively shallow depths. The fluid is normally water, either in liquid or vapour form depending on the temperature and pressure. In some cases, there is a naturally occurring aquifer where the water is continuously replenished and all we need to do is to tap into that reservoir and bring the fluid to the surface. In other cases, the water is pumped down the well, heated and then extracted. For this type of system, granite is the favoured rock system since it has low hydraulic conductivity (and so the injected water is not lost) but retains heat well.

## Electricity generation from geothermal energy

For high temperature fields, electricity can be generated in a conventional steam turbine, using steam directly from the geothermal well. In recent years, binary systems have become popular. In these systems, the steam or hot water from the well is used to heat another fluid, usually organic, with a lower boiling point and a high vapour pressure at low temperatures. This fluid is then used in the turbine. This allows geothermal fields with temperatures in the range of 85–150°C to be utilized for electricity production. Typical plant sizes are a few MW up to about 50MW, but large fields can accommodate several plants.

The utilization of the geothermal energy can be improved by using a CHP approach, where the output is both electricity and hot water, but this relies on a local need for the heat produced since it cannot be transmitted over long distances. Since geothermal power plants make use of heat stored in the Earth, the geothermal power plant has an inherent storage capacity. That allows the output to be controlled to follow demand, rather than being subject to weather conditions like many other renewable technologies.

There are, of course, environmental impacts from geothermal plants which can be summarized as follows:

- the effects of borehole drilling, construction and access roads;
- the emission of gases held in the ground water under pressure, but released when it is brought to the surface (this can include greenhouse gases in some cases, although at significantly lower amounts than for generation of the same amount of energy from fossil fuels);
- trace elements in waste water;
- the potential to trigger seismic events (although there is some debate over whether these are a result of the plant construction and operation or would have occurred in any case).

In 2005, 55TWh of electricity was produced from geothermal energy in 25 countries (WEC, 2007), with the US, Philippines, Mexico and Indonesia being the largest producers. There remains a large untapped resource of geothermal energy that could be developed over the coming years.

## Direct use of geothermal heat

For low temperature geothermal fields, the energy can be directly used for space heating, either for single buildings or in district heating schemes. It is common to use a closed loop system whereby a heat exchanger is used to transfer the heat from the geothermal water to fresh water which flows through the water in a closed circuit. Using the geothermal water directly in the radiators is only possible if the quality of the water is sufficiently high. A water temperature in the range of 60–90°C is required for the supply water, with a typical return temperature of 25–40°C. Geothermal heat can also be used for greenhouses, agricultural drying, swimming pools and other industrial uses.

In the last few years, the fastest growing sector of direct use has been geothermal (or ground source) heat pumps. The operation of heat pumps, whether using ground, water or air as the heat source, is explained in the next section. Because the heat pumps use normal ground temperatures, they can be used in a wide range of locations.

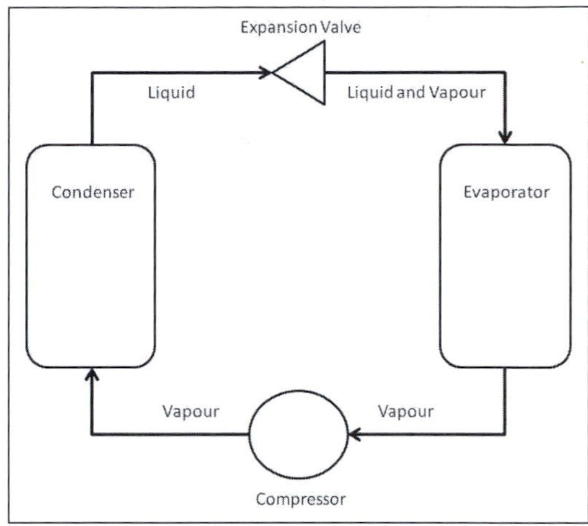


**Figure 7.47** Simplified heat pump

By 2005, geothermal heat pumps accounted for over half of the direct use capacity for geothermal energy and 32 per cent of the energy use of 87,500TJ (WEC 2007). The total number of installations was estimated to be 1.3 million, mainly in the US, Europe and China.

## Heat pumps

The heat engine is a device that transforms thermal energy into mechanical work. In the heat engine, energy flows from a hot source to a cool sink and produces work. The Second Law of Thermodynamics states that heat cannot flow spontaneously from a cold source to a hotter sink, but when external energy is applied to the system then this transfer is possible. This is the basis of the heat pump. The most common use of this cycle is the refrigerator where heat is extracted from an enclosed space and exhausted to the atmosphere. The system uses a working fluid, which can be expanded into a gas and condensed into a liquid. In its gaseous state a compressor raises the pressure of a working fluid. The output of the compressor, a hot pressurized gas is then passed through a condenser or heat exchanger, where it is cooled into a high pressure liquid state. This is then passed through an expansion valve. The expansion of the liquid back into a gas requires heat absorption from the surroundings (see Figure 7.47).

In this system the work is applied by the compressor. The service that is needed, either heating or cooling, will determine the configuration of the system. For example, if space cooling is required then the evaporator will be located in the space and the condenser in the sink. If heating is required then this will be reversed with the condenser located in the space and the evaporator in the sink. A heat pump can deliver both heating and cooling services.

Practical heat pump systems incorporate a reversing valve which allows the direction of travel of the refrigerant to be reversed. Such systems use either the atmosphere or the ground or water as the source or sink and their deployment depends upon local factors. Deployment strategies include vertical techniques that use a borehole drilled into the earth, rock or an underground water source and horizontal techniques that use coils buried about 1m below the surface of the ground or immersed in a body of water, such as a lake. These strategies rely on the fairly constant temperature found underground and within water bodies and it is the temperature between source and sink that allows heat energy transfer. In areas where it is difficult to use vertical or horizontal strategies, an air-source heat pump can be used.

Heat pumps require an external source of power to drive the compressor. Typically this is an electric motor for stationary systems (space heating and cooling) or a mechanical source for mobile systems (vehicle air conditioning). The performance of a heat pump system is called the Coefficient of Performance (COP). In terms of space heating, a heat pump can transfer up to three or four times the amount of heat when compared to the heat generated by a conventional resistance heater, for the same energy input. This does not mean that the heat pump is more efficient. Rather, it reflects that a heat pump works in a different way and is an efficient means of thermal energy transfer. A heat pump can be considered as a renewable resource as it uses heat stored in the atmosphere or ground or water. It does require an external source of power, typically electricity, and for this technology to be viewed as completely renewable this would have to be generated from other renewable technologies. There are both economic and financial benefits to be accrued by the use of heat pump technologies, although these will vary depending upon location and type of fuel currently used (EST, 2008).

Heat pump technology can make a contribution to climate goals. Within the EU heat pumps that use the ground or water as source and sink are classified as a renewable technology. The air-source heat pump requires a considerable amount of electricity to function and the EU has set a standard minimum COP of 2.9 for this technology to be considered a renewable technology (OJ, 2007). Heat pumps have been included in a proposed directive that classifies which technologies can be included in the new renewable targets set for member states (European Commission, 2008). The Directive is designed to address all sectors of the renewable energy industry with the goal of helping member states to reach the Commission's target of 20 per cent of Europe's energy being produced from renewable sources by 2020. Measures to remove barriers to growth for renewables are included.

One further issue with heat pumps concerns the refrigerant. The most common refrigerants are hydroflourocarbons (HFCs). Leakage of the refrigerant either during operation or decommissioning can contribute to global warming as many synthetic refrigerants have a high Global Warming Potential (GWP) (Forsen, 2005). EU Directive 2006/40/EC phases out the use of the refrigerant R-134a from air conditioning systems in motor vehicles by 2011 and Regulation (EC) No. 842/2006 introduces controls for HFCs by setting minimum standards for inspection and recovery. This ironically has focused interest on alternative substances and research has been undertaken into the use of carbon dioxide, a gas that was used in the early days of refrigeration but was later abandoned as it required high pressures and the development of synthetic refrigerants made the production of equipment much easier. Although carbon dioxide is a greenhouse gas, it is significantly less damaging than synthetic refrigerants.

The EU has undertaken research into alternative substances for heat pumps including carbon dioxide, ammonia and hydrocarbons through the Sherpa (Sustainable Heat and Energy Research for Heat Pump Application) project. With the market expected to double by 2010, finding an environmentally-friendly working fluid is a priority. A doubling of the market would increase the annual energy and $CO_2$ emission savings to 100TWh and 40 million tons respectively. The project, which ended in 2007, developed and tested prototype models using environmentally benign substances such as carbon dioxide and also ensured that they would be compliant with future legislation, as synthetic refrigerants such as Freon R22 will be phased out by 2010 in the EU (Thonon, 2006).

Heat pumps offer considerable environmental and energy security benefits. In many respects it is a mature technology, although the introduction of environmentally benign refrigerants is a technical challenge. The inclusion of heat pumps within the renewable target will help to ensure a more rapid take-up of the technology as its versatility, in terms of both heating and cooling and the variety of sources and sinks it can use, mean that it has a wide geographical spread.

## Implementation of renewable energy technologies

This chapter has considered the individual renewable technologies in terms of their principles of operation, the resources available and their current status, but there are some issues relating to implementation that apply to several, if not all, technologies. So it is more efficient to discuss these together here.

### Variability of output

Most renewable technologies rely on a source of energy that is climate dependent and has some variance with time, whether known (e.g. tidal) or more random (e.g. wind, solar). The energy system has to cope with that variability in some way, for example, by:

- combining sources with different and complementary variability;
- providing storage of electricity or heat to modify the resultant supply profile;
- providing sufficient back-up capacity from conventional generation to meet load requirements when the renewable output is low.

The most obvious renewable technology that does not fall into this category is bioenergy, which is similar to conventional fuels in that it is stored energy. We therefore have the possibility of combining bioenergy with other renewable sources to even out some of the variability. Some technologies, such as large hydropower schemes, concentrated solar power and geothermal, have the potential to include storage in their mode of operation. There have also been considerable improvements in the forecasting of output from both wind and solar systems as experience is gained. Furthermore, large penetrations of wind in countries such as Denmark have not shown the detrimental effects on the grid that some had predicted.

Coping with large amounts of renewable energy in the supply system may require some rethinking of our current approach to energy supply. It has been argued that having electrical generation systems with intermittent output will require this to be balanced by a large amount of conventional generation to meet the times when the renewable system output is reduced. A recent study for the UK showed that, for an assumed wind penetration of up to 20 per cent on the electricity grid, the additional balancing reserve to meet short term fluctuations was only around 5–10 per cent of the installed wind capacity (UKERC, 2006). Whilst individual systems are variable in output and individual demand is also variable, the aggregation of system outputs and demand smooth out those variations. So, the variation of output of a wind farm in Cornwall may be offset, say, by the output of a housing estate with solar photovoltaic systems in Birmingham. Or on a wider basis, wave power systems off the Norwegian coast may complement a CSP plant in Spain. We would expect more use of district based schemes, particularly for heating, so as to balance out demand. There is also an option for demand management to modify the load profile to meet the supply profile rather than the traditional approach of changing the supply to meet the load.

## Distributed generation or microgeneration

Whilst some renewable energy systems are very large, a general feature of all sources is that they can be implemented on a range of scales, some of them down to the level of an individual user. Where the resource itself is distributed, it makes little sense to adopt the old scheme of large central power stations. The most efficient way of harnessing as much energy as possible is in a distributed form, which also happens to be the way in which the energy is used. Technologies such as large hydropower, large tidal and some geothermal plants already have a concentration of resource in a specific location. Technologies such as concentrated solar power require large installations in order to reach the high temperatures required. But the other technologies (wind, solar PV, solar heating, bioenergy, wave power, tidal and marine currents, etc.) all occur in a distributed fashion. Where the implementation is at the level of the individual household, commercial building or small community and the generation technology is connected at the user level (e.g. low voltage grid), then this is often termed microgeneration.

This distributed approach to energy generation and supply has a number of advantages:

- It reduces the need for transmission of energy over long distances and therefore the accompanying losses.
- It allows the use of the most appropriate renewable technology at each site.
- Although anecdotal, there is some evidence that microgeneration, where the user is involved in the production of their own energy, encourages behavioural change in terms of energy efficiency, recycling, etc.
- A distribution of energy sources, system sizes and system locations provides a more secure system, in that loss of one system or group of systems does not impact severely on the overall energy supply.
- The variability of output can be reduced by aggregation of generation in different locations.

## Capacity factor

The capacity factor (or sometimes load factor) of an electricity generator expresses the output of the generator over a given period, usually one year, compared to the output of a generator operating at the nameplate rating throughout the same period. For a conventional generator, where it is theoretically possible to operate at full power continuously by providing the requisite amount of fuel, the capacity factor gives a measure of how reliable the generator is (since any downtime will reduce the output) and how the generator is utilized in the overall system. For example, gas fired generators which are used for rapid response to changing load demands may only be utilized for short periods when demand is high and so will have a relatively low capacity factor, even though they may be very reliable and have good conversion efficiency.

For systems based on renewable sources, the generator only runs at nameplate rating when the energy source is equal to the value at which the generator is rated. For example, a wind turbine rated at 1MW for wind speeds of 14 $ms^{-1}$ only provides 1MW of output at this wind speed and above (up to the maximum design speed of, say, 25$ms^{-1}$). For lower wind speeds, the output is lower, but this does not indicate a malfunction of the turbine or lack of utilization of the electricity generated (as would be the case for a fossil fuelled generator). Similarly, a 100kW PV system only produces 100kW when the intensity of the sunlight is 1$kWm^{-2}$ and the array temperature is

25°C (or some combination that gives the same result), but in this case we also know that there will be no output in the hours of darkness. Thus, the system will exhibit a low capacity factor, even if it is working exactly as designed.

For a renewable energy system, the capacity factor generally gives us some information about the resource at the site not the quality of the generator. For example, a site with a high average wind speed will have a higher capacity factor than one with a low average wind speed, for the same turbine design. Thus, the UK, with its better wind regime, tends to have higher capacity factors (at around 30 per cent) than those in mainland Europe. In particular, the capacity factor does not provide a measure of the efficiency of the system. It also cannot be used to predict variability. Two different systems can have the same capacity factor but widely different generation profiles. Therefore the capacity factor is not a good measure for renewable energy systems and, whilst it is necessary to take into account the generation profile, it should not be used as a direct comparison with a conventional system where fuel is constantly available.

## Costs

Costs of renewable energy technologies have not been included in this chapter, for two main reasons. First, the energy cost arising from any particular system is highly dependent on the design of the system and the quantity of the resource (i.e. the location). Second, the many market development schemes, coupled with research and development activities, mean that the costs of certain renewable technologies (particularly PV, CSP and marine) are likely to change quite significantly over the coming few years. There are several publications which address the current costs of renewable technologies in more depth than is possible in this book, for example, the *Renewables 2007 Global Status Report*, (REN21, 2007)

Nevertheless, a few general remarks about cost issues relating to renewable energy systems can be made. For most renewable technologies, with the exception of bioenergy, the main expenditure relates to capital cost since the fuel is provided by nature. There are also some ongoing operation and maintenance costs, which vary depending on the nature of the system, but these are generally much lower than the initial installation cost. This is in contrast to conventional fossil fuelled power stations where the ongoing costs of fuel and operation tend to dominate. The effect of this on the derived costs of energy are discussed in more depth in Chapter 2, but the main conclusion is that standard economic approaches favour expenditure in the future rather than now, that is technologies with future fuel costs instead of initial capital costs. By contrast, addressing the future energy needs of the planet whilst also addressing the environmental needs favours action now rather than in the future, especially where those future fuel costs are difficult to predict. We need to be able to factor in the environmental, security and social costs when making decisions on which energy sources to choose.

# Summary

This chapter has described a wide range of renewable technologies for the generation of energy. We have concentrated on existing uses of that energy, in the form of electricity or heat, and not on what might be done in the future (e.g. hydrogen production, discussed in more detail in Chapter 8). It is clear that there has been rapid progress both in the technology and in the implementation of renewable energy over the last decade and that much more is needed in the coming years. We have provided a snapshot of the current market status of many of the technologies in the period 2005–2007, since there is always some delay in publishing market numbers, but, because of the rapid changes in this sector, we encourage interested readers to seek out the latest versions of the reviews detailed in the references in order to update the numbers presented. Each different technology has different features and operates in different locations, with perhaps only solar photovoltaics, solar thermal, wind, microhydro and bioenergy being widely available around the world, but this allows the best solution to provide our energy needs in any given situation. What has become clear in the last few years is that renewable technologies should not be seen as competing with one another, rather that we will need most of these technologies to be employed to their fullest possible extent in the coming years. It is important that substantial progress in implementation is made in the period up to 2020, so setting a firm foundation for even greater implementation to meet our climate targets in 2050.

## References


COM (2008) Communication from the Commission to the European Parliament, the Council, the European Economic and Social Committee and the Committee of the Regions, 20 20 by 2020, Europe's climate change opportunity, COM(2008) 30 final, Brussels

DLR (2005) *Concentrating Solar Power for the Mediterranean Region*, Final Report, German Aerospace Center (DLR)

EIA (2007) *Annual Energy Review*, EIA. Available at: www.eia.doe.gov/emeu/aer/overview.html

EST ( 2008) *Ground Source Heat Pumps*, Energy Saving Trust (EST). Available at: www.energysavingtrust.org.uk/generate_your_own_energy/types_of_renewables/ground_source_heat_pumps

European Biofuels Technology Platform (2008) Strategic Research Agenda and Strategy Deployment. Available at: www.biofuelstp.eu/srasdd/080111_sra_sdd_web_res.pdf

European Commission (2008) Proposal for a Directive of the European Parliament and of the Council on the promotion of the use of energy from renewable sources, COM(2008) 19 final2008/0016 (COD). Available at: http://ec.europa.eu/energy/climate_actions/doc/2008_res_directive_en.pdf

Forsen, M. (2005) Heat Pumps: Technology and Environmental Impact, Report prepared by the Swedish Heat Pump Association (SVEP). Available at: http://ec.europa.eu/environment/ecolabel/pdf/heat_pumps/hp_tech_env_impact_aug2005.pdf

Fthenakis, V. and Alsema, E. (2005) 'Photovoltaics energy payback times, greenhouse gas emissions and external costs, status', *Progress in Photovoltaics: Research and Applications*, vol. 14, pp275–280

Green, M. A., Emery, K., Hishikawa, Y. and Warta, W. (2008) 'Solar cell efficiency tables (Version 32)', *Progress in Photovoltaics: Research and Applications*, vol. 16, pp435–440

Grubb, M. J. and Meyer, N. I. (1993) 'Wind energy: Resources, systems and regional strategies', in Johansson T. B. et al eds., *Renewable Energy, Sources for Fuels and Electricity*, Washington, DC: Island Press

Hill, R., O'Keefe, P. and Snape, C. (1995) *The Future of Energy Use*, London: Earthscan

IEA (2006) *World Energy Outlook*, Paris: International Energy Agency

IEA (2007) Bioenergy, Potential Contribution of Bioenergy to the World's Future Energy Demand, IEA Bioenergy Programme, Paris: IEA

IEA (undated) Bioenergy, Biomass Combustion and Co-firing: An Overview, IEA Bioenergy Programme, Task 32 Brochure, Paris: IEA

IEA-PVPS (2008) *Trends in Photovoltaic Applications*, Survey report of selected IEA countries between 1992 and 2007, International Energy Agency Photovoltaic Power Systems Programme, Report IEA-PVPS T1-17, Paris: IEA

Martinot, E. (2008) *Renewables 2007 Global Status Report*, REN21, Paris: REN21 Secretariat and Washington, DC: Worldwatch Institute

Miles, R. W., Forbes, I. and Zoppi, G. (2007) 'Inorganic photovoltaic cells', *Materials Today*, vol. 10, no. 11, pp20–27

NETBIOCOF (2006) Integrated Network for Biomass Co-firing, EC Project SES6-CT-02007, First state of the art report. Available at: www.netbiocof.net/

NREL (undated) Information pages on Ocean Thermal Energy Conversion, National Renewable Energy Laboratory, US. Available at: www.nrel.gov/otec/what.html

OJ (Official Journal) (2007) Commission Decision, 9 November 2007 establishing the ecological criteria for the award of the Community eco-label to electrically driven, gas driven or gas absorption heat pumps, OJ L 301, 20 November 2007, p14. Available at: http://ec.europa.eu/environment/ecolabel/pdf/heat_pumps/criteria_en.pdf

OkSolar.com (undated) World Solar Radiation Map. Available at: www.oksolar.com/abctech/images/world_solar_radiation_large.gif

Overend, R. P. (2007) *Survey of Energy Resources; Bioenergy*, London: World Energy Council

REN21 (2007), Paris: REN21 Secretariat and Washington, DC: Worldwatch Institute) edited by E. Martinot available at www.ren21.net/pdf/RE2007_Global_Satus_Report_pdf.

Thonon, B. (2006) The Sherpa Project: Natural Refrigerants for Heat Pumps, Conference Paper, 7th IIR Gustav Lorentzen Conference on Natural Working Fluids, Trondheim, Norway, 28–31 May 2006, www.r744.com/knowledge/papers/files/pdf/pdf_236.pdf

UKERC (2006) The Costs and Impacts of Intermittency: An assessment of the evidence on the costs and impacts of intermittent generation on the British electricity network. Available at: www.ukerc.ac.uk/Downloads/PDF/06/0604Intermittency/0604IntermittencyReport.pdf

Wavenet (2003) Results from the work of the European Thematic Network on Wave Energy, Final Report, ERK5-CT-1999-20001, European Commission, Energy, Environment and Sustainable Development Programme. Available at: www.wave-energy.net/Library/WaveNet%20Full%20Report(11.1).pdf

WEC (2007) Survey of Energy Resources, London: World Energy Council

# 8

# Energy Futures

## Introduction

In this chapter we will discuss the likely shape of the energy economy in future years. This is not a prediction as it would be impossible to accurately predict what will happen. We do know that there are extreme pressures to develop a new trajectory. Energy security concerns have seen the development of new nuclear capacity on the one hand and the push for more renewable capacity on the other. Nowhere is the debate more focused than Europe. For example, Sweden has announced that it will revoke a 30-year ban on the development of new nuclear capacity. The EU has announced the 20/20/20 goals for 2020. Clean coal technology that uses carbon capture technology is being demonstrated along with geological storage options. The EU still actively advocates carbon trading as a means of reducing emissions. In the US the election of a new President heralds a new era of re-engagement in the global climate dialogue. Additionally the US seems likely to introduce a Cap and Trade system similar to that of the EU ETS.

In summary there are considerable pressures on policy makers to shift the trajectory of energy system development to a low carbon route with little indication at present of how this will be realized. We do know that whatever decisions are made will lock the energy development trajectory for some time. For example, the proposal to develop a coal fired power station at Kingsnorth in the UK will mean that the power station will be in existence for at least 30 years and if no effective technology for carbon capture and storage is developed, then it will emit carbon into the atmosphere throughout its lifetime. More radical voices argue that we must completely change the energy economy and develop a hydrogen energy economy. There are strong reasons to support such a move as the combustion of hydrogen produces water, a harmless substance. This chapter will look at the potential for a hydrogen energy economy. It will also consider whether carbon trading can be an effective vehicle for reducing carbon and driving technological innovation.

One aspect of the energy debate that has received little attention is behaviour. We argue that the starting point for any approach to energy system development must be efficiency of both demand and supply sides. In general, the efficiency of the supply side is driven by technological developments in response to policy changes or new developments such as biofuels. Technology and new developments also influence the demand side. But lifestyles, how we view and use energy services, is an increasingly important factor implying a significant role for social learning.

## Hydrogen and the energy system

Hydrogen in its free state is an energy source. When combined with oxygen it releases energy

**Table 8.1** Hydrogen production methods

| Hydrogen production technology | Benefits | Barriers |
|---|---|---|
| *Electrolysis:* splitting water using electricity | Commercially available with proven technology; well-understood industrial process; modular; high purity hydrogen, convenient for producing $H_2$ from renewable electricity, compensates for intermittent nature of some renewables | Competition with direct use of renewable electricity |
| *Reforming (stationary and vehicle applications)*: splitting hydrocarbon fuel with heat and steam | Well-understood at large scale; widespread; low cost hydrogen from natural gas; opportunity to combine with large scale $CO_2$ sequestration ('carbon storage') | Small scale units not commercial; hydrogen contains some impurities – gas cleaning may be required for some applications; $CO_2$ emissions; $CO_2$ sequestration adds costs; primary fuel may be used directly |
| *Gasification:* splitting heavy hydrocarbons and biomass into hydrogen and gases for reforming | Well-understood for heavy hydro-carbons at large scale; can be used for solid and liquid fuels; possible synergies with synthetic fuels from biomass; biomass gasification being demonstrated | Small units very rare; hydrogen usually requires extensive cleaning before use; biomass gasification still under research; biomass has land use implications; competition with synthetic fuels from biomass |
| *Thermochemical cycles* using cheap high temperature heat from nuclear or concentrated solar energy | Potentially large scale production at low cost and without greenhouse gas emission for heavy industry or transportation; international collaboration (US, Europe and Japan) on research, development and deployment | Complex, not yet commercial, research and development needed over 10 years on the process: materials, chemistry technology; high temperature nuclear reactor (HTR) deployment needed, or solar thermal concentrators |
| *Biological production:* algae and bacteria produce hydrogen directly in some conditions | Potentially large resource | Slow hydrogen production rates; large area needed; most appropriate organisms not yet found; still under research |

*Source:* EU Commission, 2003

in the form of heat. The byproduct of this chemical reaction is water. Hydrogen offers the potential to produce carbon free energy. Some believe that hydrogen will be the fuel of the future. For example, Rifkin envisages a future world where hydrogen is the dominant fuel (Rifkin, 2002). However, the problem with the notion of a hydrogen energy economy is that while there are clear environmental benefits, hydrogen is not readily available in its free form and has to be produced. This requires energy.

Hydrogen can be produced from water by electrolysis. In this process, energy, in the form of electric power, is passed through water, transforming it into its component elements of hydrogen and oxygen. The hydrogen effectively carries part of the energy used in the process. In other words hydrogen is an energy carrier. When hydrogen is recombined with oxygen, energy is released and water formed. This seems like a virtuous circle where hydrogen is created and its recombination produces useful energy and no harmful byproducts. However, the energy input to the process has to be generated from an energy source. If, for example, fossil fuels were used to generate the power needed for electrolysis, then damaging emissions would be produced, reducing the potential benefits of the process.

Another method of producing hydrogen involves reforming natural gas by removing the carbon atom. The byproduct is carbon dioxide. Electrolysis and reformation are the most common methods of hydrogen production, but each has its drawbacks. Reformation requires methods for storing carbon dioxide, such as Carbon Capture and Storage (CCS). Electrolysis would also require CCS, or the use of renewable or nuclear sources for electricity production. As discussed later in this section it can be more efficient to use electricity generated by renewable capacity directly as opposed to using it to produce hydrogen. Other methods for hydrogen production include coal gasification which would require CCS, biological production through a bioreactor or through thermal decomposition or thermolysis. These methods for hydrogen production are shown in Table 8.1 along with a summary of benefits and barriers.

As a fuel, hydrogen is versatile. For example, it can be used in a combustion process to produce heat and mechanical work or with a fuel cell to produce electricity. Figure 8.1 shows the variety of sources that can be used to produce hydrogen and the variety of uses to which it can be put.

The use of hydrogen and fuel cells in transport systems has attracted considerable interest as it offers the opportunity to have pollution free transport systems that would bring considerable benefits to urban areas. This has been discussed in Chapter 4. Although this offers considerable environmental benefits in terms of reduced greenhouse gas emissions and urban pollution, there are questions about the efficiency of the process as energy is needed to store hydrogen in a form suitable for mobile purposes. Figure 8.2 shows the number of steps needed, each requiring an energy input, before the hydrogen is sufficiently compressed or liquefied for onboard storage in a vehicle fuel tank. This is shown on the left hand side of Figure 8.2. Each step needed to make hydrogen suitable for onboard storage requires an energy input which effectively lowers the efficiency of the overall cycle. The right hand side of Figure 8.2 shows a route using electricity

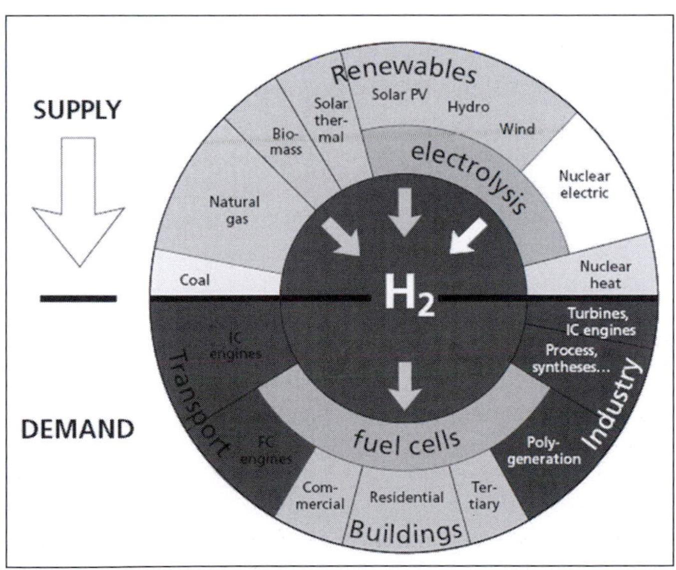


*Note:* Sector size shown has no connection with current or expected markets

*Source:* EU Commission, 2003

**Figure 8.1** Hydrogen: Primary energy sources, energy converters and applications

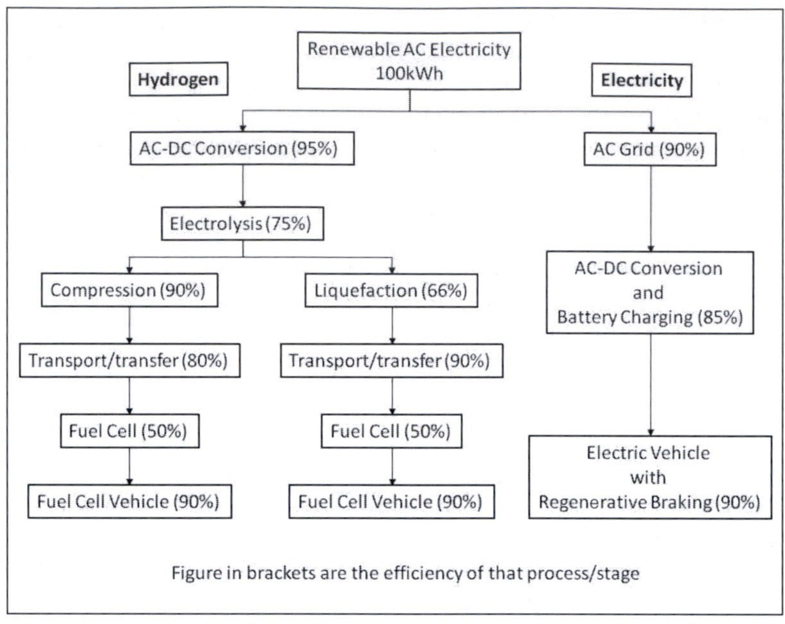


Source: Adapted from Bossel, 2006

**Figure 8.2** Hydrogen route versus electrical route for vehicles

**Table 8.2** Hydrogen storage technologies

| Hydrogen storage technology | Benefits | Barriers |
|---|---|---|
| Compressed gas cylinders | Well understood up to pressures of 200bar; generally available; can be low cost | Only relatively small amounts of $H_2$ are stored at 200bar; fuel and storage energy densities at high pressure (700bar) are comparable to liquid hydrogen, but still lower than for gasoline and diesel; high pressure storage still under development |
| Liquid tanks | Well understood technology; good storage density possible | Very low temperatures require super insulation; cost can be high; some hydrogen is lost through evaporation; energy intensity of liquid hydrogen production; energy stored still not comparable to liquid fossil fuels |
| Metal hydrides | Some technology available; solid-state storage; can be made into different shapes; thermal effects can be used in subsystems; very safe | Heavy; can degrade with time; currently expensive; filling requires cooling circuit |
| Chemical hydrides | Well-known reversible hydride formation reactions, e.g. NaBH; compact | Challenges in the logistics of handling of waste products and in infrastructure requirements |
| Carbon structures | May allow high storage density; light; may be cheap | Not fully understood or developed; early promise remains unfulfilled |

*Source:* EU Commission, 2003

directly to power a vehicle. Again there are losses associated with battery charging, but the overall loss is less than that of the hydrogen route.

Figure 8.2 only considers one aspect of a possible hydrogen economy and ignores the issue of vehicle range. If, in the future, both battery and hydrogen storage technologies were able to offer ranges equal to, or greater than, that of conventional vehicles then the battery route would be the most efficient. Although the range of both electrical and hydrogen powered vehicles has improved, more development is needed in order to match that of hydrocarbons. In addition there are issues for hydrogen related to onboard storage. Storage options for hydrogen are shown in Table 8.2. High pressure or cryogenic methods offer the most promising options at present but there are safety issues to be resolved. However, as discussed in Chapter 4, considerable investment will be needed in infrastructure.

The example of the vehicle sector does raise broader questions about an energy economy based on hydrogen. It is clear that at times where power is available from renewable capacity and there is a need then it is likely to be more efficient to use that power directly. Hydrogen can be produced and stored for use at a later time provided a method for capturing renewable resources providing the additional energy cost is recognized and integrated into system development. Hydrogen stored from excess renewable capacity can be used to generate power. There are times when the electron economy is likely to be a better option than the hydrogen economy and vice-versa. Although there are issues around the use of fuel cells and hydrogen for the vehicle sector, they are used to produce electricity for residential or commercial purposes. In this instance the need to minimize the size and weight of the equipment is not as pressing.

As discussed in Chapter 4 fuel cells are modular, meaning that they can be built to meet particular electrical loads. These constructions are known as fuel cell plants. Fuel cell plants have the advantage, when used for on-site power generation, of lower pollutant emissions, comparable efficiencies to thermal generators and are virtually silent. Fuel cell plants have been developed to provide heat and power to residential properties. Commercial systems, running on natural gas, have been developed with power ratings ranging from 1 to 4.5MW. The electrical efficiency of these systems is around 40 per cent, which is comparable to thermal plants. By using the heat generated by the fuel cell the overall system efficiency can be as high as 80 per cent (Carrette et al, 2001). If this technology were coupled to hydrogen production methods that used either renewable, nuclear or fossil fuel systems with CCS, then that approach would provide a clear environmentally benign niche role for fuel cell plants.

It is this configuration that offers a view of how hydrogen can play an effective role in the renewable energy economy. Renewables are best realized at the local level and require a development strategy that is based on the principles of 'capture/harvest-when-available' and 'store-until-required'. As discussed in Chapter 7 renewables are variable both in terms of location and time. Although some broad assumptions can be made about availability, matching the resource to real time need, the current basis of operation of the energy system, is not possible for a system that uses intermittent supplies. This implies that the structure of a renewable energy economy is local and relies on the local resource base. Figure 8.3 sets out an overview of a renewable energy economy that merges aspects of both the hydrogen and the electron economy. Note that Figure 8.3 illustrates the case for electricity. However, a system could be developed that would include heat.

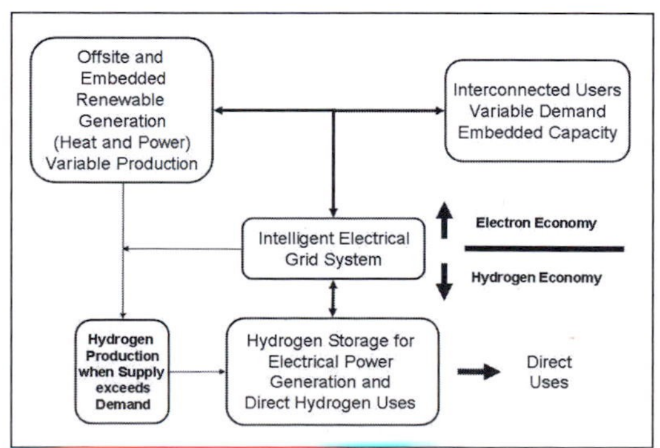


**Figure 8.3** Hydrogen and electron economies

In Figure 8.3 renewable technologies including wind, solar, photovoltaic, wave, tidal and biomass are used to collect and harvest energy. If, for example, the supply from the renewable capacity exceeds demand then the excess, both heat and power, can be used to produce hydrogen. Effectively this is a form of energy storage. When demand exceeds supply then the hydrogen can be used to generate electrical power and heat. It can also be used for direct purposes such as transportation. The management of the system would be undertaken by an intelligent grid that would manage the different resources. Although this is indicative of the role that hydrogen could play in a renewable energy economy, development work has been done in connecting together different types of renewable electricity producing technologies and managing then through a smart electrical grid system. At present these operate at a small scale, although there is potential to scale-up these technologies. This approach to intelligent management can also be applied to heat networks.

However it is unlikely that systems such as this will be developed in a finished form. The most likely development trajectory will be one that begins with smaller systems that have the potential to evolve. Experience with renewable technologies is varied. The assumption in the existing system is that the user is essentially passive, with energy services available on demand. With the system shown in Figure 8.3 we do know that there must be greater involvement in system development. Lessons learned from small scale renewable projects identify a number of points for successful deployment and development. These can be summarized as:

- *Needs assessment*: ensuring that a clear understanding of energy needs is generated.
- *Energy mapping*: knowing what local energy resources exist.
- *Support systems* (technical, human and financial) that are needed.
- *Appropriate level*: defining the entry level (O'Brien et al, 2007).

Essentially this means that learning, both of developers and users, is a key feature of successful system development. As with any technology transfer we cannot assume a single direction and system developers will have to ensure that technologies are appropriate to the capacity of the user and the available resource base and are easily expanded as capacity grows. Research into localized solutions do show the viability of smaller intelligent systems and micro-grids and Plug and Play technological approaches that allow a system to evolve as users gain confidence in both the technology and their ability to manage the system (Watson et al, 2006, Abu-Sakarkh et al, 2005).

In summary hydrogen offers the potential to produce clean energy. But this comes at a cost. In thinking about an overall approach to future energy system development then, it is clear that hydrogen has a role to play. However, because of some of the inefficiencies in producing hydrogen with existing technologies, it is likely that hydrogen will, initially, be most successfully deployed in niche and specialist applications. If an effective and safe method of onboard storage can be found, then hydrogen may have a significant role in the transportation sector. In the longer term hydrogen can play a key role of system development, not as the focus of the system, but one of a number of strategies for ensuring continuing access to a range of energy services.

## Market mechanisms and de-carbonizing the energy system

The Kyoto Protocol introduced a market-based approach to reducing greenhouse gas emissions. The thinking behind a Cap and Trade system is that it will drive innovation and those companies that can develop innovative low carbon technologies will become market leaders, not only benefitting from income derived from the sale of their products but also from income derived from selling carbon credits which they no longer need. On the face of it this would seem to be a 'no-brainer' with everyone, including the environment, winning. The argument is that trading spurs innovation and eventually this will lead to massive reductions in carbon and the gradual emergence of low carbon energy systems. This section will evaluate the effectiveness of carbon trading as it appears to be the only option on the table. At present the largest carbon trading system, the EU Emissions Trading Scheme (EU ETS), is operated within Europe. It is now clear that the US administration also favours a Cap and Trade system. This could mean that future approaches to energy will have a significant component of Cap and Trade in their architecture.

## Emission trading

Emission trading is an administrative approach to pollution control that provides an economic incentive for emission reductions. The first major emissions trading programme was adopted in 1976 by the US Environmental Protection Agency. It allowed new polluting plants to be built in exchange for 'offsets' that reduced air pollution by a greater amount than other sources in the same region. The view that markets could provide more cost effective solutions to dealing with pollution than regulation culminated in the Clean Air Act Amendments of 1990, which set up a national sulphur dioxide trading programme to save power plants money in the effort to control acid rain, as well as encouraging states to use emissions trading to reduce urban smog. The two market mechanisms that have been developed from this experience are known as Cap and Trade and Project Based or Offset mechanisms:

- *Cap and Trade* is a policy approach to controlling large amounts of emissions from a group of sources at costs that are lower than if sources were regulated individually. The approach first sets an overall cap; that is the maximum amount of emissions per compliance period that will achieve the desired environmental effects. Authorizations to emit in the form of emission allowances are then allocated to affected sources, and the total number of allowances cannot exceed the cap. Over time the level of the cap is lowered.
- *Project Based or Offsets* is where emission credits can be earned by developing projects that produced emission reductions that would not have occurred if the project had not been undertaken.

These mechanisms have become the centrepiece of the market-based approach to reducing greenhouse gas (GHG) emissions in the Kyoto Protocol. Carbon markets and trading have become an increasingly important aspect of the mitigation challenge (Lohmann, 2006).

## What is carbon trading?

Carbon trading can be defined as a transaction where one party pays another party in return for GHG emission assets that the buyer can use to meet its objectives. There are two main categories of transactions:

- *Allowance-based transactions*, in which the buyer purchases emission allowances issued by regulators under cap and trade regimes, such as AAUs under the Kyoto Protocol, or EUAs under the EU ETS. Such schemes combine environmental performance (defined by the total amount of allowances issued by the regulator, setting a cap on the global level of emissions from mandated entities) and flexibility, through trading, in order for participants to meet compliance requirements at the lowest possible cost.
- *Project-based transactions*, in which the buyer purchases ERs from a project that can verifiably demonstrate GHG emission reductions compared with what would have happened otherwise (for instance investing in wind power or other renewable energy sources instead of coal fired power generation or improving energy efficiency at a large industrial facility to reduce energy demand and hence, GHG emissions from power generation). The most prominent examples of

## Box 8.1 JI and CDM of the Kyoto Protocol

***Joint Implementation (JI)*** allows an Annex I Party to implement an emission-reducing or sink-enhancing project in the territory of another Annex I Party and count the resulting emission reduction units (ERUs) towards meeting its own Kyoto target.

***The Clean Development Mechanism (CDM)*** provides for Annex I Parties to implement project activities that reduce emissions in non-Annex I countries. CDM project activities must have the approval of all Parties involved and must reduce emissions below those emissions that would have occurred in the absence of the CDM project activity.

Carbon credits are measures of the amount of carbon dioxide or its equivalent that is removed by an action. A credit is the equivalent of one metric tonne of carbon dioxide reduction. Different schemes have different nomenclatures for carbon credits:

1 For Joint Implementation projects the credits are known as Emission Reduction Units (ERUs).
2 For the Clean Development Mechanism (CDM) the credits are known as Certified Emission Units (CERs).
3 For schemes designed at enhancing sinks the credits are known as Removal Units (RMUs).
4 For Emissions Trading schemes the credits are known as Assigned Amount Units (AAUs).

**Table 8.3** Carbon markets: Volumes and values 2006–2007

| | 2006 | | 2007 | |
|---|---|---|---|---|
| | Volume ($MtCO_2$)e | Value (MUS$) | Volume ($MtCO_2$e) | Value (MUS$) |
| Allowances | | | | |
| EU ETS | 1,104 | 24,436 | 2,061 | 50,097 |
| New South Wales | 20 | 225 | 25 | 224 |
| Chicago Climate | 10 | 38 | 23 | 72 |
| Sub total | 1,134 | 24,699 | 2,109 | 50,394 |
| Project-based | | | | |
| Primary CDM[a] | 537 | 5,804 | 551 | 7,426 |
| Secondary CDM | 25 | 445 | 240 | 5,451 |
| JI | 16 | 141 | 41 | 499 |
| Other compliance[b] | 33 | 146 | 42 | 265 |
| Sub total | 611 | 6,536 | 874 | 13,641 |
| Total | 1,745 | 31,235 | 2,983 | 64,035 |

*Notes*

[a] The primary CDM transactions refer to the first sale of CERS from the project owner to the buyer. Secondary CDM refers to the on-sale of primary CERs.

[b] Other compliance refers to the voluntary carbon market that allows institutions, companies and individual citizens in the North to offset their carbon emissions. A common example of the voluntary market is that of individuals from the North who travel by plane or cars and believe that by donating a bit of money they are 'offsetting' the emissions they have generated through their lifestyle. For instance, numerous European airlines encourage passengers to donate a certain sum of money to be used in projects that will apparently offset the emissions they generate by flying. This leads passengers to believe that by donating money, the carbon dioxide released during the flight will be automatically absorbed somewhere else, and this will compensate for the emissions involved. One of the major problems associated with this is that it is not subject to any form of regulation, making it virtually impossible to determine if the money donated has been used in an effective manner.

*Source:* Adapted from Ambrosi and Capoor, 2008, p1

such activities are under the CDM (Clean Development Mechanism) and the JI (Joint Implementation) mechanisms of the Kyoto Protocol, generating CERs and ERUs respectively. See Box 8.1 for further details on CDM, JI and emission credit nomenclatures. These are known as the Flexibility Mechanisms.

Carbon Trading should be seen as one of a number of methods for reducing GHG emissions, such as improving efficiency and developing renewable capacity. Since the introduction of the flexibility mechanisms there has been considerable growth in carbon trading and a number of markets have been established. According to the World Bank the value of carbon trading in 2007 was US$64 billion, a little over double that of 2006. The largest single market was the EU ETS with trade valued at US$50 billion, almost 80 per cent of the global market. The figures for 2006 and 2007 are shown in Table 8.3.

Estimates suggest that we are adding some 4.1 billion metric tonnes of carbon dioxide equivalents per annum to the global atmosphere (EIA, 2008). From Table 8.3 we can see that carbon credits amounting to almost 3 billion tonnes of carbon dioxide equivalents were traded in 2007. This would suggest that this approach offers considerable potential for greenhouse reductions. However, the reality is very different.

## Effectiveness of carbon trading in mitigation

One of the underpinning principles of trading is that market solutions will find the optimum cost effective solution to a problem. Effective mitigation of greenhouses gases requires a reduction in the activities that produce GHGs. Large scale reductions mean that large scale change is needed. In order to reduce GHGs, a shift in the energy systems to more efficient and low carbon technologies are needed. While there is a strong social dimension to the climate problem, there are also strong technological dimensions to the solutions. However, experience from Cap and Trade in the US Sulphur Dioxide programme shows that the market approach does not drive innovation, suggesting that regulation may be a more effective driver for technological change (Taylor et al, 2005).

This is an important issue, particularly when considering some of the largest carbon producers; the energy companies. Cap and Trade does not pay any attention to industry type. Many large electricity producers have invested heavily in fossil fuel capacity and are effectively 'locked-in' to a fossil fuel economy. As opposed to shifting to renewable capacity, there is likely to be a tendency towards seeking efficiency improvements (given the mature state of the technology, these will only be marginal) and to purchase emission credits, the cost of which will be passed on to consumers. For those industries that are not structurally locked into fossil fuels but produce greenhouse gases, Cap and Trade does provide incentives. However, these industries are, in the short term, more likely to seek quick wins so they can profit from trading their emission rights as opposed to investing in low carbon solutions that will lower their emissions and provide sellable technologies. In short Cap and Trade is more likely to encourage a focus on the cheapest methods of emission reductions as opposed to driving the radical shifts that have been argued as being necessary to avoid dangerous climate change.

Further problems have been created by the way in which the carbon market has been constructed. This is most evident in the largest carbon market, the EU ETS. This began in 2005 and covered some 10,000 energy intensive plants in the EU producing some 40 per cent of carbon emissions. The allowances (an equivalent value of €120 billion and allocated free of charge except in four member states where some were auctioned, but only in one member state was the full 5 per cent auctioned, as allowed in the first phase (Hepburn et al, 2006)) were established by National Allocations by each member state. Many of these were set too high and when the market started operation in 2005 the price of carbon fell (Grubb and Neuhoff, 2006). In fact the industries included in the EU ETS emitted 66 million tonnes less than the level set by the cap.

The generous allowance levels set by some member states militates against the scarcity principle, essential for a robust market, and casts doubt on the willingness of the EU members to meet international climate commitments. Grubb, notes:

> *If the current national allocation plans are allowed to stand, it could seriously undermine the credibility of the EU ETS and the mechanism of carbon trading as an effective way to tackle carbon emissions.* (Carbon Trust, 2006)

The second phase of ETS was launched in January 2008. For this phase the EU broadened the scope for auctioning. However, only ten EU members opted for this and of these four auctioned less than 1per cent of their total allocations. In 2006 Neuhoff et al had argued that such levels of subsidies proposed under the second phase could mean that distorted allocation decisions could lead to a situation where it was more profitable to construct a coal fired power station than it would be if the ETS did not exist (Neuhoff et al, 2006). Although tighter caps have been introduced in this phase, critics of the system argue that the Linking Directive, introduced in 2004 and implemented in 2005, which provides a link between the EU ETS and the flexibility mechanisms in the Kyoto Protocol, undermine the cap as it allows credits earned from CDM and JI projects to be offset against domestic targets. In addition critics claim that this undermines member states' domestic climate policies and damages the EU's credibility in climate negotiations (Greenpeace, 2003; Climate Action Network Europe et al, 2003). However, the ease and speed with which this measure was adopted indicates the influence of the member states within the EU negotiation processes and their desire to have a system which eases pressure on domestic activities. In effect the original view of the EU and its institutions that mixing of market mechanisms was to be avoided had shifted to a point where market mechanisms were increasingly viewed as a more effective means of delivering environmental goals. This reflects the gradual shift away from Command and Control to market-based alternatives in EU environmental policy formulation (Flåm, 2007). A study by Point Carbon Advisory Services commissioned by WWF (World Wide Fund for Nature) of five EU member states (Germany, the UK, Italy, Spain and Poland) into the scale of windfall profits that could accrue during Phase 2 (2008–2012) of the EU ETS, estimates these as being between 23 and 71 billion, in total, based on an EUA price of 21–32/t $CO_2$ (Point Carbon, 2008). The prospect of profits for nothing may also help to explain the ease with which the second phase of the scheme was agreed.

In 2008, the European Commission proposed a number of changes to the scheme, including centralized allocation (no more national allocation plans) by an EU authority, a turn to auctioning a greater share (60+ per cent) of permits rather than allocating freely, and inclusion of the greenhouse gases nitrous oxide and perfluorocarbons (MEMO 08/35, 2008). The main elements of the new system, which would enter into force in 2013 and run until 2020, are:

1 Total EU industrial emissions in 2020 capped at 21 per cent below 2005 levels; a maximum of 1720 million allowances. To achieve the total number of emissions, allowances circulating at the end of 2012 will be cut by 1.74 per cent annually.

**2** The scheme will be enlarged to include new sectors, including aviation, petrochemicals, ammonia and the aluminium sector, as well to include two new gases (nitrous oxide and perfluorocarbons), meaning that around 50 per cent of all EU emissions would be covered. Road transport and shipping remain excluded but shipping could be included at a later stage. Agriculture and forestry are not included because of the difficulties related to measuring emissions from these sectors with accuracy.

3 In order to achieve an average 10 per cent reduction of greenhouse gases from sectors not covered by the ETS, such as transport, buildings, agriculture and waste by 2020, the Commission has set national targets accord-

ing to countries' GDP. Richer countries are asked to make bigger cuts (up to 20 per cent in the case of Denmark, Ireland and Luxembourg) while poorer states (notably Portugal, as well as all of the countries that joined the EU after 2004 except Cyprus) will in fact be entitled to increase their greenhouse emissions in these sectors (by up to 19 and 20 per cent respectively for Romania and Bulgaria) in order to take into account their high expectations for GDP growth.

4 Smaller installations, emitting under 10,000 tonnes of $CO_2$ per year, will be allowed to opt out from the ETS, provided that alternative reduction measures are put in place.
5 Industrial greenhouse gases prevented from entering the atmosphere through the use of CCS technology, are to be credited as not emitted under ETS.
6 Auctioning: today, 90 per cent of pollution allowances are handed out to industrial installations for free, but a huge increase in auctioning is planned for as early as 2013. It is estimated that around 60 per cent of the total number of allowances will be auctioned in 2013. Full auctioning for the power sector is expected to be the rule from 2013 onwards; this is expected to lead to a 10–15 per cent rise in electricity prices. In other sectors, free allocations will gradually be completely phased-out on an annual basis between 2013 and 2020, although certain energy-intensive sectors could continue to get all their allowances for free in the long term if the Commission determines that they are at significant risk of carbon leakage, that is relocation to third countries with less stringent climate protection laws. The sectors affected by this measure are yet to be determined.
7 The distribution method for free allowances will be developed at a later stage by expert panels within the Commission. They are likely to be based on grandfathering or historic emissions and performance criteria.
8 Competitiveness: to minimize the risk to European competitiveness through carbon leakage (external industries exporting to the EU and gaining a competitive advantage from lax climate controls) if a global agreement is not reached, the EU will consider some compensatory measures to protect EU industries.
9 Flexibility and third countries: assuming a global climate change deal is reached, member states will continue to be entitled to meet part of their target by financing emission reduction projects in countries outside the EU, although the use of such credits will be limited to 3 per cent of member states' total emissions in 2005 – or, in other words, around one quarter of the total reduction effort.

At present these are proposals that will have to be negotiated through the EU and it is likely that they will be very different from those set out above. It is difficult to judge at present how effective ETS will be in mitigating GHGs in the future.

## CDM and JI

There is concern that credits earned through these mechanisms will be used to offset domestic targets, effectively undermining the opportunity to drive radical reform within sponsors. A JI project must have the approval of the parties involved and provide a reduction in emissions by sources, or an enhancement of removals by sinks that is additional to any that would otherwise occur. JI is overseen by a 'Supervisory Committee' and like the CDM, ERUs must be certified by 'Independent Entities'. JI Emission Reduction Units, however, can accrue only after 2008, although, several countries have invited investments into early JI emission reduction projects and grant post-2008 emission rights from their budgets for the pre-2008 reductions.

Both JI and CDM have critics. JI enables countries to purchase and sell emissions credits, which could allow countries to buy their way out of their Kyoto commitments. In the search for cost effective projects, the cheapest mode for generating carbon credits, more expensive renewable energy and energy efficiency projects would be a low priority Leakage is another problem area. Leakage refers to outside effects, for instance, a project increasing the efficiency of cars will reduce emissions, but indirectly, by lowering the cost of driving, could encourage car owners to increase their vehicular use and even purchase more cars.

## The Clean Development Mechanism (CDM)

The Clean Development Mechanism (CDM) provides for Annex I Parties (members of OECD and EIT (economies in transition) – essentially the developed industrialized countries) to implement project activities that reduce emissions in non-Annex I countries (the developing world). CDM project activities must have the approval of all Parties involved and must reduce emissions below those emissions that would have occurred in the absence of the CDM project activity. The Marrakech Accords defined an elaborate CDM 'project cycle' that is overseen by the CDM Executive Board (EB), whose 10 members are elected by the UNFCCC Conference of the Parties (COP). The Board formerly registers projects and checks whether they conform to the rules. Simplified rules for small-scale CDM projects were agreed at COP 8 (2002), and rules for sequestration CDM projects were agreed at COP 9 (2003).

CDM emission credits are added to the overall emissions budget of Annex I countries and therefore, their quality must be guaranteed. Hence, emission credits only accrue after independent verification through 'Operational Entities' (OEs), which are mainly commercial certification companies. These emission credits are referred to as Certified Emission Reductions (CERs).

In CDM projects a share of the proceeds must be used to promote sustainable development within the host country. Critics argue that many projects neither address the real problem of reducing GHG in industrialized countries nor promote sustainable development in the host nation. CDM projects a range of type and size of project. A number have focused on large hydropower and a clean coal project. Projects of this scale can generate a large amount of credits and will be attractive to investors, but might not promote sustainable development. A report by the World Commission on Dams in 2000 illustrated that large hydropower projects have seriously negative social and environmental impacts and have regularly underperformed (World Commission on Dams, 2000).

Additionally, CDM clean coal projects do not address future resource issues and hinder renewable energy projects therefore failing to promote sustainable development. Many believe that CDM is not an effective mechanism for the promotion of renewable energy projects in developing countries as the structure of CDM means the search for least-cost carbon credits is the paramount consideration. More expensive projects like renewables are sidelined as the multiple benefits they provide are not rewarded (CDM Watch, 2005).

The CDM has also proved unsuccessful at promoting projects that address energy efficiency and transport, both of which are critical in achieving sustainable development in the South and combating climate change globally. The World Bank estimated that the potential for efficiency projects is significant, however, they note in a recent paper that the limited number of projects to date suggests they face barriers not fully reflected in analyses of the achievable potential (Haites, 2004). Furthermore, a study conducted by the OECD found that:

> *...a large and rapidly growing portion of the CDM project portfolio has few direct environmental, economic or social effects other than GHG mitigation, and produces few outputs other than emissions credits* (Ellis et al, 2004, p32)

The World Bank claims that CDM projects can deliver sustainable development but it does acknowledge there is only anecdotal evidence to support this view. The Bank claims that a small, but distinct, market niche is developing, which rewards CDM projects that deliver strong sustainability benefits. In short there is little evidence that the effectiveness of carbon markets will

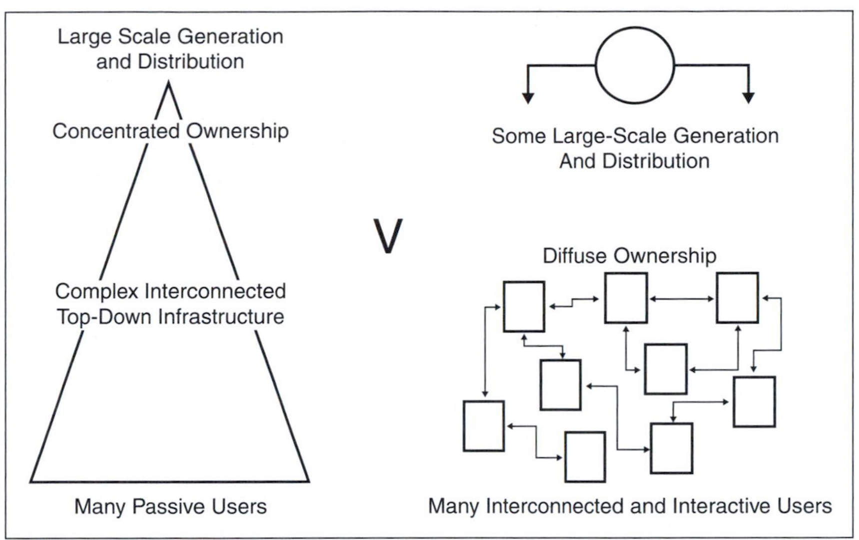


*Source:* Ambrosi and Capoor, 2008, p59

**Figure 8.4** Players and institutions in the carbon market

reduce GHGs, although a preliminary analysis undertaken by Ellerman and Buchner of the EU ETS suggests that the market has reduced emissions by 2.5–5 per cent below what they otherwise would have been (Tirpack, 2008). Whether or not, in the long term, carbon trading can make a significant contribution to the deep cuts that are needed, remains to be seen. What is alarming is that it is clear from the World Bank report, *State and Trends of the Carbon Market*, 2008, that the market is evolving into a fully-fledged system, as shown in Figure 8.4, that has many of the features associated with financial markets, such as monetization of carbon as a step toward securitization, derivatives, bond transactions linked to the future price of carbon and so on (Ambrosi and Capoor, 2008, p65). The credit crunch and bank failures of 2008 were to a large extent driven by inappropriate use of instruments of this type. Whether or not there will be sufficient regulation to ensure transparency in the carbon market and that it remains focused on delivering real cuts remains to be seen.

One final point is that during 2009 the market price of carbon fell to record low levels. This has raised questions about the need for a floor price for carbon. Effectively this would mean shoring-up a system that has not delivered on carbon reductions. There are voices now being raised about the need for carbon tax. For example, the Yale economist William Nordhaus, the respected climate scientist Jim Hansen and the Chief Executive Officer of Exxon are just a few of the voices that are calling for the introduction of a carbon tax. Resource-based taxes such as a carbon tax, are argued as being more effective in driving innovation than market mechanisms. However, the application of resource based taxes, such as a carbon tax, will need to be implemented in conjunction with other instruments such as policy, legislation, regulation and efficiency standards if they are to be effective.

## Other carbon reduction approaches

As the previous section highlighted, there is still some doubt as to whether a carbon market can be designed to generate real carbon reductions. Other alternatives to the UNFCCC and its Kyoto Protocol are proposals for Contraction and Convergence and the Greenhouse Development Rights Framework.

'Contraction and Convergence' is a proposal for an emission allowance on a capitation basis eventually driving to similar levels of carbon allowance (Meyer, 2001). It is based on national allowances and therefore does not consider broader global problems such as development but it wishes to push for similar carbon allowances for everyone. In contrast, the Greenhouse Development Rights Framework argues that poor people should not be paying for the excesses of the rich (Baer et al, 2007).

Averting climate change requires stabilizing atmospheric GHG concentrations at some 80 per cent below current levels while simultaneously, giving the poor a right to develop (Baer et al, 2007). This is based on the polluter pays principle rather than 'common but differentiated responsibilities' of the UNFCCC. The Greenhouse Development Rights Framework proposes a median global income level of US$9000 below which people would not have to curb carbon emissions. That is, they are surviving rather than polluting and therefore exempt from obligation.

The Greenhouse Development Rights Framework estimates that it would cost between 1 per cent and 3 per cent of global GDP to avert catastrophic climate change in an equitable fashion, a figure that is similar to that of Stern. Rough calculations suggest that the US would bear some 33 per cent, the EU around 25 per cent and China and India less than 1 per cent each. The drive to a global framework suggested by the Greenhouse Development Rights Framework is the most radical proposal since it places the development agenda into the global commons issue.

Global climate change must be addressed at the level of global and macro economics (Stern, 2007; UNDP, 2007). This essentially means linking it to the global poverty alleviation programme of the Millennium Development Goals. The cost of this global effort could be as much as 5.5 percent of global GDP (IPCC, 2007) but this must be considered against the possible cost of damages between 5 and 20 per cent of global GDP as stated in the Stern Report (Stern, 2007).

Technological 'leapfrogging' is needed to achieve global emission reduction of 80 per cent and stabilizing the climate below a 2°C increase. Obviously such leapfrogging can only occur if the poorest can access relevant technologies (O'Brien et al, 2007). This can only occur if richer industrialized nations simultaneously reduce their structural reliance on fossil fuels (Baer et al, 2007; Smith, 2007).

## The changing energy landscape

Developing an energy system that meets climate targets, sustainable development objectives and energy security concerns is a massive challenge. Throughout this book we have looked at a variety of technologies that can improve the efficiency of both supply and demand sides of the energy systems. In addition we have looked at technologies that can be used to capture renewable resources. Decarbonizing the energy system will be difficult. For some the continuing use of large scale technologies such as nuclear and CCS offer a method of decarbonizing the supply side. However, nuclear technology is inflexible and we would still have to find a way of dealing with the unresolved problem of nuclear waste (O'Brien and O'Keefe, 2006). CCS is still only at the development stage and raises questions about the long-term security of carbon stored in geological formations. Others advocate the use of renewable technologies and there have been suggestions that Europe could meet its energy requirements from solar technologies deployed in the Sahara desert. There will be considerable political barriers to overcome before such a scheme could be realized. The one area that most seem to agree on is that improving demand-side efficiency is vital. We have seen in earlier chapters the considerable scope for improving building and appliance efficiencies. This is an important consideration and an important first step in decarbonizing the energy system. The next is what kind of approach is needed for the supply side. For renewable technologies a strategy that recognizes the difference between concentrated energy supplies, such as fossil fuels, and the distributed nature of renewable resources, such as solar and wind, means that a renewable approach will require a 'capture and store until needed' strategy that recognizes the intermittent nature of the source. Storage is a key problem. Electricity cannot be effectively stored and this requires the development of a hydrogen economy alongside a renewable energy economy to maximize the opportunities for capture and storage.

There are many voices calling for urgent action on reducing carbon emissions, although to date an agreed stabilization level and timetable have yet to be finalized. We do know that whatever target and timetable is agreed it will take a long time to change the energy system. There is a fear that if we opt for certain technologies, such as nuclear power, we will be locked into a development trajectory for many years. This could also drain funds away from the development and deployment of alternatives. We also know that energy security concerns are focusing attention on the development of indigenous resources and ensuring a diverse energy mix. We also know that there will have to be a step change in technologies, for example, the motor car where it is likely that hybrids and then either electric or fuel cell vehicles or both will dominate the fleet. We have already seen the deployment of solar thermal water heaters and photovoltaic panels on household roofs. New standards for zero emission buildings will continue that drive. The impact on

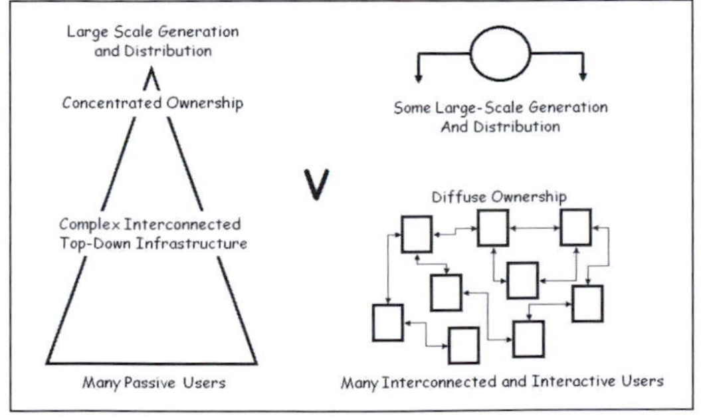


*Source:* O'Brien, 2009

**Figure 8.5** Contrasting models of energy system structure

the energy landscape of these changes is likely to be considerable. Conventional or current energy systems tend to be pyramidal in shape as shown in the drawing on the left hand side of Figure 8.5. This can be characterized as a system with a focused or concentrated ownership and many passive consumers. The only connection between the user and the energy system is the energy service provided and the utility bills. Bills typically state the amount and price of the units of energy used and there is no information on how that relates to energy services, for example, the amount of used space and water heating and for cooking and the amount of electricity used for lighting and appliances.

The introduction of more distributed technologies will change the look of the physical landscape, but the changes are likely to be subtle, for example, the deployment of solar thermal collectors and photovoltaic panels on roofs, the use of LED street lighting and the different noise emanating from electric or hydrogen vehicles. Many of the changes will be invisible, such as increased thermal efficiency of buildings. What will be different is the more distributed structure of the energy system as shown on the right hand side of Figure 8.5.

The use of Net Metering and Feed-In Tariffs has, in some countries, encouraged the participation of households in the energy system as both producer and consumer. Net metering provides an incentive for consumer investment in renewable energy generation. Net metering enables customers to use their own generation to offset their consumption by allowing their electric meters to turn backwards when they generate electricity in excess of their demand. Without net metering, a second meter is usually installed to measure the electricity that flows back into the system. The Feed-in Tariff is the mechanism where producers of power from the household upwards are paid for the power they produce. The new generation of Smart Meters planned for installation in the UK does not have this feature. It does allow communication with the utility so that billing data are more accurate and timely, and the possibility of having a real-time display of energy use in the household.

Infrastructure developments of this type do point to a different kind of energy future where there is greater involvement by the user in the energy system. However, whether such developments can help to solve the interrelated problems of energy security and climate change within a sustainable development context is debatable. Shifting the direction of development of the energy system is a complex and slow process. Decisions to develop nuclear power or to develop carbon capture and storage will take many years to realize and then will only address part of the energy system. There are no easy answers and no simple solutions. One of the problems that policy makers face is the lack of an agreed target for greenhouse gas concentration and a timetable for reaching that target. It is impossible to predict when and what agreement will be reached in terms of targets and timetables. But we do know that whatever is agreed, we cannot approach energy system development with the same mindset that has led to the current system structure. This means that we will have to learn how to do things very differently. We will need to challenge some of the assumptions that underpin the current system, for example, should we expect the system to deliver energy 24/7, should we be more self-reliant, should we change lifestyles to help meet international climate goals or is this the responsibility of government and the energy sector? There are no obvious answers to these questions. We do know that energy is absolutely essential to life itself and the ways in which we live our lives. But the production and use of energy has consequences and in the future we cannot afford to ignore these. We will have to learn from the individual through to the institutional level that decisions about the ways we produce and use energy cannot be made without regard to wider concerns. Energy system development must be an integral part of sustainable development.

## References


Abu-Sharkh, S., Li, R., Markvart, T., Ross, N., Wilson, P., Yao, R., Steemers, K., Kohler, J. and Arnold, A. (2005) *Microgrids: Distributed On-site Generation*, Technical Report 22, Tyndall Centre for Climate Research. Available at: www.tyndall.ac.uk/research/theme2/final_reports/it1_33.pdf

Ambrosi, P. and Capoor, K. (2008) *State and Trends of the Carbon Market*, World Bank. Available at: http://siteresources.worldbank.org/NEWS/Resources/State&Trendsformatted06May10pm.pdf

Baer, P., Athanasiou, T. and Kartha, S. (2007) *The Right to Development in a Climate Constrained World: The Greenhouse Development Rights Framework*, Ecoequity, Christian Aid, Heinrich Boll Foundation and Stockholm Environment Institute, Boston, MA

Bossel, U. (2006) 'Does a Hydrogen Economy Make Sense?' *Proceedings of the IEEE*, vol. 94, no. 10

Carbon Trust (2006) 'EU ETS hits crunch time: New research on National Allocation Plans shows urgent action needed'. Carbon Trust 7 November 2006. Available at: www.carbontrust.co.uk/about/presscentre/071106_euets.htm

Carrette, L., Friedrich, K. A. and Stimming, U. (2001) 'Fuel cells – Fundamentals and applications', *Fuel Cells*, vol 1, no. 1. Available at: www3.interscience.wiley.com/cgi-bin/fulltext/84502989/PDFSTART

CDM Watch (2005) 'The World Bank and the Carbon Market: Rhetoric and Reality'. Available at: http://cdmwatch.org/files/World%20Bank%20paper%20final.pdf

Climate Action Network Europe, Greenpeace, and World Wide Fund for Nature International (2003) 'Letter to Margot Wallström, Commisioner for the Environment, Re: The proposed linking of CDM/JI with the EU emission trading system'. 10 July. Available at: www.climnet.org/EUenergy/ET/ETCDMJIletter10_07_2003.pdf

Directorate-General for Research (2003) Directorate-General for Energy and Transport. Office for Official Publications of the European Communities, Luxembourg.

EIA (2008) Greenhouse Gases, Climate Change and Energy, EIA, Washington, DC. Available at: www.eia.doe.gov/bookshelf/brochures/greenhouse/greenhouse.pdf

Ellis, J. Corfee-Morlot, (OECD) and Winkler, H. (Energy Research Centre, University of Cape Town) (2004) Taking Stock of Progress under the Clean Development Mechanism (CDM), p32, OECD, Paris. Available at: www.oecd.org/dataoecd/58/58/32141417.pdf

EU Commission (2003) Hydrogen Energy and Fuel Cells: A vision of our future, Directorate-General for Research; Directorate-General for Energy and Transport. EU Commission, Brussels. Available at: http://ec.europa.eu/research/fch/pdf/hlg_vision_report_en.pdf

Flåm, K.H. (2007) 'A Multi-level Analysis of the EU Linking Directive Process: The Controversial Connection between EU and Global Climate Policy', Fridtjof Nansen Institute. Available at: www.fni.no/doc&pdf/FNI-R0807.pdf

Greenpeace (2003) 'Seven Reasons to Reject the Linking Directive', 24 October 2003. Greenpeace. Available from: www.greenpeace.eu/downloads/climate/PRon7ReasonsToRejectLinkingDir.pdf

Grubb, M. and Neuhoff, K. (2006) 'Allocation and competitiveness in the EU emissions trading scheme: Policy overview', *Climate Policy*, vol. 6, pp7–30

Haites, E. (2004) *Estimating the Market Potential for the CDM: Review of models and lessons learned*, Washington, DC: World Bank Carbon Finance Business Unit

Hepburn, C., Grubb, M., Neuhoff, K., Matthes, F. and Tse, M. (2006) 'Auctioning of EU ETS phase II allowances: How and why?' *Climate Policy*, vol. 6, pp 137–160. Available at: www.electricitypolicy.org.uk/pubs/tsec/hepburn.pdf

IPCC (2007) *Climate Change 2007: Synthesis Report.* Contribution of Working Groups I, II and III to the Fourth Assessment Report of the Intergovernmental Panel on Climate Change Core Writing Team, Pachauri, R.K. and Reisinger, A. (eds) IPCC, Geneva

Lohmann, L. (ed.) (2006), *Carbon Trading: A Critical Conversation on Climate Change, Privatisation and Power: Chapter 2* 'Made in the USA': A short history of carbon trading, Dag Hammarskjold Foundation, Durban Group for Climate Justice and The Corner House. Available at: www.thecornerhouse.org.uk/pdf/document/carbonDDch2.pdf

MEMO 08/35 (2008) Questions and Answers on the Commission's proposal to revise the EU Emissions Trading System. Available at: http://europa.eu/rapid/pressReleasesAction.do?reference=MEMO/08/35&format=HTML&aged=0&language=EN&guiLanguage=en

Meyer, A. (2001) *Contraction and Convergence: The Global Solution to Climate Change*, Totnes, Devon: Green Books, Schumacher Briefings No 5

Neuhoff, K. M., Åhman, R., Betz, J., Cludius, F., Ferrario, K., Holmgren, G., Pal, M., Grubb, F., Matthes, K., Rogge, M., Sato, J., Schleich, J., Sijm, A., Tuerk, C., Kettner, N. and Walker (2006) 'Implications of announced phase II national allocation plans for the EU ETS', *Climate Policy*, vol. 6, pp411–422

O'Brien, G. (2009) 'Resilience and vulnerability in the European energy system', *Energy and Environment*, vol. 20, no. 3, pp399–410

O'Brien, G. and O'Keefe, P. (2006) 'The future of nuclear power in Europe: A response', *International Journal of Environmental Studies*, vol 63, pp121–130

O'Brien, G., O'Keefe, P. and Rose, J. (2007) 'Energy, poverty and governance', *International Journal of Environmental Studies,* vol. 64, no. 5, pp607–618

Point Carbon (2008) EU ETS Phase II – The potential and scale of windfall profits in the power sector. A report for WWF by Point Carbon Advisory Services. Available at: *http://assets.panda.org/downloads/point_carbon_wwf_windfall_profits_mar08_final_report_1.pdf*

Rifkin, J. (2002) *The Hydrogen Economy: The Creation of the Worldwide Energy Web and the Redistribution of Power on Earth*, Oxford: Polity

Smith, K. (2007) *The Carbon Neutral Myth: Offset Indulgences for your Climate Sins*, Amsterdam: Transnational Institute,

Stern, N. (2007) *The Economics of Climate Change: The Stern Review*, Cambridge: Cambridge University Press

Taylor, M., Hounshell, M. and Rubin, D.A. (2005) 'Regulation as the Mother of Invention: The Case of SO2 Control', *Law and Policy*, vol. 27, pp348–378

Tirpack, D. (2008) The Carbon Market: IPCC III Chapter 13 – Policies, Instruments and Cooperative Arrangements, World Resources Institute, International Institute for Sustainable Development. Available at: http://unfccc.int/files/meetings/intersessional/awg-lca_1_and_awg-kp_5/presentations/application/vnd.ms-powerpoint/bkk_tirpak_emissions_trading.pps#328,16,Summary

UNDP (2007) *Human Development Report 2007/2008. Fighting Climate Change: Human Solidarity in a Divided World.* Available at http://hdr.undp.org/en/reports/global/hdr2007-2008/

Watson, J., Sauter, R., Bahaj, B. James, P.A., Myers, L. and Wing, R. (2006) Unlocking the Power House: Policy and system change for domestic micro-generation in the UK. Available at: www.sustainabletechnologies.ac.uk/PDF/project%20reports/109%20Unlocking%20Report.pdf

World Commission on Dams (2000) Dams and Development: A New Framework for Decision Making. The Report of the World Commission on Dams, 2000. Available at: www.dams.org//docs/report/wcdreport.pdf

# Appendix 1

# Global Energy Resources

The following tables have been adapted from data supplied by the World Energy Council to Show: A New Way to Look at the World. Data are for 2005. Available at: http://show.mapping-worlds.com/

**Table A1.1** Energy resources by type

| | |
|---|---|
| **Reserves** | |
| Oil | 1,215,186,000,000 barrels |
| Oil shale | 2,826,103,000,000 barrels |
| Natural gas | 176,462,000,000,000 cubic metres |
| Coal | 847,888,000,000 tons |
| Bitumen | 245,914,000,000 barrels |
| Bagasse reserves | 168,162,130 tons |
| **Capacity** | |
| Wind power | 59,335 megawatts |
| Solar electric | 3,902,290 kilowatts |
| Nuclear power | 370,576 megawatts |
| Geothermal heat use | 282,016 terajoules |
| Peat production | 13,580,000 tons |
| Hydropower potential | 16,475 terawatt-hours |

The following tables show countries that possess energy resources that are greater than 0.5 per cent of the global total.

**Table A1.2** Energy resources by country

| Country | Bagasse reserves (number of tons) | % of world total (%) |
|---|---|---|
| Argentina | 3,528,981 | 2.0986 |
| Australia | 8,790,719 | 5.2275 |
| Brazil | 45,859,978 | 27.2713 |
| China | 14,919,553 | 8.8721 |
| Colombia | 4,373,621 | 2.6008 |
| Cuba | 2,119,000 | 1.2601 |
| Egypt, Arab Rep. | 1,956,000 | 1.1632 |
| El Salvador | 1,031,195 | 0.6132 |
| Guatemala | 3,285,095 | 1.9535 |
| India | 24,801,796 | 14.7487 |
| Indonesia | 3,969,050 | 2.3603 |
| Kenya | 866,410 | 0.5152 |
| Mauritius | 854,006 | 0.5078 |
| Mexico | 9,159,426 | 5.4468 |
| Pakistan | 4,607,482 | 2.7399 |
| Panama | 256,370 | 0.1525 |
| Peru | 1,132,200 | 0.6733 |
| Philippines | 3,559,350 | 2.1166 |
| South Africa | 4,087,086 | 2.4304 |
| Sudan | 1,186,283 | 0.7054 |
| Swaziland | 1,063,960 | 0.6327 |
| Thailand | 7,479,912 | 4.4480 |
| US | 4,514,411 | 2.6846 |
| Venezuela, RB | 1,124,700 | 0.6688 |
| Vietnam | 1,425,605 | 0.8478 |
| Zambia | 403,798 | 0.2401 |
| World total | 168,162,130 | |

| Country | Bitumen reserves (number of barrels) | % of world total (%) |
|---|---|---|
| Canada | 173,605,000,000 | 70.5958 |
| Kazakhstan | 42,009,000,000 | 17.0828 |
| Russian Federation | 28,367,000,000 | 11.5353 |
| World total | 245,914,000,000 | |

| Country | Coal reserves (number of tons) | % of world total (%) |
|---|---|---|
| Australia | 76,600,000,000 | 9.0342 |
| Brazil | 7,068,000,000 | 0.8336 |
| Bulgaria | 1,996,000,000 | 0.2354 |
| Canada | 6,578,000,000 | 0.7758 |
| China | 114,500,000,000 | 13.5041 |
| Colombia | 6,959,000,000 | 0.8207 |
| Czech Republic | 4,501,000,000 | 0.5308 |
| Germany | 6,708,000,000 | 0.7911 |
| India | 56,498,000,000 | 6.6634 |
| Indonesia | 4,328,000,000 | 0.5104 |
| Kazakhstan | 31,300,000,000 | 3.6915 |
| Poland | 7,502,000,000 | 0.8848 |
| Russian Federation | 157,010,000,000 | 18.5178 |
| Serbia | 13,885,000,000 | 1.6376 |
| South Africa | 48,000,000,000 | 5.6611 |
| Ukraine | 33,873,000,000 | 3.9950 |
| US | 242,721,000,000 | 28.6265 |
| World total | 847,888,000,000 | |

| Country | Geothermal heat use (number of terajoules) | % of world total (%) |
|---|---|---|
| Algeria | 2,417 | 0.8570 |
| Australia | 2,968 | 1.0524 |
| Austria | 6,872 | 2.4367 |
| Brazil | 6,622 | 2.3481 |
| Bulgaria | 1,672 | 0.5929 |
| Canada | 2,547 | 0.9031 |
| China | 45,373 | 16.0888 |
| Denmark | 4,400 | 1.5602 |
| Finland | 1,950 | 0.6915 |
| France | 5,196 | 1.8424 |
| Georgia | 6,307 | 2.2364 |
| Germany | 3,864 | 1.3701 |
| Hungary | 7,940 | 2.8154 |
| Iceland | 24,744 | 8.7740 |
| India | 1,606 | 0.5695 |
| Israel | 2,193 | 0.7776 |
| Italy | 8,916 | 3.1615 |
| Japan | 10,301 | 3.6526 |
| Jordan | 1,540 | 0.5461 |
| Mexico | 3,628 | 1.2865 |
| New Zealand | 9,670 | 3.4289 |
| Norway | 3,085 | 1.0939 |
| Romania | 2,841 | 1.0074 |
| Russian Federation | 6,144 | 2.1786 |
| Serbia | 2,457 | 0.8712 |
| Slovak Republic | 3,034 | 1.0758 |
| Sweden | 36,000 | 12.7652 |
| Switzerland | 4,229 | 1.4996 |
| Turkey | 19,000 | 6.7372 |
| US | 34,607 | 12.2713 |
| World total | 282,016 | |

| Country | Hydropower potential (number of terawatt-hours) | % of world total (%) |
|---|---|---|
| Argentina | 130 | 0.7891 |
| Australia | 100 | 0.6070 |
| Bhutan | 99 | 0.6009 |
| Bolivia | 126 | 0.7648 |
| Brazil | 1,488 | 9.0319 |
| Cameroon | 115 | 0.6980 |
| Canada | 981 | 5.9545 |
| Chile | 162 | 0.9833 |
| China | 2,474 | 15.0167 |
| Colombia | 200 | 1.2140 |
| Congo, Dem. Rep. | 774 | 4.6980 |
| Ecuador | 134 | 0.8134 |
| Ethiopia | 260 | 1.5781 |
| France | 100 | 0.6070 |
| Greenland | 120 | 0.7284 |
| India | 660 | 4.0061 |
| Indonesia | 402 | 2.4401 |
| Iraq | 90 | 0.5463 |
| Italy | 105 | 0.6373 |
| Japan | 136 | 0.8255 |
| Kyrgyz Republic | 99 | 0.6009 |
| Madagascar | 180 | 1.0926 |

| | | |
|---|---|---|
| Malaysia | 123 | 0.7466 |
| Myanmar | 130 | 0.7891 |
| Nepal | 151 | 0.9165 |
| Norway | 200 | 1.2140 |
| Pakistan | 219 | 1.3293 |
| Paraguay | 106 | 0.6434 |
| Peru | 395 | 2.3976 |
| Russian Federation | 1,670 | 10.1366 |
| Sweden | 100 | 0.6070 |
| Tajikistan | 264 | 1.6024 |
| Turkey | 216 | 1.3111 |
| US | 1,752 | 10.6343 |
| Venezuela, RB | 246 | 1.4932 |
| World total | 16,475 | |

| Country | Natural gas reserves (number of cubic metres) | % of world total (%) |
|---|---|---|
| Algeria | 4,504,000,000,000 | 2.5524 |
| Azerbaijan | 1,350,000,000,000 | 0.7650 |
| Canada | 1,633,000,000,000 | 0.9254 |
| China | 2,350,000,000,000 | 1.3317 |
| Egypt, Arab Rep. | 1,894,000,000,000 | 1.0733 |
| India | 1,101,000,000,000 | 0.6239 |
| Indonesia | 2,754,000,000,000 | 1.5607 |
| Iran, Islamic Rep. | 26,740,000,000,000 | 15.1534 |
| Iraq | 3,170,000,000,000 | 1.7964 |
| Kazakhstan | 3,000,000,000,000 | 1.7001 |
| Kuwait | 1,586,000,000,000 | 0.8988 |
| Libya | 1,491,000,000,000 | 0.8449 |
| Malaysia | 2,480,000,000,000 | 1.4054 |
| Nigeria | 5,150,000,000,000 | 2.9185 |
| Norway | 2,358,000,000,000 | 1.3363 |
| Qatar | 25,633,000,000,000 | 14.5261 |
| Russian Federation | 47,820,000,000,000 | 27.0993 |
| Saudi Arabia | 6,848,000,000,000 | 3.8807 |
| Turkmenistan | 2,860,000,000,000 | 1.6207 |
| United Arab Emirates | 6,071,000,000,000 | 3.4404 |
| US | 5,866,000,000,000 | 3.3242 |
| Uzbekistan | 1,850,000,000,000 | 1.0484 |
| Venezuela, RB | 4,315,000,000,000 | 2.4453 |
| World total | 176,462,000,000,000 | |

| Country | Nuclear power generation (number of megawatts) | % of world total (%) |
|---|---|---|
| Belgium | 5,801 | 1.5654 |
| Brazil | 1,901 | 0.5130 |
| Bulgaria | 2,722 | 0.7345 |
| Canada | 12,500 | 3.3731 |
| China | 6,572 | 1.7735 |
| Czech Republic | 3,368 | 0.9089 |
| Finland | 2,696 | 0.7275 |
| France | 63,363 | 17.0985 |
| Germany | 20,303 | 5.4788 |
| India | 3,040 | 0.8203 |
| Japan | 47,839 | 12.9094 |
| Korea, Rep. | 16,810 | 4.5362 |
| Russian Federation | 21,743 | 5.8674 |
| Slovak Republic | 2,460 | 0.6638 |
| Spain | 7,588 | 2.0476 |
| Sweden | 8,961 | 2.4181 |
| Taiwan | 4,904 | 1.3233 |
| Ukraine | 13,107 | 3.5369 |
| UK | 12,144 | 3.2771 |
| US | 99,988 | 26.9818 |
| World total | 370,576 | |

| Country | Oil reserves (number of barrels) | % of world total (%) |
|---|---|---|
| Algeria | 23,241,000,000 | 1.9125 |
| Angola | 9,050,000,000 | 0.7447 |
| Azerbaijan | 7,000,000,000 | 0.5760 |
| Brazil | 11,772,000,000 | 0.9687 |
| Canada | 15,034,000,000 | 1.2372 |
| China | 16,189,000,000 | 1.3322 |
| India | 6,202,000,000 | 0.5104 |
| Iran, Islamic Rep. | 137,490,000,000 | 11.3143 |
| Iraq | 115,000,000,000 | 9.4636 |
| Kazakhstan | 39,600,000,000 | 3.2588 |
| Kuwait | 101,500,000,000 | 8.3526 |
| Libya | 41,464,000,000 | 3.4122 |
| Mexico | 13,671,000,000 | 1.1250 |
| Nigeria | 36,220,000,000 | 2.9806 |
| Norway | 9,547,000,000 | 0.7856 |
| Qatar | 15,207,000,000 | 1.2514 |
| Russian Federation | 74,400,000,000 | 6.1225 |

| | | |
|---|---|---|
| Saudi Arabia | 264,310,000,000 | 21.7506 |
| Sudan | 6,402,000,000 | 0.5268 |
| United Arab Emirates | 97,800,000,000 | 8.0482 |
| US | 29,922,000,000 | 2.4623 |
| Venezuela, RB | 80,012,000,000 | 6.5843 |
| World total | 1,215,186,000,000 | |

| Country | Oil shale resources (number of barrels) | % of world total (%) |
|---|---|---|
| Australia | 31,729,000,000 | 1.1227 |
| Brazil | 82,000,000,000 | 2.9015 |
| Canada | 15,241,000,000 | 0.5393 |
| China | 16,000,000,000 | 0.5662 |
| Congo, Dem. Rep. | 100,000,000,000 | 3.5384 |
| Estonia | 16,286,000,000 | 0.5763 |
| Italy | 73,000,000,000 | 2.5831 |
| Jordan | 34,172,000,000 | 1.2092 |
| Morocco | 53,381,000,000 | 1.8889 |
| Russian Federation | 247,883,000,000 | 8.7712 |
| US | 2,085,228,000,000 | 73.7846 |
| World total | 2,826,103,000,000 | |

| Country | Peat production (number of tons) | % of world total (%) |
|---|---|---|
| Belarus | 1,993,000 | 14.6760 |
| Estonia | 279,000 | 2.0545 |
| Finland | 3,200,000 | 23.5641 |
| Ireland | 4,395,000 | 32.3638 |
| Russian Federation | 1,487,000 | 10.9499 |
| Sweden | 1,276,000 | 9.3962 |
| Ukraine | 707,000 | 5.2062 |
| World total | 13,580,000 | |

| Country | Solar electric capacity (number of kilowatts) | % of world total (%) |
|---|---|---|
| Austria | 24,000 | 0.6150 |
| Bangladesh | 3,500 | 0.0897 |
| China | 70,000 | 1.7938 |
| France | 33,570 | 0.8603 |
| Gabon | 148 | 0.0038 |
| Germany | 1,429,000 | 36.6195 |
| India | 85,000 | 2.1782 |
| Italy | 34,000 | 0.8713 |
| Japan | 1,421,908 | 36.4378 |
| Luxembourg | 23,600 | 0.6048 |
| Netherlands | 50,776 | 1.3012 |
| Spain | 51,900 | 1.3300 |
| Switzerland | 26,300 | 0.6740 |
| Thailand | 23,700 | 0.6073 |
| US | 496,000 | 12.7105 |
| World total | 3,902,290 | |

| Country | Wind power capacity (number of megawatts) | % of world total (%) |
|---|---|---|
| Australia | 708 | 1.1932 |
| Austria | 819 | 1.3803 |
| Canada | 683 | 1.1511 |
| China | 1,266 | 2.1336 |
| Denmark | 3,129 | 5.2734 |
| France | 723 | 1.2185 |
| Germany | 18,428 | 31.0576 |
| Greece | 573 | 0.9657 |
| India | 4,434 | 7.4728 |
| Ireland | 496 | 0.8359 |
| Italy | 1,639 | 2.7623 |
| Japan | 1,078 | 1.8168 |
| Netherlands | 1,224 | 2.0629 |
| Portugal | 1,063 | 1.7915 |
| Spain | 10,028 | 16.9006 |
| Sweden | 493 | 0.8309 |
| UK | 1,565 | 2.6376 |
| US | 9,149 | 15.4192 |
| World total | 59,335 | |

# Appendix 2

# Global Carbon Dioxide Emissions

The following table has been adapted from data supplied by the World Energy Council to Show: A New Way to Look at the World. Data are for 2005. Available at http://show.mappingworlds.com/

| Country | $CO_2$ emissions (metric tons of carbon) | $CO_2$ emissions (% world total) |
|---|---|---|
| Afghanistan | 189,000 | 0.0025 |
| Albania | 1,002,000 | 0.0134 |
| Algeria | 52,915,000 | 0.7058 |
| Angola | 2,154,000 | 0.0287 |
| Antigua and Barbuda | 113,000 | 0.0015 |
| Argentina | 38,673,000 | 0.5158 |
| Armenia | 995,000 | 0.0133 |
| Aruba | 588,000 | 0.0078 |
| Australia | 89,125,000 | 1.1888 |
| Austria | 19,051,000 | 0.2541 |
| Azerbaijan | 8,555,000 | 0.1141 |
| Bahamas, The | 548,000 | 0.0073 |
| Bahrain | 4,623,000 | 0.0617 |
| Bangladesh | 10,137,000 | 0.1352 |
| Barbados | 346,000 | 0.0046 |
| Belarus | 17,699,000 | 0.2361 |
| Belgium | 27,471,000 | 0.3664 |
| Belize | 216,000 | 0.0029 |
| Benin | 651,000 | 0.0087 |
| Bermuda | 150,000 | 0.0020 |
| Bhutan | 113,000 | 0.0015 |
| Bolivia | 1,902,000 | 0.0254 |
| Bosnia and Herzegovina | 4,254,000 | 0.0567 |
| Botswana | 1,173,000 | 0.0156 |
| Brazil | 90,499,000 | 1.2071 |
| British Virgin Islands | 23,000 | 0.0003 |
| Brunei Darussalam | 2,403,000 | 0.0321 |
| Bulgaria | 11,608,000 | 0.1548 |
| Burkina Faso | 299,000 | 0.0040 |
| Burundi | 60,000 | 0.0008 |
| Cambodia | 146,000 | 0.0019 |
| Cameroon | 1,047,000 | 0.0140 |
| Canada | 174,401,000 | 2.3262 |
| Cape Verde | 75,000 | 0.0010 |
| Cayman Islands | 85,000 | 0.0011 |
| Central African Republic | 69,000 | 0.0009 |
| Chad | 34,000 | 0.0005 |
| Chile | 17,025,000 | 0.2271 |
| China | 1,366,554,000 | 18.2274 |
| Colombia | 14,629,000 | 0.1951 |
| Comoros | 24,000 | 0.0003 |
| Congo, Dem. Rep. | 574,000 | 0.0077 |
| Congo, Rep. | 966,000 | 0.0129 |
| Cook Islands | 8,000 | 0.0001 |
| Costa Rica | 1,747,000 | 0.0233 |
| Croatia | 6,410,000 | 0.0855 |
| Cuba | 7,042,000 | 0.0939 |
| Cyprus | 1,841,000 | 0.0246 |
| Czech Republic | 31,910,000 | 0.4256 |
| Côte d'Ivoire | 1,408,000 | 0.0188 |
| Denmark | 14,444,000 | 0.1927 |
| Djibouti | 100,000 | 0.0013 |
| Dominica | 29,000 | 0.0004 |
| Dominican Republic | 5,357,000 | 0.0715 |
| Ecuador | 7,983,000 | 0.1065 |
| Egypt, Arab Rep. | 43,160,000 | 0.5757 |
| El Salvador | 1,682,000 | 0.0224 |
| Equatorial Guinea | 1,480,000 | 0.0197 |

| Country | $CO_2$ emissions (metric tons of carbon) | $CO_2$ emissions (% world total) |
|---|---|---|
| Eritrea | 206,000 | 0.0027 |
| Estonia | 5,167,000 | 0.0689 |
| Ethiopia | 2,177,000 | 0.0290 |
| Faeroe Islands | 180,000 | 0.0024 |
| Falkland Islands | 12,000 | 0.0002 |
| Fiji | 292,000 | 0.0039 |
| Finland | 17,947,000 | 0.2394 |
| France | 101,927,000 | 1.3595 |
| French Guiana | 274,000 | 0.0037 |
| French Polynesia | 183,000 | 0.0024 |
| Gabon | 374,000 | 0.0050 |
| Gambia, The | 78,000 | 0.0010 |
| Georgia | 1,067,000 | 0.0142 |
| Germany | 220,596,000 | 2.9424 |
| Ghana | 1,961,000 | 0.0262 |
| Gibraltar | 102,000 | 0.0014 |
| Greece | 26,374,000 | 0.3518 |
| Greenland | 156,000 | 0.0021 |
| Grenada | 59,000 | 0.0008 |
| Guadeloupe | 473,000 | 0.0063 |
| Guatemala | 3,333,000 | 0.0445 |
| Guinea | 365,000 | 0.0049 |
| Guinea-Bissau | 74,000 | 0.0010 |
| Guyana | 394,000 | 0.0053 |
| Haiti | 479,000 | 0.0064 |
| Honduras | 2,077,000 | 0.0277 |
| Hong Kong, China | 10,204,000 | 0.1361 |
| Hungary | 15,597,000 | 0.2080 |
| Iceland | 608,000 | 0.0081 |
| India | 366,301,000 | 4.8858 |
| Indonesia | 103,170,000 | 1.3761 |
| Iran, Islamic Rep. | 118,259,000 | 1.5774 |
| Iraq | 22,271,000 | 0.2971 |
| Ireland | 11,552,000 | 0.1541 |
| Israel | 19,433,000 | 0.2592 |
| Italy | 122,726,000 | 1.6369 |
| Jamaica | 2,889,000 | 0.0385 |
| Japan | 343,117,000 | 4.5766 |
| Jordan | 4,491,000 | 0.0599 |
| Kazakhstan | 54,627,000 | 0.7286 |
| Kenya | 2,888,000 | 0.0385 |
| Kiribati | 8,000 | 0.0001 |
| Korea, Dem. Rep. | 21,578,000 | 0.2878 |
| Korea, Rep. | 127,007,000 | 1.6940 |
| Kuwait | 27,102,000 | 0.3615 |
| Kyrgyz Republic | 1,562,000 | 0.0208 |
| Lao PDR | 349,000 | 0.0047 |
| Latvia | 1,936,000 | 0.0258 |
| Lebanon | 4,436,000 | 0.0592 |
| Liberia | 128,000 | 0.0017 |
| Libya | 16,342,000 | 0.2180 |
| Lithuania | 3,630,000 | 0.0484 |
| Luxembourg | 3,076,000 | 0.0410 |
| Macao, China | 602,000 | 0.0080 |
| Macedonia, FYR | 2,842,000 | 0.0379 |
| Madagascar | 745,000 | 0.0099 |
| Malawi | 285,000 | 0.0038 |
| Malaysia | 48,437,000 | 0.6461 |
| Maldives | 198,000 | 0.0026 |
| Mali | 154,000 | 0.0021 |
| Malta | 669,000 | 0.0089 |
| Martinique | 352,000 | 0.0047 |
| Mauritania | 697,000 | 0.0093 |
| Mauritius | 872,000 | 0.0116 |
| Mexico | 119,473,000 | 1.5936 |
| Moldova | 2,096,000 | 0.0280 |
| Mongolia | 2,333,000 | 0.0311 |
| Montserrat | 17,000 | 0.0002 |
| Morocco | 11,229,000 | 0.1498 |
| Mozambique | 591,000 | 0.0079 |
| Myanmar | 2,662,000 | 0.0355 |
| Namibia | 674,000 | 0.0090 |
| Nauru | 39,000 | 0.0005 |
| Nepal | 830,000 | 0.0111 |
| Netherlands | 38,748,000 | 0.5168 |
| Netherlands Antilles | 1,115,000 | 0.0149 |
| New Caledonia | 703,000 | 0.0094 |
| New Zealand | 8,611,000 | 0.1149 |
| Nicaragua | 1,093,000 | 0.0146 |
| Niger | 331,000 | 0.0044 |
| Nigeria | 31,101,000 | 0.4148 |
| Niue Islands | 1,000 | 0.0000 |
| Norway | 23,894,000 | 0.3187 |
| Oman | 8,428,000 | 0.1124 |
| Pakistan | 34,277,000 | 0.4572 |
| Palau | 65,000 | 0.0009 |
| Palestinian Authority | 177,000 | 0.0024 |
| Panama | 1,544,000 | 0.0206 |
| Papua New Guinea | 668,000 | 0.0089 |
| Paraguay | 1,140,000 | 0.0152 |
| Peru | 8,590,000 | 0.1146 |
| Philippines | 21,960,000 | 0.2929 |
| Poland | 83,801,000 | 1.1178 |
| Portugal | 16,067,000 | 0.2143 |
| Qatar | 14,430,000 | 0.1925 |

| | | | | | |
|---|---|---|---|---|---|
| Reunion | 621,000 | 0.0083 | Syrian Arab Republic | 18,662,000 | 0.2489 |
| Romania | 24,664,000 | 0.3290 | São Tomé and Principe | 25,000 | 0.0003 |
| Russian Federation | 415,951,000 | 5.5480 | Taiwan | 65,807,000 | 0.8777 |
| Rwanda | 156,000 | 0.0021 | Tajikistan | 1,365,000 | 0.0182 |
| Samoa | 41,000 | 0.0005 | Tanzania | 1,187,000 | 0.0158 |
| Saudi Arabia | 84,116,000 | 1.1220 | Thailand | 73,121,000 | 0.9753 |
| Senegal | 1,362,000 | 0.0182 | Timor-Leste | 48,000 | 0.0006 |
| Serbia | 14,544,000 | 0.1940 | Togo | 630,000 | 0.0084 |
| Seychelles | 149,000 | 0.0020 | Tonga | 32,000 | 0.0004 |
| Sierra Leone | 271,000 | 0.0036 | Trinidad and Tobago | 8,880,000 | 0.1184 |
| Singapore | 14,252,000 | 0.1901 | Tunisia | 6,242,000 | 0.0833 |
| Slovak Republic | 9,898,000 | 0.1320 | Turkey | 61,677,000 | 0.8227 |
| Slovenia | 4,422,000 | 0.0590 | Turkmenistan | 11,381,000 | 0.1518 |
| Solomon Islands | 48,000 | 0.0006 | Uganda | 498,000 | 0.0066 |
| South Africa | 119,203,000 | 1.5900 | Ukraine | 90,020,000 | 1.2007 |
| Spain | 90,145,000 | 1.2024 | United Arab Emirates | 40,692,000 | 0.5428 |
| Sri Lanka | 3,146,000 | 0.0420 | UK | 160,179,000 | 2.1365 |
| St. Helena | 3,000 | 0.0000 | US | 1,650,020,000 | 22.0083 |
| St. Kitts and Nevis | 34,000 | 0.0005 | Uruguay | 1,494,000 | 0.0199 |
| St. Lucia | 100,000 | 0.0013 | Uzbekistan | 37,615,000 | 0.5017 |
| St. Pierre and Miquelon | 17,000 | 0.0002 | Vanuatu | 24,000 | 0.0003 |
| St. Vincent & the Grenadines | 54,000 | 0.0007 | Venezuela, RB | 47,084,000 | 0.6280 |
| | | | Vietnam | 26,911,000 | 0.3589 |
| Sudan | 2,829,000 | 0.0377 | Western Sahara | 65,000 | 0.0009 |
| Suriname | 623,000 | 0.0083 | Yemen, Rep. | 5,759,000 | 0.0768 |
| Swaziland | 261,000 | 0.0035 | Zambia | 624,000 | 0.0083 |
| Sweden | 14,465,000 | 0.1929 | Zimbabwe | 2,880,000 | 0.0384 |
| Switzerland | 11,035,000 | 0.1472 | World total | 7,497,252,000 | 100.00 |

# Appendix 3

# Global Warming Potential (GWP)

UNFCCC and its Kyoto Protocol are responsible for regulating a basket of six greenhouse gases. Three of these occur naturally (note they are also produced by human action, for example, combustion); namely carbon dioxide, methane and nitrous oxide and three are manufactured, namely hydrofluorocarbons (HFCs), perfluorocarbons (PFCs) and sulphur hexafluoride ($SF_6$). HFC and PFC represent a family of different chemicals that have been manufactured for different purposes. There are other gases that exacerbate the greenhouse effect such as Ozone Depleting Substances (ODS) that are regulated by the Montreal Protocol.

It is known that different gases have different impacts on the greenhouse effect. The greenhouse effect is primarily a function of the concentration of water vapour, carbon dioxide and other trace gases in the atmosphere that absorb the terrestrial radiation leaving the surface of the Earth. Changes in the atmospheric concentrations of these greenhouse gases can alter the balance of energy transfers between the atmosphere, space, land and the oceans. A gauge of these changes is called radiative forcing, which is a simple measure of changes in the energy available to the Earth-atmosphere system.

Radiative forcing is a measure of how the energy balance of the Earth-atmosphere system is influenced when factors that affect climate are altered. The word radiative arises because these factors change the balance between incoming solar radiation and outgoing infrared radiation within the Earth's atmosphere. This radiative balance controls the Earth's surface temperature. The term forcing is used to indicate that the Earth's radiative balance is being pushed away from its normal state. Radiative forcing is usually quantified as the 'rate of energy change per unit area of the globe as measured at the top of the atmosphere', and is expressed in units of 'Watts per square metre' (Forster et al, 2007). From Table A3.1 we can see that the single largest contributor to the greenhouse effect is carbon dioxide.

To compare the relative climate effects of greenhouse gases, it is necessary to assess their contribution to changes in the net downward infra-red radiation flux at the tropopause (the top of the lower atmosphere) over a period of

**Table A3.1** Global atmospheric concentration, concentration rates and residence times of selected greenhouse gases

| Atmospheric variable | $CO_2$ | $CH_4$ | $N_2O$ | $SF_6$ |
|---|---|---|---|---|
| Pre-industrial concentration | 278 | 0.700 | 0.270 | 0 |
| Atmospheric concentration (1998) | 365 | 1.745 | 0.314 | 4.2 |
| Rate of concentration change | 1.5 | 0.007 | 0.0008 | 0.24 |
| Atmospheric lifetime | 50–200 | 12 | 114 | 3200 |

*Notes:*
1 $SF_6$ specified in parts per trillion (ppt), all others in parts per million (ppm)
2 No single lifetime can be specified for $CO_2$ because of the different rates of uptake by different removal processes.

*Source:* Adapted from IPCC, 2001

**Table A3.2** GWP and residence times of selected greenhouse gases

| Gas | 100 year GWP | Atmospheric lifetime |
|---|---|---|
| Carbon Dioxide ($CO_2$) | 1 | 50–200 |
| Methane ($CH_4$) | 21 | 12 +/–3 |
| Nitrous Oxide ($N_2O$) | 310 | 120 |
| Hydrofluorocarbons (HFCs) | 140–11,700 | 1.5–264 |
| Perfluorocarbons (PFCs) | 6,500–9,200 | 3,200–50,000 |
| Sulphur Hexafluoride ($SF_6$) | 23,900 | 3,200 |

*Source:* Adapted from IPCC, 1996

time. Ultimately the best way to do this is by comparing different emission scenarios in climate models, but a simple working method has been derived for use by Parties to the UNFCCC. This provides the relative contribution of a unit emission of each gas, relative to the effect of a unit emission of carbon dioxide integrated over a fixed time period. A 100-year time horizon has been chosen by the Convention in view of the relatively long time scale for addressing climate change. The factor is known as the Global Warming Potential (GWP). GWPs are intended as a quantified measure of the globally averaged relative radiative forcing impacts of a particular greenhouse gas. It is defined as the cumulative radiative forcing, both direct and indirect effects, integrated over a period of time from the emission of a unit mass of gas relative to some reference gas (IPCC, 1996). Carbon dioxide was chosen as this reference gas. Direct effects occur when the gas itself is a greenhouse gas. Indirect radiative forcing occurs when chemical transformations involving the original gas produce a gas or gases that are greenhouse gases, or when a gas influences other radiatively important processes such as the atmospheric lifetimes of other gases. The relationship between gigagrams (Gg) of a gas and Tg $CO_2$ Eq. can be expressed as follows:

Tg $CO_2$ Eq = (Gg of gas) × (GWP) × (Tg/1,000 Gg)

where:

Tg $CO_2$ Eq = Teragrams of carbon dioxide equivalents

Gg = Gigagrams (equivalent to a thousand metric tons)

GWP = Global Warming Potential

Tg = Teragrams

GWP values allow policy makers to compare the impacts of emissions and reductions of different gases. According to the IPCC, GWPs typically have an uncertainty of roughly 35 per cent, although some GWPs have larger uncertainty than others, especially those in which lifetimes have not yet been ascertained.

IPCC has used the values for GWP calculated for the Second Assessment Report (SAR) as the basis for estimating greenhouse gas inventories. The Third Assessment Report did update some of these values. Table A3.2 gives an overview of the GWP and residence times used in the inventory.

The Third Assessment Report (TAR) of the GWPs of several gases were revised relative to the IPCC's Second Assessment Report (SAR) and new GWPs were calculated for an expanded set of gases. This also included an improved calculation of $CO_2$ radiative forcing and an improved $CO_2$ response function. The atmospheric lifetimes of some gases have been recalculated. Because the revised radiative forcing of $CO_2$ was found to be about 12 per cent lower than that in the SAR, the GWPs of the other gases relative to $CO_2$ tend to be larger, taking into account revisions in lifetimes. However, there were some instances in which other variables, such as the radiative efficiency or the chemical lifetime, were altered that resulted in further increases or decreases in particular GWP values. In addition, the values for

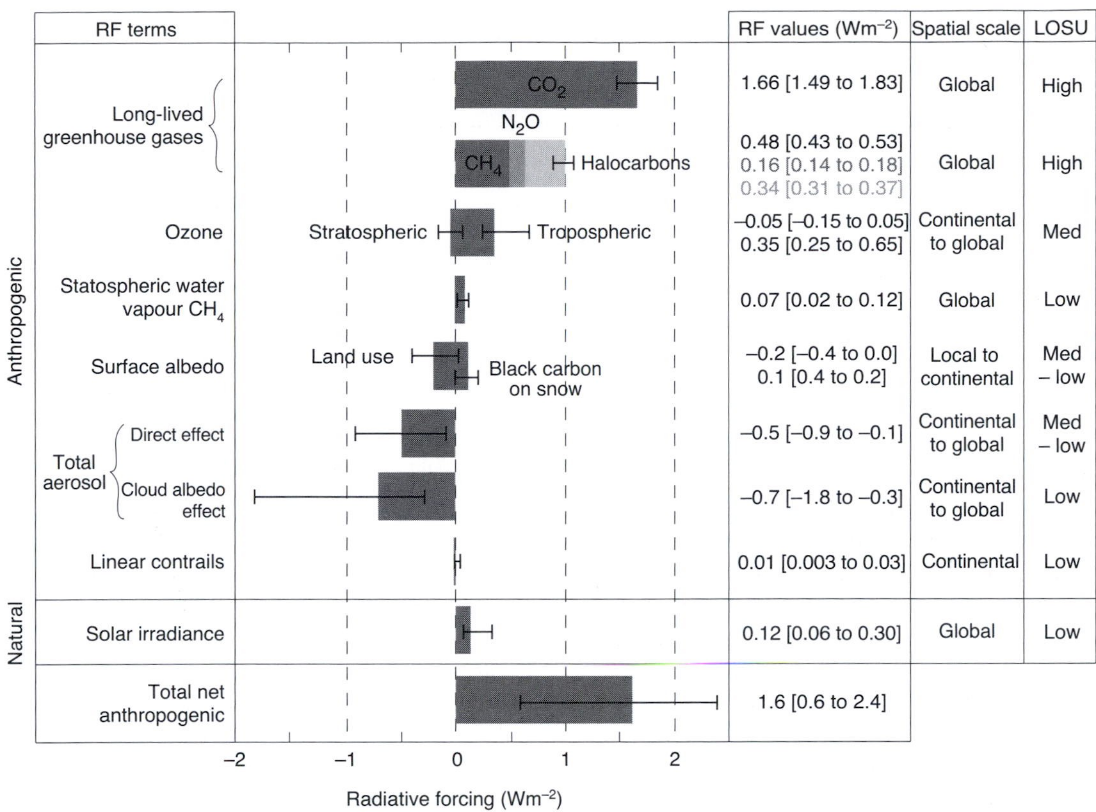


*Source:* IPCC, 2007

**Figure A3.1** Radiative forcing components

radiative forcing and lifetimes were calculated for a variety of halocarbons, which were not presented in the SAR.

The Fourth Assessment Report presents a more comprehensive assessment of the radiative potential of a larger number of gases. In addition IPCC combined data from other sources that lower radiative forcing such as albedo, which tends to reflect solar away from the planet, and Figure A3.1 shows the combined effects. This shows that there is a discernible increase in radiative forcing due, in the main, to the emission of anthropogenic greenhouse gases.

IPCC reports a number of changes to the GWP and residence time in the atmosphere, for example, methane is given values of 12 and 21 respectively and nitrous oxide 114 and 310 respectively. A full listing can be found in Table 2.14 in Forster et al, 2007, available at: http://www.ipcc.ch/pdf/assessment-report/ar4/wg1/ar4-wg1-chapter2.pdf

Finally, the report concludes that the atmospheric concentration of carbon dioxide was 380ppm with a growth rate of 1.9ppm for the period 1995–2005. The average growth rate between 1960 and 2005 was 1.4ppm. The primary source of the increased atmospheric concentration of carbon dioxide since the pre-industrial period results from fossil fuel use, with land use change providing another significant but smaller contribution. Annual fossil carbon dioxide emissions increased from an average of 6.4 (6.0–6.80GtC (equivalent to 23.5 [22.0–25.0]GtCO[2] as 1GTC = 3.67 $GtCO_2$) per year in the 1990s to 7.2 [6.9–7.5]GtC (26.4 [25.3–27.5]$GtCO_2$) per year in 2000–2005 (2004 and 2005 data are interim estimates). Carbon dioxide emissions associated with land use change are estimated to be 1.6 (0.5–2.7)GtC (5.9 [1.8–9.9]Gt$CO_2$) per year over the 1990s, although these estimates have a large uncertainty.

The global atmospheric concentration of methane has increased from a pre-industrial value of about 715ppb to 1732ppb in the early 1990s, and was 1774ppb in 2005. The atmospheric concentration of methane in 2005 exceeds by far the natural range of the last 650,000 years (320–790ppb) as determined from ice cores. Growth rates have declined since the early 1990s, consistent with total emissions (sum of anthropogenic and natural sources) being nearly constant during this period. IPPC concludes that it is *very likely* that the observed increase in methane concentration is due to anthropogenic activities, predominantly agriculture and fossil fuel use, but relative contributions from different source types are not well determined.

The global atmospheric nitrous oxide concentration increased from a pre-industrial value of about 270ppb to 319ppb in 2005. The growth rate has been approximately constant since 1980. More than a third of all nitrous oxide emissions are anthropogenic and are primarily due to agriculture (Forster et al, 2007).

## References

Forster, P.V., Ramaswamy, P., Artaxo, T., Berntsen, R., Betts, D.W., Fahey, J., Haywood, J., Lean, D.C., Lowe, G., Myhre, J., Nganga, R., Prinn, G., Raga, M., Schulz and Van Dorland, R. (2007) 'Changes in Atmospheric Constituents and in Radiative Forcing', in Solomon, S., D. Qin, M. Manning, Z. Chen, M. Marquis, K. B. Averyt, M. Tignor and H. L. Miller (eds) *Climate Change 2007: The Physical Science Basis. Contribution of Working Group I to the Fourth Assessment Report of the Intergovernmental Panel on Climate Change*, Cambridge University Press, Cambridge, UK and New York

IPCC (1996) *Climate Change 1995: The Science of Climate Change*, Intergovernmental Panel on Climate Change; J. T. Houghton, L. G. Meira Filho, B. A. Callander, N. Harris, A. Kattenberg, and K. Maskell (eds); Cambridge University Press, Cambridge, UK

IPCC (2001) *Climate Change 2001: A Scientific Basis*, Intergovernmental Panel on Climate Change; J.T. Houghton, Y. Ding, D.J. Griggs, M. Noguer, P.J. van der Linden, X. Dai, C.A. Johnson, and K. Maskell, eds.; Cambridge University Press. Cambridge, U.K.

IPCC (2007) 'Summary for Policymakers', in Solomon, S., D. Qin, M. Manning, Z. Chen, M. Marquis, K. B. Averyt, M. Tignor and H. L. Miller (eds) *Climate Change 2007: The Physical Science Basis. Contribution of Working Group I to the Fourth Assessment Report of the Intergovernmental Panel on Climate Change,* Cambridge University Press, Cambridge, UK and New York

# Appendix 4

# Measurements and Conversion Tables

**Table A4.1** Conversion factors

| | MJ | GJ | kWh | toe | tce |
|---|---|---|---|---|---|
| 1MJ = | 1 | 0.001 | 0.2778 | $2.4 \times 10^{-15}$ | 3.6 10–5 |
| 1GJ = | 1000 | 1 | 277.8 | 0.024 | 0.036 |
| 1kWh = | 3.60 | 0.0036 | 1 | $8.6 \times 10^{-15}$ | $1.3 \times 10^{-14}$ |
| 1toe = | 42000 | 42 | 12000 | 1 | 1.5 |
| 1tce = | 28000 | 28 | 7800 | 0.67 | 1 |
| | **PJ** | **EJ** | **TWh** | **Mtoe** | **Mtce** |
| 1PJ = | 1 | 0.001 | 0.2778 | 0.024 | 0.036 |
| 1EJ = | 1000 | 1 | 277.8 | 24 | 36 |
| 1TWh = | 3.60 | 0.0036 | 1 | 0.086 | 0.13 |
| 1Mtoe = | 42 | 0.042 | 12 | 1 | 1.5 |
| 1Mtce = | 28 | 0.028 | 7.8 | 0.67 | 1 |

**Table A4.2** Energy rates for different power outputs

| Rate | Joules per hour | Joules per year | Kilowatt-hours per year | Oil equivalent per year | Coal equivalent per year |
|---|---|---|---|---|---|
| 1W | 3600J | 31.54MJ | 8.76 | $0.75 \times 10^{-3}$ toe* | $1.1 \times 10^{-3}$ tce* |
| 1kW | 3.6MJ | 31.54GJ | 8760 | 0.75toe | 1.1tce |
| 1MW | 3.6GJ | 31.54TJ | $8.76 \times 10^{6}$ | 750toe | 1100tce |
| 1GW | 3.6PJ | 31.54PJ | $8.76 \times 10^{9}$ | 0.75Mtoe | 1.1Mtce |
| 1TW | 3.6TJ | 31.54EJ | $8.76 \times 10^{12}$ | 750Mtoe | 1100Mtce |

* the energy equivalent of 0.75 kg of oil or 1.1 kg of coal

**Table A4.3** Measurements and SI and other equivalents

| Quantity | Unit | Standard International (SI) equivalent |
|---|---|---|
| Mass | 1oz (ounce) | $=2.834 \times 10^{-2}$ kg |
| | 1lb (pound) | =0.4536kg |
| | 1 ton | =1016kg |
| | 1 short ton | =972kg |
| | 1t (tonne) | =1000kg |
| | 1u (unified mass unit) | $=1.660 \times 10^{-27}$ kg |
| Length | 1in (inch) | $=2.540 \times 10^{-2}$ m |
| | 1ft (foot) | =0.3048m |
| | 1yd (yard) | =0.9144m |
| | 1mi (mile) | =1609m |
| Speed | 1km $hr^{-1}$ (kph) | $=0.2778m/s^{-1}$ |
| | 1mi $kr^{-1}$ (mph) | $=0.4470m\ s^{-1}$ |
| Area | $1in^2$ | $=6.452 \times 10^{-4}m^2$ |
| | $1ft^2$ | $=9.290 \times 10^{-2}m^2$ |
| | $1yd^2$ | $=0.8361m^2$ |
| | 1 acre | $=4047m^2$ |
| | 1ha (hectare) | $=10^4m^2$ |
| | $1mi^2$ | $=2.590 \times 10^6m^2$ |
| Volume | $1in^3$ | $=1.639 \times 10^{-5}m^3$ |
| | $1ft^3$ | $=2.832 \times 10^{-2}m^3$ |
| | $1yd^3$ | $=0.7646m^3$ |
| | 1 litre | $=10^{-3}m^3$ |
| | 1gal (UK) | $=4.546 \times 10^{-3}m^3$ |
| | 1gal (US) | $=3.785 \times 10^{-3}m^3$ |
| | 1 bushel | $=3.637 \times 10^{-2}m^3$ |
| Force | 1lbf (weight of 1 lb mass) | =4.448N |
| Pressure | 1lbf $in^{-2}$ (of psi) | =6895Pa |
| | 1bar | $=10^5Pa$ |
| Energy | 1ft lb (foot-pound) | =1.356J |
| | 1eV (electron volt) | $=1.602 \times 10^{-19}$ J |
| | 1MeV | $=1.602 \times 10^{-13}$ J |
| Power | 1HP (horse power) | =745.7W |

**Table A4.4** Fossil fuel equivalents

| Petroleum (1 million tonnes) | Coal (1 million tonnes) | Natural gas (1 million therms) |
|---|---|---|
| 7.5 million barrels | 600,000 tonnes of oil | 100 million cubic feet |
| 425 million therms | 250 million therms | 2.75 million cubic meters |
| 1.7 million tonnes of coal | 7500GWh of energy | 4000 tonnes of coal |
| 12,500GWh of energy | | 2400 tonnes of oil |
| | | 29.3GWh of energy |

**Table A4.5** Multipliers

| Symbol | Prefix | Multiply by | Description |
|---|---|---|---|
| E | exa | $10^{18}$ | One quintillion |
| P | peta | $10^{15}$ | One quadrillion |
| T | tera | $10^{12}$ | One trillion |
| G | giga | $10^{9}$ | One billion |
| M | mega | $10^{6}$ | One million |
| k | kilo | $10^{3}$ | One thousand |
| h | hecto | $10^{2}$ | One hundred |
| da | deca | 10 | Ten |
| d | deci | $10^{-1}$ | One tenth |
| c | centi | $10^{-2}$ | One hundredth |
| m | milli | $10^{-3}$ | One thousandth |
| u | micro | $10^{-6}$ | One millionth |
| n | nano | $10^{-9}$ | One billionth |
| p | pico | $10^{-12}$ | One trillionth |
| f | femto | $10^{-15}$ | One quadrillionth |
| a | atto | $10^{-18}$ | One quintillionth |

# Appendix 5

# Costing Energy Projects

Investing in an energy project, like any other project, requires a method for assessing if the project is value for money. From a commercial perspective, if a project pays back less than the initial cost then it does not make any sense to go ahead with the project, no matter how laudable the aims of the project may be. There are two methods of estimating the costs and benefits of an action; Discounted Cash Flow (DCF) and Cost–Benefit Analysis (CBA). Typically DCF is used in investment finance, property development and corporate financial management and is based on the concept of the time value of money. Future cash flows are estimated and discounted to give their present values. CBA is typically used by governments to evaluate the desirability of a given intervention in markets. The aim is to gauge the efficiency of the intervention relative to the status quo. The costs and benefits of the impacts of an intervention are evaluated in terms of the public's willingness to pay for them (benefits) or willingness to pay to avoid them (costs). Inputs are typically measured in terms of opportunity costs – the value in their best alternative use. The guiding principle is to list all of the parties affected by an intervention, and place a monetary value of the effect it has on their welfare as it would be valued by them.

## Discounted Cash Flow (DCF)

This is a valuation method that is used to estimate the attractiveness of an investment opportunity. DCF analysis uses future free cash flow projections and discounts them (most often using the Weighted Average Cost of Capital or WACC) to arrive at a present value, which is used to evaluate the potential for investment. This is known as the Net Present Value (NPV) which is the difference between the present value of cash inflows and the present value of cash outflows. If the value arrived at through DCF analysis is higher than the current cost of the investment, the opportunity may be a good one.

The WACC is a calculation of an organization's cost of capital in which each category of capital is proportionately weighted. All capital sources – common stock, preferred stock, bonds and any other long-term debt – are included in a WACC calculation. WACC is calculated by multiplying the cost of each capital component by its proportional weight and then summing:

$$\text{WACC} = (E/V) \times Re + (D/V) \times Rd \times (1 - Tc)$$

where:

| | | |
|---|---|---|
| Re | = | cost of equity |
| Rd | = | cost of debt |
| E | = | market value of the organization's equity |
| D | = | market value of the organization's debt |
| V | = | E + D |
| E/V | = | percentage of financing that is equity |
| D/V | = | percentage of financing that is debt |
| Tc | = | corporate tax rate |

## Cost–Benefit Analysis (CBA)

CBA is a tool for assessing the viability of different investments that considers the future realization of costs and benefits. CBA is a process by which decisions are analysed. The benefits of a given situation or action are summed and then the costs associated with taking that action are subtracted. Non-monetized items such as global warming or the effects of pollution are assigned a monetary value so that they can be included in the analysis. In general there are five-steps in conducting a CBA:

1. Listing the candidate projects to be assessed.
2. Listing the social costs and benefits for each project.
3. Quantifying each of these costs and benefits with supporting technical evidence.
4. Calculating a money value for each cost and benefit.
5. Arriving at a final evaluation.

In this sense CBA deals with the economic costs and benefits to society. As discussed earlier the WACC refers specifically to an organization and this determines the discount rate. In CBA the discount rate is the 'social rate of time preference' (SRTP). In theory, the SRTP is the mean discount rate expressed by each individual in the population affected by the project in question. Usually, the SRTP is determined by reference to the market rate of interest, but that rate is subject to central bank influence. Overall, establishing an SRTP is difficult, so some government agencies simply apply a given rate. Often, a range of discount rates are applied to a project as sensitivity analysis to test for robustness. Selection of the SRTP requires careful consideration, as results tend to be very sensitive to changes in the given rate. In the UK, for example, this rate, termed the 'social time preference rate' (STPR), is defined by the UK Treasury Green Book as:

> *the value society attaches to present, as opposed to future, consumption.*

The STPR has two components:-

- The rate at which individuals discount future consumption over present consumption, on the assumption that no change in per capita consumption is expected, represented by $\rho$; and,
- An additional element, if per capita consumption is expected to grow over time, reflecting the fact that these circumstances imply future consumption will be plentiful relative to the current position and thus have lower marginal utility. This effect is represented by the product of the annual growth in per capita consumption (g) and the elasticity of marginal utility of consumption ($\mu$) with respect to utility.

Estimates of $\rho$ comprise two elements: catastrophic risk and assessing the value of the non-monetized aspects.

The first component, catastrophe risk, is the likelihood that there will be some event so devastating that all returns from policies, programmes or projects are eliminated, or at least radically and unpredictably altered, for example, technological advancements that lead to premature obsolescence, or natural disasters, major wars, etc. The scale of this risk is, by its nature, hard to quantify. The second component, pure time preference, reflects individuals' preference for consumption now, rather than later, with an unchanging level of consumption per capita over time. Evidence seems to suggest a value of 1.5 per cent a year for the near future.

Estimates of $\mu$ and g suggest a value of 1 and 2 per cent per year.

This gives an STPR of 3.5 per cent. However it should be noted that there are many uncertainties and discount rates will vary. For example over time this value will decline with the Treasury estimating it at 1 per cent after 30 years (HM Treasury, 2007).

The next difficulty is assessing the value of the non-monetized aspects of a project. For example investment in a wind farm may bring public goods in terms of reducing the production of greenhouse gases. However, for some it may have an adverse visual impact that detracts from its amenity value. This is usually determined through Willingness to Pay or Willingness to Accept techniques. These are discussed in Chapter 2.

## Investing in energy projects

On the face of it, investing in energy projects is no different to any other kind of investment. It follows that it should be relatively straightforward to decide on the merits of a particular energy investment. There are a number of points to consider:

- The capital costs of land, buildings and equipment. These should be relatively easy to determine for a particular type of plant such as a coal, gas or nuclear power station.
- The costs of capital reflect the fact that typically an organization would have to borrow all or some of the capital needed to build the plant. This will have to be repaid and those that invested in the project will expect to generate a return. Generally, it is expected that this would have a higher rate of return than could be earned simply by depositing the cash in a bank, otherwise there would be no economic incentive for investors.
- The operating costs of the plant over its lifetime. Technologies will have a rated lifetime and an associated budget requirement for maintenance and upkeep. However, operating costs include labour and fuel costs. Labour costs may be reasonably easy to estimate as the number and type of personnel should be known. A factor for wage inflation can be estimated. Fuel costs are more problematic. Oil and gas has seen sharp fluctuations recently and there is great uncertainty surrounding future energy prices. For renewable systems that use wind, wave, tide, water or sunlight, there are no fuel costs but the intermittency for some renewable resources will need to be factored in. Renewable resources such as biomass will have on-going fuel costs.
- The income of the plant can be determined from the rated output over its lifetime and the income streams that it will generate in the future. The cost of electricity for the consumer, for example, is well known. However, estimating what consumers will pay in the future is problematic. Factors influencing the price of electricity are inflation and variations in the fuel costs and the availability of other alternative sources that may be developed during the lifetime of the plant. This means that the further into the future that estimates are made, the greater the discount rate will have to be, as risk increases. That means income streams near to the end of the lifetime will need to be discounted at a higher rate than those generated just after the plant goes online.
- The residual value of the plant at the end of its service life is the value of the plant or of its individual components and is determined from the possibilities of its alternative use. On the assumption that equipment or parts of a plant can be sold, the expected liquidation yield from the sale is usually taken as the residual value. Often one of the most valuable residual assets could be the land as this may be sold for another purpose.
- Depreciation can be defined as the decrease in value of an asset due to use and/or time. A very simple operational definition of depreciation can be given as:

  Depreciation = (cost of asset – residual value)/service life

Although estimating some of these values is problematic there are other factors that need to be considered. For example, during the lifetime of a plant new regulations could be implemented that would require the plant to be upgraded. This would typically entail new investment. An example of a possible uncertainty is whether or not a decision may be taken in the EU to require new coal burning plants to be Carbon Capture ready. This implies that the design would have to be such that carbon capture equipment could be fitted if that becomes a requirement. At present this is being discussed at EU level and there is a political aspiration for this approach. Some contingency will need to be factored into the investment decision in anticipation of such likely changes.

The area where there is great uncertainty is the external costs of energy production. We do know that the use of fossil fuels generates greenhouse gases and pollution. The costs associated with such emissions are usually not carried by the energy producer. Typically, they are paid for from public funds, for example, impacts on health will be paid through the health system. There is a more detailed discussion on external costs in Chapter 2.

A further problematic area when costing energy systems are subsidies. Energy subsidies are measures that, for consumers, keep prices below market levels or, for producers, above market levels, or reduce costs for consumers and producers. Energy subsidies may be direct cash transfers to producers, consumers or related bodies, as well as indirect support mechanisms, tax exemptions and rebates, price controls, trade restrictions, planning consent and limits on market access. They may also include energy conservation subsidies. In historical terms, subsidies for renewable energy have been lower in comparison to other forms of energy (EEA, 2004). There are arguments for and against energy subsidies.

The main arguments for subsidies are:

- *Security of supply* – subsidies are used to ensure adequate domestic supply by supporting indigenous fuel production in order to reduce import dependency, or supporting overseas activities of national energy companies.
- *Environmental improvement* – subsidies are used to reduce pollution and fulfil international obligations such as the Kyoto Protocol.
- *Economic benefits* – subsidies in the form of reduced prices are used to stimulate particular economic sectors or segments of the population, e.g. alleviating poverty and increasing access to energy in developing countries.
- *Employment and social benefits* – subsidies are used to maintain employment, especially in periods of economic transition.

The main arguments against subsidies are:

- Some energy subsidies are counter to the goal of sustainable development, as they may lead to higher consumption and waste, exacerbating the harmful effects of energy use on the environment, creating a heavy burden on government finances and weakening the potential for economies to grow, undermining private and public investment in the energy sector.
- Subsidies can impede the expansion of distribution networks and the development of more environmentally benign energy technologies, and do not always help the people that need them most.
- Energy subsidies often go to capital intensive projects at the expense of smaller or distributed alternatives (Mackenzie and Pershing, 2004).

Globally, subsidies for oil, coal, gas and nuclear power have totalled in the tens of billions of dollars annually and, although this may help some to access energy systems, it can drain resources away from the development of alternative and renewable approaches. Koplow et al (2007) discuss ten ways in which subsidies distort energy markets. In summary these are:

1 *Absence of charges on GHG emissions* – lack of realistic price for carbon in carbon trading effectively distorts this market approach.
2 *Oil security* – the price paid to police supply routes is a form of subsidy.
3 *Cap on liability for accidents in nuclear power* – in the event of a major nuclear accident, the majority of the costs will be met from public funds.
4 *Tax credits and exemptions for biofuels* – subsidies can distort food markets and lead to significant environmental damage as land is turned to their production.
5 *Cross-subsidies in electricity markets* – distortions in price when electricity supplies are switched between suppliers to meet consumer needs.
6 *Domestic subsidies to energy consumption* – tend to militate against new technologies and efficiency and conservation strategies.

7 *Subsidies for nuclear waste disposal* – not factored into the costs and funded by public money.
8 *Tax exemptions for petroleum use in international transportation* – this is a major issues for the aviation sector
9 *Tax credits for US alternative coal production* – slight changes in product development and clean coal technologies can hamper development of alternatives.
10 *Coal subsidies in Germany* – have been subsidised substantially for many years.

## Summary

Costing energy projects can be problematic. Using the straightforward method with discounted cash flow can give an indication of the economic viability of an investment. But when external costs and subsidies are added to the equation, then the decision becomes more complex. Cost–benefit analysis can help in evaluating a number of different approaches to meeting energy needs through different approaches and has the capacity to include external costs and subsidies as well as be viewed from a social welfare perspective.

## References

EEA (2004) Energy subsidies in the European Union: A brief overview, EEA Technical report 1/2004. Available at: http://reports.eea.europa.eu/technical_report_2004_1/en/Energy_FINAL_web.pdf

Koplow, D. (Lead Author), Earth Track (Content Partner), Cleveland, Cutler J. (Topic Editor) (2007) 'Ten most distortionary energy subsidies', in Cutler J. Cleveland (ed.), *Encyclopedia of Earth*, Washington, DC: Environmental Information Coalition, National Council for Science and the Environment. Available at: http://www.eoearth.org/article/Ten_most_distortionary_energy_subsidies

Mackenzie, J. and Pershing, J. (2004) Removing Subsidies: Levelling the Playing Field for Renewable Energy Technologies, World Resources Institute, Thematic Background Paper. Available at: http://www.renewables2004.de/pdf/tbp/TBP04-LevelField.pdf

HM Treasury (2007) Green Book: Appraisal and Evaluation in Central Government, HM Treasury. Available at: http://www.hm-treasury.gov.uk/d/2(4).pdf

# Index